Fundamentals of Crime Mapping

Rebecca Paynich, PhD
Director, Master of Arts in Criminal Justice
Associate Professor of Criminal Justice
Curry College
Milton, Massachusetts
Member, Academy of Criminal Justice Sciences
Member, American Society of Criminology

Bryan Hill
Crime Analyst
Glendale Police Department
Glendale, Arizona
Member, International Association of Crime Analysts
Member, Arizona Association of Crime Analysts

JONES AND BARTLETT PUBLISHERS
Sudbury, Massachusetts
BOSTON   TORONTO   LONDON   SINGAPORE

World Headquarters

Jones and Bartlett Publishers
40 Tall Pine Drive
Sudbury, MA 01776
978-443-5000
info@jbpub.com
www.jbpub.com

Jones and Bartlett Publishers
Canada
6339 Ormindale Way
Mississauga, Ontario L5V 1J2
Canada

Jones and Bartlett Publishers
International
Barb House, Barb Mews
London W6 7PA
United Kingdom

Jones and Bartlett's books and products are available through most bookstores and online booksellers. To contact Jones and Bartlett Publishers directly, call 800-832-0034, fax 978-443-8000, or visit our website www.jbpub.com.

Substantial discounts on bulk quantities of Jones and Bartlett's publications are available to corporations, professional associations, and other qualified organizations. For details and specific discount information, contact the special sales department at Jones and Bartlett via the above contact information or send an email to specialsales@jbpub.com.

Copyright © 2010 by Jones and Bartlett Publishers, LLC

All rights reserved. No part of the material protected by this copyright may be reproduced or utilized in any form, electronic or mechanical, including photocopying, recording, or by any information storage and retrieval system, without written permission from the copyright owner.

This publication is designed to provide accurate and authoritative information in regard to the Subject Matter covered. It is sold with the understanding that the publisher is not engaged in rendering legal, accounting, or other professional service. If legal advice or other expert assistance is required, the service of a competent professional person should be sought.

ArcGIS® and ArcMap™ graphical user interfaces are the intellectual property of ESRI and are used herein by permission. Copyright © ESRI. All rights reserved.

CrimeStat® is a registered trademark of Ned Levine & Associates. Ned Levine, CrimeStat III: A Spatial Statistics Program for the Analysis of Crime Incident Locations. Ned Levine & Associates, Houston, TX, and the National Institute of Justice, Washington, DC. November 2004.

Production Credits

Acquisitions Editor: Jeremy Spiegel
Associate Editor: Megan R. Turner
Production Director: Amy Rose
Production Assistant: Julia Waugaman
Marketing Manager: Jessica Faucher
Manufacturing and Inventory Control Supervisor:
　Amy Bacus

Composition: Shepherd Incorporated
Cover Design: Kristin E. Parker
Cover and Title Page Image:
　© Miflippo/Dreamstime.com
Printing and Binding: Malloy Incorporated
Cover Printing: Malloy Incorporated

Library of Congress Cataloging-in-Publication Data

Paynich, Rebecca.
　Fundamentals of crime mapping / by Rebecca Paynich and Bryan Hill.
　　p. cm.
　Includes bibliographical references and index.
　ISBN 978-0-7637-5575-1 (pbk.)
　1. Crime analysis. 2. Geographic information systems. 3. Digital mapping. I. Hill, Bryan. II. Title.
　HV7936.C88P39 2009
　363.250285--dc22
　　　　　　　　　　　　2008055383

6048

Printed in the United States of America
13 12 11 10 09　10 9 8 7 6 5 4 3 2 1

Contents

Foreword xi
Preface xiii
Use of ArcGIS and CrimeStat With This Book xv
Acknowledgments xvii

I Foundations and Theoretical Perspectives 1

1 Introduction to Crime Mapping 3
By Paul Cooke

Learning Objectives 3
Introduction 4
Theoretical Explanations of Crime and Place 4
    Crime as a Criminal Event 7
Crime Analysis 9
A Short History of Crime Mapping 12
Mapping Basics 13
    Map Information 13
    Types of Maps Used By Crime Analysts 14
    Map Data 15
Mapping Software and Resources 16
    Software 16
    Resources 17
Online Resources 17
Conclusion 18
Recommended Reading 18
Questions for Review 19
Chapter Glossary 19
Chapter Exercises 20
    Exercise 1a: The Data DVD 22
    Exercise 1b: ArcMap—A Trip Around Glendale, Arizona 23
    Exercise 1c: Arc Catalog Part I 35
    Exercise 1d: Arc Catalog Part II 39
    Exercise 2: Graphics 43
References 50

2 Social Disorganization and Social Efficacy 53

Learning Objectives   53
Introduction   53
The Chicago School of Criminology   56
Social Efficacy   58
Broken Windows Theory: Can It Be That Simple?   59
Online Resources   64
Conclusion   64
Recommended Reading   65
Questions for Review   65
Chapter Glossary   66
Chapter Exercises   67
    Exercise 3: Tables, Addresses, and Maps   67
    Exercise 4a: Address Locators and Find Addresses   77
    Exercise 4b: Adding Tables Part I   82
    Exercise 4c: Adding Tables Part II   91
References   95

3 Environmental Criminology 97

Learning Objectives   97
Introduction   97
The CPTED Approach   100
    Newman's Defensible Space   100
    CPTED: Theory and Research   102
The Environment, Opportunity, and Decision Making   107
    Rational Choice Theory   107
    Routine Activities Theory   109
    Crime Pattern Theory   111
Lifestyle Exposure Approaches   113
Displacement and Diffusion   115
    Types of Displacement   116
    Diffusion of Benefits   117
Online Resources   118
Conclusion   118
Recommended Reading   118
Questions for Review   119
Chapter Glossary   119
Chapter Exercises   120
    Exercise 5a: Selecting Features Part I   120
    Exercise 5b: Selecting Features Part II   124
    Exercise 5c: Selecting Features Part III   127
    Exercise 5d: Selecting Features Part IV   133
    Exercise 6: Buffers   136
References   145

4 Geography and Individual Decision Making: Victims and Offenders 149
By Lorie Velarde

Learning Objectives   149
Introduction   149
Mental Maps and Awareness Space   150
Target Backcloth   152
Journey to Crime   153
Geographic Profiling   155
   Hunting Methods   156
   Investigative Strategies   157
   Computerized Geographic Profiling   158
   Training in Geographic Profiling   160
Geographic Profiling Case Example   160
Online Resources   164
Conclusion   164
Recommended Reading   165
Questions for Review   165
Chapter Glossary   166
Chapter Exercises   166
   Exercise 7: Journey to Crime Basics   166
   Exercise 8: CrimeStat III: Journey to Crime Techniques   170
References   182

II Research and Crime Data 185

5 Research and Applications in Crime Mapping 187

Learning Objectives   187
Introduction   187
Research   188
   Public Housing   188
   Businesses   190
   Schools   191
   Transit Stops   192
   Hot Spots   193
   Drugs   195
Applications   195
   Investigation and Prosecution   196
   Proactive Policing   199
Online Resources   204
Conclusion   204
Recommended Reading   205
Questions for Review   205

Chapter Exercises   205
   Exercise 9: Spatial Deviation   205
   Exercise 10: Time to Practice on Your Own   215
   Exercise 11: A Discussion of Victimology   216
References   219

6 Crime Mapping and Analysis Data 225

Learning Objectives   225
Introduction   226
Measuring Crime   230
   Official Crime Data   231
   Self-Report   236
   The Crime Funnel   237
Contextual Data   238
   US Census Bureau   239
   Local Social Agencies   240
Tools for Crime Analysts and Mappers   241
   Microsoft   241
   Statistical Package for the Social Sciences   242
   Other Software Tools   243
Data Sharing Issues   243
Conclusion   244
Recommended Reading   245
Questions for Review   245
Chapter Glossary   246
Chapter Exercises   247
   Exercise 12: Joins and Relates   247
References   252

7 People and Places: Current Crime Trends 253

Learning Objectives   253
Introduction   253
People   254
   Victims   254
   Offenders   259
   Motivations for Offending   260
   Victim–Offender Relationships   262
   Group and Other Miscellaneous Patterns   263
   Victim Precipitation   266
   Alcohol and Drug Use   267
Places   268
   Location   268
   Urban Versus Rural: Is There Really a Difference?   271
Online Resources   272
Conclusion   272

Questions for Review   272
Chapter Exercises   273
    Exercise 13: Practical Administrative Exercise 1—
        The Media   273
References   279

III Statistics and Analyses 281

8 A Brief Review of Statistics 283

Learning Objectives   283
Introduction   283
A Crash Course in Statistics   284
    Types of Data   284
    Descriptive Statistics   292
    Inferential Statistics   297
Classification in Mapping   309
Online Resources   313
Conclusion   313
Recommended Reading   317
Questions for Review   317
Chapter Glossary   317
Chapter Exercises   320
    Exercise 14: Aggregating Data for Shaded Grid Maps   320
    Exercise 15: Hot Spots, Hot Areas, Choropleth Maps, and Problem
        Areas   325
References   333

9 Distance Analysis 335

Learning Objectives   335
Introduction   335
Distance Analyses   337
    Mean Center Analysis   337
    Journey to Crime Analysis   339
    Spider Distance Analysis   341
    Buffer Analysis and Queries (Theme Selection)   341
    Distance Between Hits Analysis   342
    Distance and Time Analysis   344
Online Resources   345
Conclusion   346
Recommended Reading   347
Questions for Review   347
Chapter Glossary   347

Chapter Exercises   347
    Exercise 16: Count or Density   347
    Exercise 17: Distance Between Hits and the Mean Center   351
    Exercise 18: Spatial Autocorrelation and Distance Decay   359
References   368

10 Hot Spot Analysis 371

Learning Objectives   371
Introduction   371
Types of Hot Spot Analyses   373
    Manual or Eyeball Analysis   373
    Graduated Color Map or Choropleth Map Analysis   375
    Grid Cell Mapping Analysis   375
    Point Pattern Analysis or Cluster Analysis   379
    Standard Deviation Analysis   383
Online Resources   384
Conclusion   386
Recommended Reading   387
Questions for Review   387
Chapter Exercises   387
    Exercise 19: Graduated Points   387
    Exercise 20: CrimeStat III Hot Spot Analysis
      Techniques I   390
    Exercise 21: CrimeStat III Hot Spot Analysis
      Techniques II   402
    Exercise 22: Density as a Predictor   406
    Exercise 23: A Probability Grid Concept   412
References   417

IV Mapping for Your Audience and Future Issues 419

11 Mapping for Your Audience 421

Learning Objectives   421
Introduction   421
GIS in Criminal Justice   422
Maps for Specific Audiences   428
    Patrol Officers   428
    Investigators   429
    Community   440
    Courts   442
    Corrections   444
Conclusion   447
Questions for Review   447

Chapter Exercises   447
  Exercise 24: Data Frames   447
  Exercise 25: Putting It All Together for One Project   452
  Exercise 26: Practical Administrative Exercise 1   466
  Exercise 27: Practical Administrative Exercise 2   467
References   468

12 Future Issues in Crime Mapping 469

Learning Objectives   469
Introduction   469
General Topics   469
Education and Crime Mapping   471
Crime Mapping Research   472
Geographic Profiling and Forecasting Crime   473
Technology   473
Other Issues   474
Recommended Reading   476
Chapter Exercises   476
  Exercise 28: Practical Strategic Exercise: Vandalism Unveiled   476
  Exercise 29: Practical Strategic Exercise 2:
      Intelligence-Led Policing   477
  Exercise 30: Practical Tactical Exercise 1: Robbery 101   479
  Exercise 31: Practical Tactical Exercise 2: Burglary 101   480
  Exercise 32: Practical Tactical Exercise 3: Pick a Crime   481
References   481

Appendix   Microsoft Excel: A Generic Primer for Crime Analysis 483

Introduction   483
Section 1   484
  Introduction to Excel and Some General Information   484
Section 2   487
  Working with Standard Deviation, Average,
    and Other Statistics   487
Section 3   494
  Converting Dates and Times   494
Converting Data into Date Format   497
Section 4   502
  Notes on Working with ArcView and Excel   502
Section 5   507
  Tactical Prediction   507
Decimal Time Predictions   507

Simple Graph Predictions 509
Split Time or Midpoint and Weighted Averaging Predictions 511
CrimeStat's Correlated Walk Prediction 515
Why Do So Many Calculations? 516
Section 6 518
    Pivot Tables and Administrative, or Strategic, Analysis 518
Pivot Table Web Resources 518
Pivot Table Example 518
Section 7 520
References 520

Index **521**

Foreword

Crime Mapping: Cutting to the Chase could be another name for the *Fundamentals of Crime Mapping* because this work takes the basics of theory, technique, and methodology and shows how they are essential for the complementary fields of crime analysis and crime mapping. The authors have laid out a straight course; thus, there are no off-ramps leading to distracting theoretical discussions yielding estimates of how many police calls for service can fit on the head of a pin. A popular line used by the mythical Sgt. Joe Friday in the TV series *Dragnet* was "just the facts ma'am." In this book it is "just the maps ma'am."

One of the many assets of this book is that some of the contributors are active crime analysts in police departments. Therefore, these individuals have designed many of the exercises and tutorials and have provided the data for analyzing specific problems and policies confronting an operational agency. Students using this book will undoubtedly gain valuable insights into how crime analysis professionals go about practicing their craft.

Acquiring the skills to become competent in crime mapping is going to be very hard work. The authors provide very detailed step-by-step instructions on the procedures for using database, statistical, and GIS software. They have plotted the course for you and even provide helpful advice emanating from their experiences, like to "periodically save your work." I would like to augment this sage advice with three things I tell my students. The first is "develop a thick skin." A student's initial maps are going to be awful. Therefore, the student has to learn to distinguish between personal and constructive criticism. The second is "use it or lose it." The point here is that the processes and procedures involved in crime mapping are quite complex and dynamic. Thus, like perfecting the playing of a musical instrument, one has to practice, practice, and practice. A fairly common problem among many criminology or criminal justice PhD students is that they take one GIS mapping course which they can list on their resume or CV as one of their specialties. Yet when they show up for job interviews and are quizzed about it or asked to demonstrate their skill, they fall flat on their face. Finally, I tell my students, "You have worked very hard in acquiring a very valuable skill—don't give it away!" The point here is that people that can do mapping very well are relatively rare and valuable. They should not sell their skills to the lowest bidder, but should be respected and treated like professionals. You would be surprised at the number of academics and bureaucrats who profess all kinds of expertise about crime mapping, but if they had to make a crime map they too would fall flat on their face.

Fundamentals of Crime Mapping is an excellent start down the path to becoming a crime analyst and cartographer.

James L. LeBeau, PhD
Southern Illinois University–Carbondale
Carbondale, IL

Preface

Have you ever faced a situation where you have copies of ArcGIS installed on computers, but no one in the office knows how to use it? We wrote this book as a teaching tool for our students who are a mix of full-time college students and practitioners currently working in the field of law enforcement. We have found in our teaching experiences that other books available do not suit the needs of these beginning students and analysts who want to gain introductory skills and knowledge about crime mapping, practical uses of the software, and some of the basics about the theory of crime. Some students need a precursor in criminological theory; some students need a discussion of crime statistics; some students need a discussion of statistical analyses that are commonly used in crime analysis; some students need a review of the existing research; and all students need a workbook that contains exercises utilizing a range of data that they could complete to begin acquiring skills in crime analysis and mapping that make sense and directly apply to what they have to do back at the office.

Section I contains a brief introduction to crime mapping and several chapters on the theoretical approaches to the study of crime and place. The discussions in these chapters are meant to be concise but comprehensive, and provide students with enough information to understand the importance of theory in creating crime maps and performing crime analyses. Section II focuses more narrowly on the existing research in the field and the types of data that are used in crime analysis and mapping. This section ends with a discussion of current crime trends. Section III discusses and explains common types of statistical analyses that are employed in crime mapping. In addition, this final section of the book provides a chapter that gives useful suggestions about tailoring map presentations based on the needs of particular audiences. The book ends with a brief chapter on future issues in crime mapping.

Mapping is a tool, and it is a powerful one when used appropriately and correctly. This book will teach you how to use that tool with skill and knowledge so that your maps will connect you and others to the geography and crime stories they depict. With knowledge and understanding, you have the opportunity to find answers to your questions about crime. We hope you enjoy the chapters and exercises and that we have helped turn law enforcement data into information for you, and with your diligence in practice with the software, knowledge you can use over and over again.

Use of ArcGIS and CrimeStat With This Book

Fundamentals of Crime Mapping contains a number of practice exercises that utilize the software programs ArcGIS and CrimeStat. While the Publisher makes every effort to direct users of this book to resources that will allow the user to download the most up-to-date versions of these programs, the Publisher is not responsible for providing the software directly. Any issues regarding technical support or assistance regarding usage of ArcGIS or CrimeStat should be made directly to the software provider.

Acknowledgments

Rebecca would like to acknowledge her family for all of the help and support they gave her throughout the process. To Jason, for all of his editing and words of encouragement; to Spencer and Dylan, for being patient; to Bonnie and Dan, and Mom and Dad for coming out to take care of the house and kids; to Ashley Bechtel for driving all the way to Pennsylvania so I could type on the laptop and make my deadline; to Julie Bissett, for going through the exercises and all of her helpful feedback; to Deanna Gordon and Pete McKeen for answering all of my Microsoft questions; to Melissa McKeen for watching the kids; and to the reviewers for their helpful suggestions. And last, but certainly not least, to my professors at WSU, my students, and to all of my colleagues in the field from whom I continue to learn on a daily basis. Without them, none of this would be possible.

Bryan would like to acknowledge his beautiful wife, Barbara, for having patience while he spent hours away from her to create these exercises and for her insistence that he write a book someday. He would also like to thank the staff of the Administration of Justice section of the Chandler Gilbert Community College who helped the Arizona Association of Crime Analysts (AACA) create and teach the courses for the occupational certificate for crime and intelligence analysis at their college. I would also like to thank my many colleagues in the field who have contributed greatly to my knowledge, and especially those analysts who belong to the AACA who educate me on a daily basis. I owe special thanks to Dr. James LeBeau for being the satellite crime analysis department for the City of Glendale, Dr. Ned Levine for being patient with my many questions, and the hundreds of other outstanding academics in this field who have furthered my ability to put theory into practice.

Foundations and Theoretical Perspectives

SECTION I

CHAPTER 1 Introduction to Crime Mapping
CHAPTER 2 Social Disorganization and Social Efficacy
CHAPTER 3 Environmental Criminology
CHAPTER 4 Geography and Individual Decision Making: Victims and Offenders

Introduction to Crime Mapping

CHAPTER 1

By Paul Cooke

▸ ▸ LEARNING OBJECTIVES

This book is written for individuals who are interested in the study and practice of crime mapping. Whether you are a practitioner or a college student, we hope you will find this book useful in your endeavor to learn more about crime mapping. This book, however, is not meant to provide you with an in depth analysis of theory or an extensive discussion of methods in crime mapping. Rather, it is meant to convey the academic and practical skills the beginning student and analyst need to create crime maps that will be useful and accurate. For students who desire more than a cursory approach, there is a plethora of existing literature, both practical and academic, that is more appropriate for advanced studies. We have listed some of these sources in the Recommended Reading section at the end of each chapter.

As a person interested in the field of crime mapping and analysis, you should understand that it is not a simple field of study. The technology is expanding rapidly, and you will need to work vigilantly to stay current in the field.

This chapter introduces the notion of crime as a "criminal event," which stresses the importance of understanding how both victims and offenders interact with their environments. This chapter also contains a brief history of crime mapping and introduces the student to the basics of using mapping tools in the analysis of crime. After studying this chapter, you should be able to:

- Define and explain the *criminal event* and understand the basic theoretical explanations offered for the relationship between crime and place.
- Define and explain the concepts of crime mapping and analysis.
- Identify key research that is rooted in crime mapping and analysis.
- Identify and explain basic map information, data, and various types of maps used by crime analysts.
- List key software and resources used by crime mappers and analysts.
- List the resources available to crime mappers and analysts.

▸ ▸ KEY TERMS

Administrative Crime Analysis
Cartography
CPTED
Crime Analysis
Criminal Intelligence Analysis
Criminal Investigative Analysis
Criminology
Geocoding
Geographic Profiling
Hypotheses
Macro
Micro
Police Operational Analysis
Projection
Scale
Strategic Crime Analysis
Tactical Crime Analysis

■ Introduction

This chapter begins with a discussion of crime as a criminal event. This view differs from other approaches to the study of crime in that it not only seeks to understand the motivations and behaviors of offenders or victims, but it seeks to explain how both offenders and victims interact with each other and their environment. Thus, we are studying crime from multiple dimensions (the dimensions of the offenders, victims, and environment) rather than simply trying to understand it from one point of view (for example, trying to understand the motivation of the offender without factoring in victim or environmental characteristics). Viewing crime as a criminal event (Sacco & Kennedy, 2002) allows us to understand crime in a spatial context in that the environment provides varying opportunities for crime by providing cues to both offenders and victims that impact their decision making. Thus, properties of the immediate space in and around criminal events are contributing factors to victimization and cannot be excluded from the study of crime.

Also included in this chapter is a brief discussion of the history of crime mapping. Understanding the history of this field (or any field, for that matter) is important to the discussion of present and future issues in crime mapping. An introduction of basic map terminology and a brief examination of the different types of maps that crime analysts produce will also be included in this chapter. A discussion and listing of resources, including mapping and analysis software options, is also provided.

■ Theoretical Explanations of Crime and Place

Crime has been a part of life for as long as people have gathered into social groups. A great deal of time and effort has been invested in trying to understand crime, most notably examining why some people engage in criminal behavior and others do not. As a result, there is a plethora of theories under the larger heading **criminology** (the study of crime and criminal behavior) that attempt to answer these questions. However, thus far, we have no generally accepted theory that explains the existence of crime in a society. For example, consensus perspectives of criminology approach crime as a normal and healthy part of any society. Conflict perspectives, on the other hand, argue that crime is the result of group conflict and unequal distributions of power.

Criminological theories also differ in their level of application. **Macro**-level theories make assumptions about societal-level variables, including the structure of government and the economy and how these variables impact crime rates within a society (which could be a city, state, country, or even the world). **Micro**-level theories make assumptions about individual characteristics (IQ, mental state, temperament, biological characteristics, and personal finances, for example) and how they influence a person's decision to commit a crime.

Unfortunately, many theories of crime focus solely on the individual or group that commits a crime and ignore other contributors to crime, including the environment. Recently, theorists have begun to broaden their approach from simple explanations of the criminal and his act to include other variables, such as victims' behaviors (Clarke & Felson, 1993; Cohen & Felson, 1979) and the physical environment (Brantingham & Brantingham, 1981, 1993). Thus, theorists began to examine crime as an "event" that was not simply a product of an interaction between persons but an interaction between victims, offenders, and their environments. When researchers began to examine the contributions of time and space on various criminal events, crime mapping became an important tool in crime analysis. In addition, with the introduction of widespread geographic information systems (GIS) data collection efforts and improved technology, the importance of crime mapping has grown exponentially over recent years to the development of theory and to the development of policy aimed at understanding and preventing crime.

Shaw and McKay (1942) put forth their social disorganization theory, which suggests, in part, that the economic composition of a community contributes to crime by affecting neighborhood order. Building upon the earlier works of Park and Burgess, they observed that higher juvenile delinquency rates tended to cluster in certain neighborhoods within urban areas. Through their research, they determined that various factors about these neighborhoods contributed to higher levels of crime. The neighborhoods suffered from the effects of poverty and residential instability, which impacted both the physical appearance and the social structure of the neighborhood itself. In these socially disorganized neighborhoods, poverty, high population density, and high population mobility created an atmosphere where higher numbers of suitable targets and motivated offenders coexisted with little or no guardianship. Thus, the clustering of motivated offenders and a lack of guardianship in certain areas of a city, often in socially disorganized neighborhoods, produces higher rates of crime. Shaw and McKay made an important discovery that helped guide the development of more recent theoretical approaches; they found that socially disorganized neighborhoods suffered from higher rates of crime regardless of who lived there. That is, whether the neighborhood was inhabited by Italians, Russians, Cape Verdeans, or Cubans, for example, the crime rate remained high. Thus, this discovery suggests that the area, not the people, is criminogenic.

Three primary theoretical perspectives have built upon Shaw and McKay's discovery and have made enormous contributions to the study and understanding of crime in a spatial context: Routine Activities Theory, Rational Choice Theory, and Crime Pattern Theory, all of which are housed within a larger theoretical framework called Environmental Criminology. Each theory provides assumptions about empirical observations of the environment, victims, and/or offenders, allowing predictions to be made about the criminal event. In turn, this provides insight into possible solutions to combat crime, which incorporates environmental factors such as crime prevention through environmental design (**CPTED**) and other situational crime prevention strategies (Crowe, 2000; Rosenbaum, Lurigio, & Davis, 1998). For example, simply improving lighting or limiting access to a parking lot may reduce thefts from automobiles in that lot, or installing emergency phones and well-lit walkways may reduce personal crimes on a college campus.

Environmental Criminology suggests that we analyze a variety of characteristics about the physical landscape, such as land use, access, and visibility, to determine likely areas that are conducive to crime. Brantingham, Dyreson, and Brantingham (1976), in their "cone of resolution," examined why there were regional differences in crime rates. In their subsequent work, *Environmental Criminology,* they suggested that to have a better understanding of crime, theorists must examine four key elements to crime including the law, offenders, targets, and place (Brantingham & Brantingham, 1981). In 1993, they proposed the groundwork for crime pattern theory, which combines Environmental Criminology with rational choice and routine activities perspectives in understanding crime (Brantingham & Brantingham, 1993).

Routine Activities Theory (Cohen & Felson, 1979) offers that three elements, when coexisting in time and space, will produce a greater likelihood that crime will occur: these include a motivated offender, a suitable target, and the absence of a capable guardian. When all three elements come together in time and space, the opportunity for crime is greater than when one or two of these elements are missing. Essentially, Routine Activities Theory suggests at the micro level that individuals are more likely to be victimized in some situations than others. For example, a person walking alone in a park at night listening to an expensive iPod is more likely to be robbed than a couple in the same park walking their large dog, not listening to expensive iPods.

At the macro level, this theory suggests that patterns in victimization have changed, in part, due to patterns in the routine activities of society in general. For example, in the last 3 or 4 decades, the number of women who have decided to work outside of the home has dramatically increased. According to the Bureau of Labor Statistics (2007), 40.8% of civilian, noninstitutional women aged 16 years and older were employed in 1970. In 2006, this percentage increased to 56.6%. This jump left an increased number of homes during the daytime hours unguarded, which may have contributed to an increase in daytime residential burglaries. In addition, this theory explains that crime rates will vary spatially due to increased opportunities caused by higher numbers of suitable targets and motivated offenders and lower levels of guardianship. Neighborhoods that are characterized by physical and social disorder have more opportunities for crime because they tend to hold more motivated offenders and have a greater lack of guardianship.

Crime as a Criminal Event

The criminal event, under Routine Activities Theory, requires that a suitable target, a motivated offender, and the absence of a capable defender or guardian come together at the same time and place. That place might be private, such as at a home (burglary or domestic violence, for example), a more public place than the home, such as work or school (workplace assaults or embezzlement), or in public spaces, such as public roads, paths, parks, or entertainment spots (robbery, assault, and riots, for example). The specifics of who is assaulted, what is stolen, or who commits the crime may change from place to place, but whenever the three elements of the criminal event converge in time and space, the likelihood of crime is increased. By the same token, if one or more of the elements is missing from the equation, a crime is less likely to occur. Clarke and Eck (2005) expanded Routine Activities Theory and developed the problem analysis triangle (or crime triangle) to include the "controllers," who are *handlers*, *managers*, and *guardians* (see **Figure 1–1**).

Figure 1–1 Problem Analysis Triangle

In this model, guardians are those persons who keep a watchful eye on people and property, including people themselves and police or private security personnel. Handlers are people who know and interact with motivated offenders. They include agents of informal social control, such as family members, friends, or teachers, and agents of formal social control, such as probation or parole officers. Last, managers are those persons who have some level of responsibility for the behaviors of persons within a specific location. For example, home owners are responsible for controlling behavior in their homes, teachers are responsible for maintaining order in their classrooms, and bar managers are responsible for controlling behavior in their bars. Home owners, teachers, and bar managers, in this sense, are managers.

In studying and mapping the criminal event, it is possible to study not just the individuals involved but also to map crime and the spatial and temporal variables needed to understand how the specific event occurred, how the elements came together, what actually transpired, and how the event was concluded. When it is understood how a specific event occurred, it becomes possible to examine similar events in an attempt to identify similarities, patterns, and ways to keep that type of event from occurring again.

It is also essential to realize that mapping criminal events is only the first step in the crime fighting process. The criminal event and the circumstances surrounding the criminal event must be analyzed to find meaning in the data. Facts do not speak for themselves! They must be critically analyzed. **Hypotheses** (statements derived from theory that can be tested to either support or disprove a theory or its assumptions) must be generated about individual criminal events and criminal events that take place in combination with other, similar events. Could the perpetrator be the same in a series of crimes? Is the method of committing the crime similar? Are times and locations connected in some way? Examining criminal events to identify relationships is the role of a crime analyst, and data collection, analysis, and mapping are the analyst's method of fighting and preventing crime.

■ Crime Analysis

There are three broad categories of analysis that fall under law enforcement analysis: crime analysis, criminal intelligence analysis, and investigative analysis (Bruce, 2004; Gottlieb, Arenberg, & Singh, 1994). There is a wide array of definitions for crime analysis in the existing literature. **Crime analysis**, as Boba defines it, "is the systematic study of crime and disorder problems as well as other police-related issues—including sociodemographic, spatial, and temporal factors—to assist the police in criminal apprehension, crime and disorder reduction, crime prevention, and evaluation" (Boba, 2005, p. 6). Bruce defines crime analysis as "focused on the study of criminal incidents; the identification of patterns, trends, and problems; and the dissemination of information that helps a police agency develop tactics and strategies to solve patterns, trends, and problems" (Bruce, 2004, p. 15). The International Association of Crime Analysts (IACA) defines crime analysis as "a type of law enforcement analysis that is focused on the study of criminal incidents; the identification and analysis of patterns, trends, and problems; and the dissemination of information that helps a police agency develop tactics and strategies to solve patterns, trends, and problems" (2005). Essentially, crime analysis draws on a variety of different types of data, including crime data, to gain a better understanding of criminal activity and the root causes of criminal activity, and to develop more effective means of combating crime and preventing victimization.

Analysts look at crime and other law enforcement data using formal analytical and statistical techniques and research methods that have been developed in the social sciences. They study arrests, crime reports, offender and victim characteristics, and crime scene evidence. They also examine other types of data collected, such as calls for service, traffic citations and accidents, census data, weather and traffic patterns, and data from other criminal justice agencies, including probation and parole reports. By analyzing numerous sources of information, the crime analyst can provide useful information to assist decision makers in a police department to fight and prevent crime.

Crime analysis includes tactical and strategic analysis (focused on criminal activity), and administrative analysis (focused on police activity). **Tactical crime analysis** examines recent criminal events and potential criminal activity by analyzing how, when, and where the events occur to establish patterns and series, identify leads or suspects, and to clear cases (Boba, 2005). Tactical analysis focuses on specific information about each crime, such as method of entry, point of entry, suspects' actions, victims' characteristics, type of weapon used, and the date, time, and location of the crime. It also considers information developed in the field from patrol officers, such as suspicious activity, calls for service, criminal trespass warnings, field interrogation cards, and personal identifying marks, such as scars and tattoos (Boba, 2005). Usually, crimes examined under tactical crime analysis are those where the victim and offender are unknown to one another. The focus of tactical analysis is the daily examination of data to identify trends and patterns concerning recent criminal activity. When a crime pattern, suspect, or investigative lead is identified, the information is compiled and disseminated to patrol officers and detectives. Crime mapping is used in tactical analysis to reveal clusters of criminal activity and to identify spatial relationships between crime and various geographic variables (Boba, 2005).

Strategic crime analysis involves the study of crime and other law enforcement issues to identify long-standing patterns of crime and other problems and to assess police responses to these problems (Boba, 2005). Typically, this analysis involves collecting a great deal of information about criminal events. In addition, "helping agencies to identify root causes of crime problems and develop creative problem-solving strategies to reduce crime" is a key goal in strategic crime analysis (IACA, 2005).

Administrative crime analysis involves the presentation of key findings of crime research and analysis to audiences within law enforcement, local government, and citizenry based on legal, political, and practical concerns. Bruce (2004, p. 22) provides several examples of administrative crime analysis, including:

- A report on demographic changes in the jurisdiction
- Miscellaneous crime statistics to support grant applications
- Preparation of Uniform Crime Report (UCR) or Incident-Based Reporting System (IBRS) reports

Another example of administrative crime analysis is the establishment of a department Web site to inform the general public about public safety issues. **Police operational analysis** is another type of analysis that is also conducted. Its focus is on the operations of police agencies, including staffing and resource deployment. Operations analysis seeks to identify how to better organize the internal operations of a police department to minimize inefficiency and maximize effectiveness (Bruce, 2004).

Crime analysis is different from **criminal intelligence analysis** in that criminal intelligence analysis typically looks at organized criminal activity and seeks to link people, events, and property. Much of the information that is analyzed is obtained by law enforcement efforts, including surveillance, informants, and undercover operations (Boba, 2005). Information that is analyzed is not necessarily criminal information and can include telephone taps, travel information, financial and tax records, and family and business relationships of the person(s) being investigated (Boba, 2005).

Crime analysis is also separate from **criminal investigative analysis** in that with criminal investigative analysis, the focus is on serial criminals (Boba, 2005). Victim characteristics and elements of crime scenes are studied to discover patterns that link related crimes together. Sometimes, in **geographic profiling**, a profile of the offender, based on the nature of the crime and the facts of the case, is developed. Often these types of serial crimes cross jurisdictional lines, and many different law enforcement agencies may become involved in high-profile cases. (See Rossmo, 2000 for a more in-depth discussion of geographic profiling).

Analysis has been utilized for many years on an informal basis but with newer technology, especially in the last 2 decades, and its prominence and utility in law enforcement applications has grown. After the events surrounding 9/11, the need and demand for good analysis has exploded. Good analysis and presentation is essential. This book is dedicated to providing you with the tools necessary to complete useful and accurate analysis that, when properly presented, can make a difference in fighting crime.

Jack Dangermond, the president and cofounder of ESRI (the company that created the ArcGIS software that is used to complete the exercises contained in this textbook) suggests that: "To gain a greater understanding of our world we need a framework that I call the 'geographic approach.' The geographic approach uses geographic science supported by geographic information system (GIS) technology as a framework for understanding our world and applying geographic knowledge to solve problems and guide human behavior" (Dangermond, 2007).

The ability to utilize such an approach has not always been an option for law enforcement. However, with the arrival of GIS and inexpensive desktop computers and printers with speed and large memory/storage capacity, even the smallest agency can use analysis and crime mapping in its daily battle against crime and disorder.

■ A Short History of Crime Mapping

This chapter contains only a brief discussion of the history of crime mapping. For a deeper understanding of its history, there are several sources you may want to look at, including Weisburd & McEwen (1988), Harries (1999), Boba (2001), and Chamard (2006). Mapping itself has a long history, but crime mapping, the subject of our study, has been traced to the early 1800s when social scientists began creating maps to illustrate their crime theories and research. Social scientists Adolphe Quetelet and A. M. Guerry are credited as being two of the first people to use spatial analysis to research crime. Using crime and other social data from France, they determined that crime was not evenly distributed across space and that it also clustered geographically with other observable social level variables, such as socioeconomic status and population density.

In the early 1900s, the New York Police Department and other large departments began to utilize single symbol point maps (pin maps) to illustrate crime locations. In the 1920s and 1930s, sociologists at the University of Chicago used graduated area maps of crime and delinquency to show the relationships between crime and social variables, such as poverty. These maps were drawn by hand and were very labor intensive.

It was not until the 1960s and 1970s that crime maps were created with the use of computers. These early computers were large, mainframe computers that were affordable only to the largest agencies. They were very expensive to operate, and producing maps was still very labor intensive. In addition, the maps that were produced were of poor quality and unsuitable for many law enforcement purposes.

Desktop computers that were suitable for mapping (still with limited quality) became available in the 1980s, but they had slow processing speeds, limited memory, and poor printer quality. In the 1990s, desktop computers with GIS capabilities and the ability to be integrated with law enforcement record management systems and other data made mapping possible for many law enforcement agencies. The additional contributions of government funding, expanded training, a more tech-savvy workforce, and higher quality computers and printers further enhanced this ability. Studies that examined the use of crime mapping in law enforcement agencies suggest that the utilization of crime mapping software, the program(s) being used, and how crime mapping is applied within an agency varies significantly (Chamard, 2006; Mamalian, et al., 1999; Paynich, Cooke, & Mathews, 2007). In general, large departments with an expansive data collection effort that serve a diverse and geographically dispersed population and that provide adequate funding and personnel to accommodate training and equipment expenses have been the most successful in adopting mapping as a strategy and have maintained their involvement through networking, conferences, and the Internet. Midsized and smaller agencies have been slower to adopt a crime mapping policy, primarily due to a lack of data, a small service population, a low crime rate, and a lack of resources and personnel. However, with examples of successful implementation, decreasing equipment costs, rising workforce capabilities, and increased training and funding available in the post-9/11 era, crime analysis and mapping should continue to grow and improve law enforcement agencies' analytic capabilities.

■ Mapping Basics

A map is a two-dimensional or flat-scale model of the Earth's surface or a portion thereof (Rubenstein, 2003). Maps portray a portion of the real world in a form we can use to find our way or find answers to questions we may have about an area we are interested in. They may be rough sketches of the way to Aunt Trudy's house or a precise computer map of a community that details every important landmark. The science of mapmaking is called **cartography** and is centuries old.

Map Information

For our purposes, we will focus on geographic information systems (GIS)—high-performance computer software that allows users to process geographically-related data. This data (topography, political boundaries, population density, calls for service, crimes, etc.) is stored as information in virtual layers (one layer for each variable) to be displayed on the computer in the form of a multilayer, virtual map. A single layer may be displayed, or several layers can be combined to show relationships among the layers (types of information). In crime mapping, the layers sit atop one another, resulting in an overall picture of crime and its spatial context. This is much like an anatomy textbook that contains different transparency sheets with drawings of the skeletal, muscular, cardiovascular, central nervous, and digestive systems. You may choose to look at the systems individually or layer them atop one another to see how they overlay and get a better picture of the entire human anatomy. In crime mapping, you may have a map with layers representing police boundaries, streets, key landmarks, and various crimes. Using GIS, you can select various layers to use in your analysis. The use of GIS to analyze environmental and social phenomena, such as crime, is what crime analysis and mapping is all about.

The Earth is very nearly a sphere, and while it may be accurately portrayed as a globe, that is not practical for most uses. Problems arise when you try to draw a sphere on a flat surface because the image is distorted. The method used to transfer locations on the Earth's surface to a flat map is called **projection**. Mapmakers have devised numerous ways of producing flat maps, but none are without flaw. There are cylindrical and conical projections, but each has distortions. The distortion produced is more severe when you attempt to map the entire world and less so as the mapped area becomes smaller. For most law enforcement purposes, the geographic area of interest is small enough that projection does not cause serious problems and can be ignored (Harries, 1999). Projection systems will be discussed in more detail in the Chapter Exercises within this textbook, but for our purposes here, you should know that there are many different projection systems to choose from, and it is important to know what system your data is projected with. Nevertheless, you should be aware of the problem and work with local experts to address this issue and, if necessary, create solutions for your purposes. The problem becomes more important as the **scale** of the map increases (from municipal to regional level, for example) and when importing and exporting data between applications. This is especially important in **geocoding** events onto a map. In the geocoding process, street addresses are positioned onto a map using latitude and longitude coordinates. "*Projections* determine how the latitude and longitude grid of the earth is represented on paper. *Coordinate systems* provide the x-y reference system to describe locations in two-dimensional space" (Harries, 1999, p. 14). Different states use different projections, and it is important for the analyst to understand which projections are used by the map layers he or she is working with.

Types of Maps Used By Crime Analysts

There are many different types of maps that might be constructed depending on what information the analyst wants to present and what audience will receive the information. He or she might map a simple, single event (such as that depicted by traditional pin maps) or show the distribution of crime across a particular area (a choropleth map, such as a police beat or district or a census block), multiple hot spots located across a jurisdiction that are related to multiple criminal events (which typically include liquor stores, pawn shops, high schools, drug houses, shopping malls, etc.), or a series of connected criminal events (such as a child molester or serial murderer). An analyst might also utilize the statistical capabilities provided in GIS software and other programs, which expand the programs' capabilities to attempt to predict an offender's next target (such was the case in the Washington sniper case) or areas where offenders might live, work, play, or retreat to after performing a criminal event.

Map Data

There are many different maps that can be created. However, most crime maps include a limited number of map components. The five major components of a GIS include hardware, software, data, personnel, and methods. Real-world data are represented by four features in a GIS:

- Point features: A discrete location usually depicted by a symbol or label. It is like a pin placed on a paper wall map. Different symbols are used to depict the location of crimes, motor vehicle crashes, traffic signs, buildings, police stations, cell towers, etc.
- Line features: A geographic feature that can be represented by a line or set of lines, such as railways, streets, rivers, power lines, bus routes, etc.
- Polygon features: A multisided figure represented by a closed set of lines, such as city boundaries, census tracts or blocks, patrol beats, neighborhood boundaries, and gang turf. Polygon features may be as large as countries or as small as a single place (like a park or cemetery) or even a single building (such as a school).
- Image features: A vertical photo (usually taken from an airplane or satellite) that is digitized and placed within the geographic information system coordinates that are associated with it.

Each type of feature has "attributes" or a table of data that describes it. The ability to view, query, relate, and manipulate the data behind these features is the real power of a GIS. Simply clicking on a point, line, polygon, or image can produce the data associated with that particular feature. For example, you can very easily identify a specific point on the map and be instantly given detailed information (as long as it is contained in the attribute table). In crime data, this information may include victim and offender characteristics, motives for the crime, and temporal (date and time of incident) and spatial (address and name of location) details.

■ Mapping Software and Resources

In today's world of crime analysis, the beginner analyst is fortunate to have a plethora of crime mapping tools, software, and resources at his or her disposal from a variety of sources. This section provides a brief overview of some of the most known and commonly used programs and resources. It is by no means a complete listing of the wonderful resources available to analysts. The field is rapidly changing and growing and these changes bring new software and resources on an almost daily basis.

Software

The wide use of crime mapping and analysis has greatly increased due to the development and availability of relatively inexpensive computers, printers, and analysis software that are adequate to the task. Today there are many mapping and GIS programs available on the market, and more are being developed each year. For the purposes of this introduction, we will look at four of the more common programs. Two are professional, commercial systems, and two are programs that were developed for special use under government funding and are available free online. The first two programs, currently the leaders in GIS systems used by law enforcement in the United States, are MapInfo (www.mapinfo.com) and ArcGIS (www.esri.com). ArcGIS is the most commonly cited program used by police departments that utilize crime mapping applications (Paynich, et al., 2007). GeoDa is a standalone spatial analysis program developed by Dr. Luc Anselin at the University of Illinois for use as an introduction to spatial analysis. CrimeStat is a standalone spatial statistics program for the analysis of incident locations. It was developed by Ned Levine & Associates under research grants from the National Institute of Justice (NIJ), who is the sole distributor of CrimeStat and makes it available for free to law enforcement and criminal justice analysts and researchers (Levine, 2003). You will use CrimeStat in several of the exercises contained in this book. CrimeStat must be used in conjunction with a mapping program, such as ArcView or MapInfo, to project the analysis onto a map. Without it, the analyses are simply numerical outputs to be interpreted by analysts.

Resources

There are also a variety of online and other resources for beginning analysts to use in their quest for information about crime analysis and mapping. Many of these sites have links to free software, with accompanying workbooks and tutorials, and publications written about topics pertinent to the analysis of crime and related social problems. New sources are always being developed and older sites abandoned. For example, the Mapping and Analysis for Public Safety (MAPS) program at the NIJ, along with the Community Oriented Policing Services (COPS) office, publishes a quarterly bulletin on geography and public safety. Topics related to crime analysis and mapping are covered in this bulletin along with discussions about various tricks of the trade. Interested persons can sign up for either an e-mail or printed version of this bulletin at http://www.ojp.usdoj.gov/nij/maps/bulletin.htm. In addition to this bulletin, students may want to join an e-mail group list, such as the CRIMEMAP listserv (http://www.ojp.usdoj.gov/nij/maps/listserv.htm), to learn and share information about crime mapping and analysis. The e-mail content of this group ranges from discussions about the best software to buy, how to use various applications, where to find data, and recommendations of various publications. Other online sources worth looking at are shown in the next section.

■ Online Resources

A list of online resources is as follows. Please note that this is far from a complete list.

- International Association of Crime Analysts: http://www.iaca.net
- Massachusetts Association of Crime Analysts: http://www.macrimeanalysts.com
- International Association of Law Enforcement Intelligence Analysts: http://www.ialeia.org
- Police Foundation: http://www.policefoundation.org
- Crime Mapping and Analysis Program: http://www.crimeanalysts.net
- Mapping and Analysis for Public Safety (MAPS): http://www.ojp.usdoj.gov/nij/maps
- Office of Community Oriented Policing Services: http://cops.usdoj.gov
- ESRI Mapping Center: http://mappingcenter.esri.com
- The Omega Group: http://www.crimemapping.com and http://www.theomegagroup.com

■ Conclusion

This chapter introduced you to the world of crime mapping and analysis. A discussion of the basics of crime mapping, along with a brief history of how crime mapping has evolved, was provided. Various types of analyses, data, and maps were explained, and outside sources were recommended. The following chapters in this section provide discussions of criminology and the explanations of crime which incorporate or are founded upon spatial aspects. Section Two examines the past and present research in crime mapping and discusses the common types of data that are used in crime analysis and mapping. Section Three contains chapters focused on the types of analyses that analysts perform in a variety of situations. Section Four, the final section of this book, discusses key points to consider in creating crime maps and performing crime analysis and also addresses future issues in crime mapping and analysis.

■ Recommended Reading

Charles Swartz at the Center for Applied Studies of the Environment, City University of New York, maintains a rather extensive list of references related to crime mapping and analysis. To view this list, visit: http://www.geo.hunter.cuny.edu/capse/projects/nij/crime_bib1.html

For a good discussion of map projections, visit: http://www.colorado.edu/geography/gcraft/notes/mapproj/mapproj_f.html

Johnson, M. R. (2000). Applying theory to crime mapping. *Crime Mapping News, 2*(4), 5–7. Retrieved January 2008, from http://www.policefoundation.org/

Kelly, J. (1999). MapInfo helps take a byte out of crime. *Crime Mapping News, 1*(4), 5–7. Retrieved January 2008, from http://www.policefoundation.org/

Levine, N. (2000). CrimeStat: A spatial statistics program for the analysis of crime incident locations. *Crime Mapping News, 2*(1), 8–9. Retrieved January 2008, from http://www.policefoundation.org/

Levine, N. (2005). CrimeStat III. *Crime Mapping News, 7*(2), 8–10. Retrieved January 2008, from http://www.policefoundation.org/

Nelson, L. (1999). Crime mapping and ESRI. *Crime Mapping News, 1*(4), 1–4, 8. Retrieved January 2008, from http://www.policefoundation.org/

Osborne, D. A., & Wernicke, S. C. (2003). *Introduction to crime analysis: Basic resources for criminal justice practice.* New York: The Haworth Press.

Questions for Review

1. What is the difference between micro- and macro-level theories of crime?
2. What is the problem analysis triangle?
3. What is crime analysis?
4. How does crime analysis differ from criminal intelligence analysis? From criminal investigative analysis?
5. Briefly define and provide an example for strategic crime analysis, tactical crime analysis, and administrative crime analysis.
6. What are projections and coordinate systems?
7. What are the four types of map data? Provide an example of each.

Chapter Glossary

Administrative Crime Analysis Administrative Crime Analysis involves the presentation of key findings of crime research and analysis to audiences within law enforcement, local government, and citizenry based on legal, political, and practical concerns.

Cartography The science of mapmaking.

CPTED The acronym for crime prevention through environmental design. Refers to strategies to reduce crime that incorporate making changes to the physical environment to limit the opportunity for crime to occur.

Crime Analysis Several good definitions exist for crime analysis, but essentially, crime analysis is the study and analysis of crime and crime-related factors in efforts to inform and develop strategies to reduce crime and the fear of crime.

Criminal Intelligence Analysis Criminal intelligence analysis typically looks at organized criminal activity and seeks to link people, events, and property.

Criminal Investigative Analysis With criminal investigative analysis, the focus is on serial criminals. Victim characteristics and elements of crime scenes are studied to discover patterns that link related crimes together.

Criminology The study of crime and criminal behavior.

Geocoding In geocoding, street addresses and other geographic reference points are positioned onto a map using latitude and longitude coordinates for computer mapping and analysis.

Geographic Profiling In geographic profiling, a profile of the offender, based on the nature of the crime and the facts of the case, is developed.

Hypotheses Statements derived from theory that can be tested to either support or disprove a theory or its assumptions.

Macro Macro-level theories of crime focus on societal-level variables, including the structure of government and the economy and how these variables impact crime rates within a society. Macro-level theories could focus on city-, state-, country-, or global-level influences.

Micro Micro-level theories of crime focus on individual characteristics, including (but not limited to) IQ, mental state, temperament, biological characteristics, and personal finances, and how they influence a person's decision to commit a crime.

Police Operational Analysis Police operational analysis focuses on the operations of police agencies, including staffing and resource deployment.

Projection The method used to transfer locations on the Earth's surface to a flat map.

Scale The scale of a map indicates how miniature the representation is. The larger the scale, the smaller the area shown on a map.

Strategic Crime Analysis Strategic crime analysis involves the study of crime and other law enforcement issues to identify long-standing patterns of crime and other problems and to assess police responses to these problems.

Tactical Crime Analysis Tactical crime analysis examines recent criminal events and potential criminal activity by analyzing how, when, and where the events occur to establish patterns and series, identify leads or suspects, and to clear cases.

■ Chapter Exercises

All of these exercises were developed using Microsoft Excel 2000/2003 and ArcGIS 9.2 with Service Pack 6 installed. Although many of these exercises will work correctly with ArcGIS 9.3 and Excel 2007, you may need to look in a different location for tools, buttons, or functions described in the exercises.

The chapter exercises created for this textbook were developed by Bryan Hill, a crime analyst for the Glendale Police Department in Glendale, Arizona. The ArcMap 9.2 exercises in this section are basic, practical exercises for flexing your GIS muscles in a practical and operational setting. The exercises will take you through administrative, strategic, and tactical exercises that many crime analysts are faced with during their careers. Hopefully, with these exercises, *and lots of practice*, you will find ways to accomplish GIS crime analysis tasks through some standard geoprocessing tactics. The format of the lessons assumes that you are a current, working analyst or are trying to become one, and you have received a request for GIS products. The data used comes from the City of Glendale, Arizona, but it has been sanitized. Any names or addresses found in the data resemble real data, but they have been created to avoid any relationship to the real world. With some intuitive thinking, you should be able to substitute your own data in these exercises where needed.

Each exercise follows what you are often asked to do as a crime analyst in the field. Enjoy—be patient! It can take several months to several years to become a proficient GIS analyst of crime. Studying, practicing, and reading about crime mapping will help you attain great skills. They won't get there by osmosis; it takes work. Think of it as trying to learn four new languages all at the same time: (1) geography, (2) GIS and spatial thinking, (3) crime theory, and (4) the ArcGIS software itself. These are all sciences and arts among themselves, and so the world of crime mapping does not come into view quickly; it takes patience and, at times, stubbornness to become proficient.

The texts in the following list contain working copies of ArcGIS that you may find useful for additional practice and reference. The first one contains a 180-day trial version of ArcMap, which is needed to complete these exercises. You may also order an evaluation copy (which usually expires 60 days after it is installed onto your computer) from the ESRI Web site (http://www.esri.com/software/arcgis/arcview/eval/evalcd.html). In addition, ESRI offers a 1-year student license for about $100.

- *Getting to Know ArcGIS Desktop, Second Edition Updated for ArcGIS 9*. ESRI Press, 380 New York Street, Redlands, CA 92373-8100; Copyright 2001, 2004; ISBN: 1-58948-083-X.
- *GIS Tutorial, Second Edition: Workbook for ArcView 9*. ESRI Press, 380 New York Street, Redlands, CA 92373-8100; Copyright 2007; ISBN: 978-1-58948-178-7.

If you get stuck on any of the exercises in this tutorial, refer to either of these two manuals and work out the problems. These exercises will assume you have some basic knowledge of crime analysis (found throughout the text of this book) and a small degree of familiarity with ArcMap (the exercises do contain very basic, introductory instructions to familiarize the student with ArcMap). If you can create a new map from the data you have available to you, you should do very well. These exercises should be simple enough for any new user to develop skills in GIS with practice. For extra help, various screen shots and other figures have been created. They also include examples of finished maps for comparison to see if you have correctly completed the exercise. They can be found on the DVD that came with this book.

Exercise 1a: The Data DVD

Data Needed for This Exercise
- ArcGIS desktop software (evaluation copy is fine, see the previous introduction to the chapter exercises).
- Everything else is included on the DVD and is a self-extracting zip file.

Lesson Objectives
- Remove the DVD from the sleeve and install the course exercises and data as needed.

Task Description
- Place the DVD in the DVD drive on the PC that has ArcGIS installed.
- Run the CIA.exe file. This will create a folder called C:\CIA on your PC and copy all of the data for the class into the appropriate folders.
- Browse and begin using this data.

1. Install ArcGIS 9x onto your computer. Make sure to also install the most recent patch for the version of ArcGIS that you own.
2. To begin using this data, copy it from the DVD to the local C:\ drive of the computer where ArcGIS is installed and where you are going to do the exercises. Remove the DVD from its sleeve and insert it into the DVD drive in your personal computer (PC).
3. If the DVD does not automatically run when inserted, open Windows Explorer and double click the CIA.exe file in the DVD directory.
4. It will take several minutes to install and extract the files needed for these exercises, so please be patient, and do not use your PC while the installation is being performed.
5. When the self-extracting process is completed, you will need to open Windows Explorer to browse the data in the folder. After the data is completely loaded onto the C:\ drive, right click on the C:\ CIA folder, choose **properties**, and in the dialog menu, make sure that all of the data in the folder *is not* read only.
6. This is accomplished by **unchecking** the box next to read only and *apply it to all subfolders*. Data stored on CDs and DVDs will

always be converted to read only, so to make edits to this data you will need to make sure you do not forget this step.
7. You are ready to start working with this data.

Exercise 1b: ArcMap—A Trip Around Glendale, Arizona

Data Needed for This Exercise
- All data is included in the ArcMap document (EX1b.mxd).

Lesson Objectives
- Open an ArcMap document.
- Use the Standard toolbar to browse data in ArcMap.
- Discover the basic elements of an ArcMap window.

Task Description
- The purpose of this exercise is to introduce you to the ArcMap interface and the Standard toolbar. You will learn how to zoom in and out on various regions within the document. You will also learn about some settings for layers that allow them to be turned on and off automatically, selecting records, measuring distances, and using the Identify tool to view data behind the points, lines, and polygons on the map display.
- You will discover where the table of contents, map display, toolbars, status bar, and other parts of the ArcMap interface are located.

1. To begin this exercise, open ArcView/ArcMap and a premade ArcMap document.
2. To accomplish this, click on the **Start menu → Programs** and find the ESRI or ArcGIS program folder. Double click on the icon for **ArcMap** to open the program. When ArcMap opens, choose **An existing map** and **Browse for maps**.
3. Find the exercise 1 file at C:\CIA\Exercises\Exercise_1\EX1b.mxd.
4. The file extension for ArcMap documents is .mxd. Like Microsoft Word (.doc) or Microsoft Excel (.xls), the .mxd extension stores all of the settings, legend items, location of windows, source locations of data (*but not the data itself*), and toolbars for the mapping project. It saves these settings when it is closed and will reopen and appear as it was when last saved and exited. This means that whatever queries, joins, and window positions the project was set to when they were closed will reappear exactly as before. The MXD file does not save the actual data for each theme in the

table of contents within the project, *but it does store the path to that data.* When the EX1b.mxd file opens, you will see an ArcMap document window that has the table of contents on the left, map display on the right, status bar at the bottom, and toolbars docked to the top, bottom, center, and possibly floating or in other locations (see DVD Figure 1–1).

Do not worry if it does not exactly match or if your toolbars are not located where they are shown in this image. Toolbars can be moved at any time, and their positions are saved when ArcMap closes on your PC. The table of contents (TOC) should look very similar to the table of contents in this image because the TOC is saved for each project.

5. The table of contents displays the themes or layers of data that are linked for this project. It also shows the types of markers, line symbols, and polygon outlines and fills that will be shown in the map display.
6. The map display is the graphic representation of the data that is listed in the table of contents. This can also be called the "drawing canvas" and basically shows you what is visible and how it would look if you printed this project "map."
7. The Standard Menu toolbar has commonly seen functions in all Windows types of applications and programs. This includes such things as File, Edit, View, etc. It also includes such items as adding data and layers, viewing the attribute table of a theme, or adding additional toolbars.
8. The Standard Display toolbar contains tools that you can use to browse around the map display and allows you to do such things as zoom in, zoom out, pan, identify an item on the map display, measure distances, and show hyperlinks.
9. The Draw toolbar gives you some tools that allow you to draw rectangles, circles, points, odd-shaped polygons, and add text as graphics to the screen. Think of these tools as the representative of the old standard wall map. They are simply graphics and contain no data or attributes behind the graphic item, unlike a theme or layer in the table of contents, which can conceivably store everything there is to know about an incident. Graphics can be useful for quickly portraying graphic information on a map and allowing a lot of freedom in adding individual pieces of annotation (text) to the map display. In an emergency response situation, the Draw toolbar is a lifesaver and allows you to draw a lot of things on the

map display without having to create a sophisticated attribute table behind it. Later, when you have more time, you can create the shapefile or data that makes the information more permanent.

10. The View Choice buttons at the bottom of the map display window will allow us to move back and forth from the map display view and the layout view of the data.

11. The Display view can be thought of as the place where we get everything to look like it needs to look to tell the story that we wish to tell.

12. The Layout view is where we create a printable map that has a legend, scale bar, north arrow, title bar, and other items that make it a good cartographic representation of the world and story we wish to tell. We generally print the map from the Layout view. The Layout view has a different set of standard tools than does the Display view, and you need to get used to recognizing the difference between the two toolbars (see DVD Figures 1–2 and 1–3).

13. Both tools control what you see; however, the Display toolbar allows the user to browse, select, zoom in, and identify various parts of the geography, or surface of the Earth, that is being examined. The Layout toolbar allows you to zoom in and out of the map you will eventually want to print and change the layout template being used. You could think of the Display toolbar as a device that allows you to interact with the map, or the graphic representation of the Earth's surface, and the Layout toolbar allows you to move around the printable "copy" of the map just to enhance the ultimate, printed version.

14. The Selection tab at the bottom of the TOC allows us to set various themes of data as either selectable or not selectable. When the items in the Selection TOC are bolded, we know that records within that theme are selected.

15. There are four buttons on the Display toolbar that allow you to zoom in and out on the map display area. These are the first four buttons on the toolbar. The ones that look like small magnifying glasses are Interactive Zoom tools, and the ones that have arrows moving inward or outward are the Standard or Static Zoom buttons. When you click on one of the Standard Zoom buttons, the map display will either zoom in or out a standard distance with each click. This is normally 200% of the current view. It also zooms in and out from the center of the current map display or where

you clicked. The Interactive Zoom (magnifying glass with + sign) tools allow the user to either click on the map display to zoom in or out 200% on the point clicked, or the user can click and hold the left mouse button down and draw a box. When you let up the left mouse button, the map will zoom to the area within or outside the box (depending on whether you are zooming in or out; see DVD Figure 1–4).

16. Use the Interactive Zoom In tool (magnifying glass with "+" sign) to draw a box around the Gateway patrol division that is bounded by the labels on the map between Glendale Av to Campbell Av and N 43rd Av to N 71st Av. Notice that as you zoom in, new data and detail is presented. This is due to some scale settings that were set for these layers when the project was saved. Continue zooming in on the police station, which is located just south of Glendale Av and N 57th Dr. Then use the Interactive Zoom Out tool (magnifying glass with "−" sign) and zoom out to the city limits again.

17. Practice zooming in and out with the four different tools until you are comfortable with their use and function.

18. There are also three buttons on the toolbar that will help if for some reason you either zoom in too far or out too far when utilizing the zoom tools (see DVD Figure 1–5).

    Often, the student will zoom in so far that there are no graphics in sight (kind of like having your nose stuck in the dirt). Occasionally they will also zoom so far out that they see a blank white screen and no map (kind of like flying to the center of the universe and trying to look back at the Earth from 900 million miles away). The tiny world symbol will zoom into the extents of all of the layers in the table of contents. "Extents" is just a fancy name for the limits the data extends across the Earth's surface or the outer regions of the data. Push the Extents button to see how it zooms out so that all the data in the table of contents is visible in the map display window.

19. Near or below the world symbol is a blue arrow pointing left and a blue arrow pointing right. These arrows allow you to go back to the previous zoom level within the current ArcMap session or forward to a zoom level already viewed, respectively. Click on these two arrows once or twice to see how you can go back to the zoom level you were at.

20. In case you make a mistake and zoom in too far, you can use the Zoom to Previous Extents (left arrow) to get back where you were in one or more steps. The right arrow allows us to zoom forward

through the steps. Practice with the World, Left Arrow, and Right Arrow extents tools and see that they let you return to previous views or go back to the entire extents of the map as needed. If either of these tools is grayed out, it means that you have not zoomed in and out on the map in this session of ArcMap. Generally, ArcMap keeps about 16 levels of zoom in memory during the current work session with the ArcMap project. This is more dependent on the amount of memory your PC has than a limitation within ArcMap. It *does not* preserve these zoom levels when you close the project.

21. Bookmarks can be used to keep zoom levels that you often need to go to time and time again.

    Go to the **View** menu item, find **Bookmarks**, and click on **Home**, **Foothills**, **Downtown**, and **Gateway** views one at a time. Notice how these bookmarks take you to saved extents.

22. To create new bookmarks, you simply zoom into an area, get it exactly as you wish, and then click **View → Bookmarks → Create** and save the current view as a new bookmark. Now you are an accomplished ArcMap analyst. Zoom back into the area directly around the Glendale Main Police Station and then create a new bookmark called "Main Police Station" (see DVD Figure 1–6).

23. Another way to zoom to a specific area is to simply change the map scale. The map scale is shown in a small white box on the menu bar that displays something like 1:245,789 (your scale may show something entirely different). By clicking in this window and typing *1:24,000* you will zoom into that scale on the center of the map display. You can change this interactively or click on the drop-down arrow to the right of the box and pick a scale from the list. You can also add scales to this list by clicking on **Customize This List** if you find that a certain scale always works well with your data. Again, these settings are only preserved in this MXD file and will not be there if you create a brand new MXD file.

24. The Pan tool is a small, white hand found between the four zoom tools and the extents tool (see DVD Figure 1–7). The Pan tool allows you to drag the map left, right, up, or down to center the parts that you wish to be in your final map. Practice moving around with the Pan tool. You should also have noticed by now that the tool you choose remains the active tool until you change it. If you have selected the Interactive Zoom tool and keep clicking on the map, you keep zooming. This is a common mistake among new users of ArcMap. After using a specific tool, you need to go

back to the default Black Arrow (graphic select or select elements) tool when browsing the data with your mouse. This keeps you from zooming in or out, panning, or doing something you really had not intended to do. *Another great piece of advice is when you are working in ArcMap, it is always wise to stop your work and save your project often. Because we are not making any changes to this project, we don't need to worry about that, but when you work on your own projects, you can save yourself hours of misery by saving your project frequently.*

25. The Select Features tool (the white arrow with the blue box by it, approximately the ninth tool in the toolbar) allows us to select graphics on the map by drawing a box around the area we are interested in (see DVD Figure 1–8). Before we use this tool, however, we need to tell ArcMap which layers or themes of data located in the table of contents we wish to select from. At the bottom of the table of contents you will see three tabs (Display, Source, and Selection). Click on the **Selection** tab and scroll through the list of themes listed. You need to make sure that both "Grids" layers are the only themes with a check box in the box next to their names. This means that when you do select functions on the map display, only the Grids layers will be affected. You will notice that there are two Grids layers listed under the boundaries grouping. These Grid layers are set up to display differently depending on the scale.

26. Go back to the Display tab and expand the Boundaries group by clicking on the **plus sign** next to Boundaries. You will see that the second Grids layer listed is grayed out in the table of contents. This is because the project has been set up so that this Grid layer only turns on when you zoom in to 1:99,999 map scale. Go up to the map scale or use the zoom tools to zoom in to any area of the map that is at the 1:99,999 scale or less. You should now see thicker red lines, which represent the quarter square mile grid boundaries. You should also observe that in the table of contents, the top Grids layer is now grayed out, and the bottom one is now visible. Now right click on the second **Grids** layer, and choose **Properties** from the pop-up menu. Now click on the **General** tab, and at the bottom you will see where the turn on/off scales are set for each theme.

27. The Properties dialog allows us to set how transparent the theme is, what label properties and symbol properties we want to use to display the data with, and a variety of other things. We will learn more about these in later lessons.

28. Close the Properties dialog box and then click on the **Select Features** tool. By holding the left mouse button, draw a box around several of the grids in the current map display and then let go of the left mouse button. You should see light blue lines around these grids where the light red lines used to be. The only grids selected will be the ones you drew your box around (see DVD Figure 1–9).
29. Now right click on the second **Grids** (bottom) theme again, and this time choose **Open Attribute Table**. You will now see a table of all of the grids in this layer. At the bottom you will see that it says "Records (*n* out of 658 selected)." Click on the button that says **Selected**, and you will now see only those records that you selected with the Select tool. The attribute data and the map display are dynamically linked. If you choose records within the table, they will show with the blue selection graphics on the map display, and if you choose graphics on the map display, they will show on the attribute table as blue, or selected, as well.
30. Close the attribute table, and then go up to the menu bar.
31. Find the **Selection** menu bar choice and click on it.
32. Find **Clear Selected Features** in the new dialog menu. Click this item. The records are no longer selected in any theme in the project.
33. Practice selecting other grids in this theme, and notice that if you zoom out past where this layer is set to display (1:99,999), you cannot select any grids in it. Not only can you control which layers are selectable in ArcMap, but with the scale settings in the Properties dialog, you can turn layers on and off automatically at different scale levels and control what can be selected at those scales.
34. You can also see this effect with several of the other layers in the table of contents, including the street themes and many others. To view the data under the group headings, click on the plus sign to the left of the item in the table of contents, and it will reveal everything below it (similar to Windows Explorer).
35. The black arrow is the Graphic Select tool and is the default tool you should remember to return to when you are done using one of the other tools. This tool allows you to select graphics and delete or move them around the map display. It has no affect on themes or layers in the table of contents. We will learn more about this tool and how to use graphics in a future exercise.

36. The Identify tool allows us to click on an item within the map display and view the attribute information for that one item in a small window. When you click this tool, a dialog menu pops up, which first asks you to choose which layer contains the information you wish to identify.
37. In the menu bar choose **View → Bookmarks → Home**.
38. Turn on the **Miscellaneous Point Data** grouping in the table of contents by checking the box next to the theme. Expand the data tree for this theme in the TOC (click plus sign). You should see that there are data layers for hospitals and multihousing locations under this theme grouping, and both of them are turned on. Click on the **Identify tool**, and you should see the Identify window (see DVD Figure 1–10). Click the down arrow on the **Identify From** dialog section, and choose **hospitals** from the drop-down list. Now find one of the little red crosses on the map display, and click on it with the Identify tool. Find which hospital is covering the Utopia road label on the map display, and then find the Boswell Hospital in Sun City around 103rd Av and Sweetwater Rd. (*Hint: Use the street labels on the outside of the map display to help find these locations.*) After you have identified both of these hospitals, change the **identify from** layer to the **multihousing** layer, and find the northernmost multihousing facility in Glendale. What is the name of this building? _____. Right click on the name of a multihousing facility you have identified, and choose **Flash** from the pop-up menu. You may have to move the Identify window around to see the point on the map flash. Right click again and choose **Zoom to** from the pop-up menu. Using what you know about the Identify tool, find the name of the public facility that is due west of this alternative living complex. (*Hint: You will need to change the identify from layer to Public_Facilties and zoom into the Claire Bridge multihousing location first.*) The name of this facility is _____.
39. *Remember to change back to the Black Arrow tool when you are done.* Go to **View → Bookmarks → Home** to return to our Standard view of Glendale.
40. Turn off the miscellaneous point data theme grouping.
41. We are now going to measure the approximate length of the Agua Fria Freeway (Loop 101) as it traverses its way through Glendale and the surrounding region. We will measure the freeway from where it comes into Glendale around N 51st Av and Beardsley Rd down to where it ends south of Camelback Rd. (*Hint: Zoom out to

full extents and use the labels to help you and the table of contents to see what a freeway looks like on the map display, and use the Agua Fria Fwy View bookmark.) The Measure tool is the second from the bottom on the Display toolbar (see DVD Figure 1–11). If you look closely, you will see that it resembles a ruler with arrows pointing left and right on the top of it. We will use the default to measure a line (left icon in the pop-up menu). We need to set the distance units we want to display in the Measure tool's menu. To do this we click on the down arrow (far right) to the right of the pop-up menu, choose **Distance**, and change this to **Miles** (see DVD Figure 1–12).

42. Choose the Measure tool and move your cursor over the approximate intersection of N 51st Av and the Agua Fria Fwy. (This will give you practice finding a location you are not familiar with. However, in the real world, an analyst should be intimately familiar with the area he or she analyzes.) Click once with the left mouse button to anchor the starting point. Move along the freeway and click the left mouse button everywhere you see a possible direction change, and try to follow the freeway down to where it ends. You want to keep the measuring line as close to the freeway as possible. When you get to the end, you should have measured approximately 15–17 miles. To end the measurement, double click the left mouse button. You should have noticed that you had a segment length and a total length result showing in the measure tool window. Each time you click once, you anchor the Measure tool on the map, and it then measures the length of the distance from that anchor to wherever your mouse is (segment). When you click again, it adds that length to *the total length* and gives you a new value for the segment length.

43. There are many other ways you can use the Measure tool, including measuring the area of a polygon you draw or measuring actual line features in a layer. You can experiment with these other options on your own as you practice to become more proficient with ArcMap through daily use. There are many emergency response uses for the Measure tool, and we will learn about some tactical uses in future exercises.

44. The last tool we will discuss on the Display toolbar is the Hyperlink tool. When you set up the attribute information correctly when developing your GIS layer, you can use this Hyperlink tool to display Word documents, PDF documents, photographs, or even open other ArcMap projects and software. The most common use is to simply open pictures. Many analysts have used this function

to display graffiti to neighborhood groups for an area and even for court testimony or crime scene explanations. If you are asked to do this, you may want to invest in a software product that can make movies of your ArcMap screens and such as you click on the various hyperlinks instead of putting them on your computer for court. If you bring your PC with ArcMap installed to court on a gang trial, you may find it is tied up in court for a very long time, and you are out of work! A crime analysis unit should have a laptop available with all this software installed for this very purpose and for presentations you may be asked to make as well. *An alternative is to use the free ArcReader or ArcExplorer software from ESRI (http://www.esri.com) with a canned project for those applications in court, or use the Animation tools in ArcMap to create a movie of the information needed.*

45. Click on the Hyperlink tool (it looks like a lightning bolt in the Standard toolbar; see DVD Figure 1–13), and then move it over the Foothills patrol division. Click the **Foothills** division to see a photo. Do the same for the **Gateway** division to see a photo of the front of the station. The photos will open in another window to be viewed and closed independently of the ArcMap window. There are no limits to what you can do with hyperlinks; however, if you link to other programs, you will likely need to create a Visual Basic for Applications (VBA) or Visual Basic (VB) macro, which requires some programming skill and familiarity with one of those programming languages.

46. Now let's put some of these skills to good use with a real-life project. The chief has asked you for some information on a subdivision called Arrowhead by the Lakes. Citizens in this very upper-class neighborhood have been complaining to the city council that cars are being stolen from their neighborhood frequently. The chief wants you to take a preliminary look at the area to see if this could be true and then advise him. To perform this analysis, we need to add in some data.

47. To find the subdivision, we need to add the C:\CIA\geodatabase\Glendale_Data.mdb\Features\SubDivisions data into our ArcMap project. We want to find the button with the image of a yellow square with a plus sign in it (or go to File → Add Data) in the menu bar and click it to bring up a Browse dialog window. Find the folder in the drop down menu and the data file listed, and add it to the ArcMap TOC by clicking on it once and clicking the **Add** button.

48. The Subdivisions theme data has been added. Turn this on and open the attribute table for this theme (right click the theme, open the attribute table).

49. Find the Arrowhead by the Lakes subdivision and click on the gray box to the far left of it in the table to select this record. Minimize the table. Look in the map display, and you should see a bright blue line around a subdivision near Union Hills Dr and N 75th Av (DJ11). Right click the Subdivisions theme, and choose **zoom to layer** in the pop-up menu. As you can see, we zoomed in a little bit but not far enough. We want to zoom into that selected record. Take your Interactive Zoom tool and draw a box just barely around the Arrowhead by the Lakes subdivision to zoom into it (see DVD Figure 1–14). You may need to scroll vertically or horizontally to make the complete subdivision visible.

50. Now we need to add in the stolen vehicle events and maybe a theme that shows the auto theft suspects who live in this neighborhood.

51. We will add two more themes as follows: (Again, do this by using the Add button. Notice that upon clicking the Add button, in the dialog box ArcMap takes you to the last place you added data from. If you do not need data from this place, simply use the drop-down enu or the icons to take you to the folder you need.)

    C:\CIA\geodatabase\Glendale_Data.MDB\Offense and CFS Data\Stolen_Vehicles

    C:\CIA\geodatabase\Glendale_Data.MDB\Persons_Data\ AutoTheft_Persons

52. Turn on one of these themes at a time. Count the number of points manually within the Arrowhead by the Lakes subdivision (if you can), or open the attributes tables for each theme to determine the number of records.

53. There are 29 stolen vehicles within the subdivision and 50 records of auto-theft-related persons living in the area. This does not mean there were 50 different subjects because this data comes from records of stolen vehicles and includes witnesses, victims, suspects, etc. We want to see if any suspects live in this area, however. Make sure that the AutoTheft_Persons theme is turned on (check the box next to the theme in the TOC) and the Stolen_Vehicles theme is not. Right click the **AutoTheft_Persons** theme, and choose **Properties** from the dialog box. Choose the **Definition Query** tab. Inside the Definition Query box type this exactly as shown: *[INVCODE] = 'S'* (the "S" in this table stands for "suspects").

54. Click the **Apply** and **OK** buttons to make this definition query active. Now you should see only two points inside the Arrowhead by the Lakes subdivision.

    Click on the **Identify** tool, select the **AutoTheft_Persons** theme as the "identify from" layer, and write the addresses of the two suspects here _____ and here _____. Close the table.

55. Select those two suspects by choosing the **Select** tool and drawing a box around just those two points at the same time.

56. You should have noticed that when you let go of the mouse it selected those two points and made them bright blue.

57. Open the attribute table again for the **AutoTheft_Persons** theme, and click on the **selected button** to view the data about these subjects.

58. Turn off the **AutoTheft_Persons** theme and turn on the **Stolen_Vehicles** theme. Use the **Select** tool to select the stolen vehicles within that subdivision (as close as you can get). View the attribute table and sort several of the fields to get an idea of the type of cars being stolen in the subdivision. It appears that Hondas and Fords are stolen most often, but what is the most frequent color of stolen vehicles?_____. What year of vehicles are stolen?_____. What style of vehicles are stolen? _____.

59. Click on the **Selection** tab at the bottom of the TOC, and you will find that a few layers are bolded and in parentheses. After the bolded theme name, it shows the number of records currently selected within those themes. Right clicking within the white portion of the Select TOC, next to the group or theme, gives you a pop-up dialog that contains a **Clear Selected Features** option.

60. Click this and see that it clears all selected features in that theme area or group. You will need to clear features from every group and area, or you can go to Selection in the menu bar and choose **Clear Selected Features** from that pop-up menu, which will clear all themes. View the map display and see that the bright blue lines are now gone.

61. In the next exercise we will learn about ArcCatalog and how to add themes from it to ArcMap.

62. This concludes Exercise 1b.

63. Please close the project and *do not save* edits.

Exercise 1c: Arc Catalog Part I

Data Needed for This Exercise
- All data is included in Arc Catalog.

Lesson Objectives
- Open ArcCatalog.
- Browse data in ArcCatalog.
- Discover the basic elements of ArcCatalog.

Task Description
- This exercise introduces you to ArcCatalog and shows you how to recognize different types of data that can be used in a GIS software application and how to arrange, view, and find data for your crime analysis project.

1. ArcCatalog is a companion software product that comes with ArcMap/ArcView. It allows you to organize, create, and manage your GIS datasets. You can copy, move, delete, search for specific data, or create shapefiles or personal geodatabases using ArcCatalog. ArcCatalog is made up of three general sections:
   - Catalog display: For seeing the data itself and browsing it
   - Catalog tree: For viewing what data exists on a PC or network
   - Toolbars: For performing various functions and procedures on your data

2. GIS data comes in a wide variety of formats, and ArcCatalog displays these data formats by using different icons or symbols in the catalog tree (see DVD Figure 1–15). The small green rectangles represent shapefiles, and different symbols on the green rectangle represent points, lines, and polygons, which are the main three types of GIS data used in ArcMap.

3. We can also use raster data (like aerial photos), ArcInfo coverages, tables, and a lot of other types of data in ArcMap. Each item has its own icon. In addition to different formats for data, ArcCatalog also has a different icon to represent layer files, which don't actually contain data but contain instructions for how data from a shapefile or geodatabase should display and what markers, line styles, polygon fill colors, etc., should be used when viewing that data.

4. Open ArcCatalog by going to the **Start menu → Programs → ArcGIS → ArcCatalog** and launching it from there.

5. ArcCatalog should open, and you should be looking at the application window that resembles the following. Your actual application window may have totally different folders and connections depending on the computer that you are using, so if it is not exactly like the one in DVD Figure 1–16, do not panic. The default in ArcCatalog is to return to the last used location, and that will affect the connections that are visible in your application window (see DVD Figure 1–16).

6. In the catalog tree at the left you should see the local drives, any network drives that have been connected on this PC, as well as databases and Internet servers or other services this PC is connected to. To make it easier to get to the data in the CIA folder, we are going to create a specific connection to that folder so we can spend less time hunting down our data for future projects. Go to the **Tools** menu in the main menu bar and choose **Options** from the drop-down dialog (see DVD Figure 1–17).

7. Uncheck the **Hide File Extensions** box if it is not already unchecked, and click **OK**. This Options dialog allows you to change how any data viewed with ArcCatalog will appear, what data will show up, and several other settings that will affect how data is viewed in ArcCatalog.

8. Now, using the Standard Catalog toolbar, choose the **Connect to Folder Button** to create a new connection to the CIA folder (see DVD Figure 1–18). When you click this button, a Connect to Folder dialog box opens (see DVD Figure 1–19), and you will need to browse the folder structure to find the C:\CIA folder. Keep in mind that your display may be different due to the computer you are using, but you should still have a C:\ drive listed. Browse through the root directory of C:\ until you find the **CIA folder**, and click on it once. After you have clicked the **OK** button, the dialog disappears and a new connection to the C:\CIA folder is created in the catalog tree (it shows up on the left side).

9. Browse through the data with ArcCatalog and look at the different icons that are used to display GIS data in ArcCatalog. A bumpy yellow rectangle is used for aerial images; a green rectangle with dots is used for point shapefiles; a green rectangle with lines is used for line shapefiles; and a green rectangle with white shapes is used for polygon shapefiles. You should also see small table icons under the lookup table's folder as well as many others.

10. Now find the folder called **Geodatabase**, and double click it to show all the personal geodatabase features that are in the CIA

folder. You can also click the plus sign in the catalog tree to show all of the feature classes within the geodatabase. You will notice that many of the symbols are like the shapefile symbols, except the rectangle behind them is gray instead of green (this signifies a geodatabase).

11. In the Standard toolbar (see DVD Figure 1–20) there are several tools to display the data with details, as icons, or a combination of both. Experiment with these four methods of displaying information about your GIS data. The last tool on the right will display large icons of the various GIS datasets. If a thumbnail has already been created for that GIS data, an image of what the data actually looks like will be visible on the icon shown in DVD Figure 1–21. To add a thumbnail for a GIS layer that does not already have one, you need to select the layer in the catalog tree on the left. For this exercise, let's choose the CIA\GIS_DATA\Calls4Service\ CALLS_FOR_SERVICE.shp data in the catalog tree. Now make sure the large icons tool in the Standard toolbar has been selected. You should now see a green rectangle with dots in it that represents this shapefile. With this shapefile still selected, click on the **Preview** tab in the catalog display window to the right to show a preview of the points in the file. In the Geography toolbar, click the **Create Thumbnail** tool (see DVD Figure 1–22). When this tool is clicked, a thumbnail is created for this shapefile. Now return to the **Contents** tab to see the results of your work. If this layer already has a thumbnail created for it, just pick a different file/layer to create one for.

12. In some cases the thumbnail may be enough to select the data you wish to use in ArcMap. In other cases you need to know a little more about the data or be able to see more of it to be able to choose the right GIS data to use. Find the C:\CIA\GIS_DATA\ BOUNDARIES\ folder and pick the AZ_CITIES.shp file. Now click the **Preview** tab in the display section again to see the full image of the Arizona cities polygon theme. Use the Zoom tool (magnifying glass with plus sign) to draw a box around the general vicinity of Glendale to zoom into the data layer. Now use the Identify tool and click on **Glendale** as well as several surrounding cities to see that you can get a great deal of data from ArcCatalog before using it in ArcMap (see DVD Figure 1–23). Now click the little green world to zoom back out to the full extents of the Arizona cities shapefile. The Pan (little hand) and Zoom Out tools on the Geography toolbar are also useful for browsing the available data and is similar to those we experimented with in ArcMap.

13. As a side note for future exercises, to create a new shapefile or geodatabase for use in ArcMap, you have to create the structure of it in ArcCatalog first. If you are simply making a copy of selected records, this can be done from inside ArcMap or in ArcCatalog; however, if you are creating a brand new theme, layer, shapefile, or geodatabase file, you will need to first create the structure and the fields it will contain in ArcCatalog before you can add anything to it in ArcMap. In future exercises we will experiment with this. The majority of us will not have to regularly create new GIS data; however, at some point in your career, you will likely run into this problem.

14. You can also browse the attribute data behind the graphics within ArcCatalog. Find the Geodatabase folder again and choose the **GLENDALE_DATA.mdb** geodatabase. Find the Features group, and within that data category, find the **Neighborhoods** feature class. Click this layer and view the thumbnail of it under the Contents tab. Now click the **Preview** tab and view the polygon layer graphic. At the bottom of the preview section, click the down arrow near the Preview field and choose **Table** rather than Geography. You will now see the attribute table data behind the polygons (see DVD Figure 1–24).

15. Browse the data in the table view and then click on the **Metadata** tab. This will bring you to a new section where the information about the data should be collected on all of your commonly used GIS layers. This not only helps you to organize and find your data when you need it, but it advises you of such things as when it was last updated and what projection it currently has. This data is vital for sharing crime information among users within and outside of your department. You should seriously consider creating metadata for all of your GIS and related data so that it will be easier to maintain, share, and develop new uses for the data.

16. Browse the three tabs under the Metadata section. The blue section headings specify the information that has been entered about this data. The green sections are areas that can be clicked on and expanded to show more data about the information element (like field names and what can be contained within them, etc.; see DVD Figure 1–25).

17. If you are done with this exercise, you can close ArcCatalog. If you are going on to the next exercise right away, leave it open and proceed.

Exercise 1d: Arc Catalog Part II

Data Needed for This Exercise
- All data is included in ArcCatalog.
- C:\CIA\Exercise\Exercise_1\EX1d.mxd (blank project).

Lesson Objectives
- Open ArcCatalog.
- Browse data in ArcCatalog.
- Find data.

Task Description
- This exercise introduces you to ArcCatalog and shows you how to find data for your crime analysis project.

1. In addition to browsing your data in ArcCatalog, you can also search for data that is specific to the project you are currently working on. In this exercise we will learn how to search for data that matches a certain set of criteria that we will establish, and we will also search for data that covers the same geographic area as an area we are interested in. If you do not have ArcCatalog already open, please open ArcCatalog.

2. If you are moving forward from the last exercise, you should be on the C:\CIA connection you created in that exercise. If you opened a new ArcCatalog session, go to the C:\CIA folder connection.

    You will go to the Standard toolbar again, but this time click the **Search** icon (it looks like a magnifying glass over a file cabinet, see DVD Figure 1–26). The Search dialog menu should appear and will now allow you to search the connections for data that matches one of four criteria:
    - Name and/or location tab: You can search for a file name or file type
    - Geography search tab: You can search for data by its location on the Earth's surface
    - Date search tab: You can search for data based on the date it was last accessed or by other dates within the metadata
    - Advanced Search tab: Search for key words or other metadata contained with the file

3. When you do a search in ArcCatalog, it will create a brand new folder for each search you perform. You will also notice that when we are done, it creates a new connection to the data in the catalog tree. ArcCatalog does not replicate the data but just creates links to where the data is located on the local, network, or Internet

connections you have on your PC and, of course, where you told it to search.

4. For our first exercise, we will look for a photograph of one of our more active gang members for a Compstat meeting. A few minutes prior to the meeting the chief tells you that you will need to pull up his photo during the meeting, and you do not remember where it is located. You are pretty sure it is a JPG image, and the suspect's name is Jerry Aquilera; however, you think his first name could actually be spelled differently in the files and are not quite sure of the last name spelling. To find the file, you will search ArcCatalog for it. Click the tab that says **Name and Location**.

5. In the search window, rename the search name to EX1d (under **Save As** on the right). Type in the suspect's name under the name field using wild cards as follows—*A*U*JER*—and make sure the Look in field is C:\CIA, and then click the **Find Now** button.

6. It will take a few seconds for it to run and search the entire CIA folder, but in less than a minute it returns two photos of two different "Jerry" Aquileras. We can view the two files by going to the ArcCatalog tree and clicking in the EX1d search under the Search Results portion of the tree (at the bottom). To actually see the two photos, we can click the **Preview** tab on the catalog display area to view the two photos one at a time (see DVD Figure 1–27). When we have found the one that we need, we can simply bring it up again whenever we want to by opening ArcCatalog and clicking on the results of the search folder called EX1d.

7. Now we will see if we can find data that pertains to a specific geographic area within the City of Glendale that we can use. We want to limit where ArcCatalog goes to search for this data, and we want to specify beat 32 for the geography of the search. In the catalog tree, move up to the CIA folder and click on it once and then click the Search button again. In the Search dialog enter *Glendale*, and on the Name and Location tab and the Look In section enter *C:\CIA*.

8. This search gives us back every file in the C:\CIA folder that had "Glendale" in the name. Name this new search EX1d2 in the Save As section. That isn't exactly what we want.

9. Now click on the Geography tab.

10. We want to use one of our layers to do the search, so at the very bottom of the Geography search tab, we see MAP. In this field, click the down arrow to the right and choose **browse**. Find the

C:\CIA\Geodatabase\Glendale_Data.mdb\Features\Beats data and choose it (see DVD Figure 1–28).

11. You will now have the beats showing and a big red box around the entire city. Use the Draw tool to draw a box around just the lower right beat in the city (beat 32).

12. You should have noticed that we changed the default of "Find data entirely within location" to "Find data overlapping location" so that we would get results back. This search would not have provided any results if we had left it to within location because all of our data in the geodatabase location that would be searched would be for the entire City of Glendale and not just beat 32.

13. If you do not get any results, go back to the Name and Location tab and make sure it indicates you want to search the entire C:\CIA folder.

14. You should have numerous overlapping datasets for this area that now appear in the EX1d2 search folder. Close the search dialog.

15. To modify an existing search, right click the **EX1d2** search under Search Results, and then pick **Properties**. Change the Look In field to a new area you wish to search. Check the box that says **Find data entirely within location** and rerun the search query. You will probably not get back any results. This is because the data layer has to exist *only* within beat 32 or the box you drew and cannot be partially within it as most of our layers are.

16. Another useful piece of information is the difference between a shapefile or geodatabase and a layer file. A layer file is simply a file that you can save after you have set the display properties for a theme of some sort. A layer file contains what symbology should be used to display that data the next time you bring it into ArcMap or view it in ArcCatalog. Browse to the GIS_DATA folder and then to the Boundaries folder within it. Preview the GLN_BEATS.shp file and the GLN_BEATS.lyr file, and you can quickly see the difference. The layer file saves a quick link to the SHP file but retains the way in which it will be displayed in ArcMap when it is used next. Layer files do not store the data or duplicate it; they only store how to display the data it is linked to and how it will be drawn. This is a very easy way to standardize your maps and save time.

17. Now we will learn about how to drag and drop data layers and information from ArcCatalog onto the ArcMap application to create a new map. We will also learn the difference between Data Display mode, Layout mode, and data frames. This will also lead

us to the definition and description of the virtual page and how data frames are used with it or as part of it.

18. The next assignment will begin from ArcCatalog. Click on the **Open ArcMap** icon in the Standard toolbar (see DVD Figure 1–29).

19. When ArcMap opens, click the **An Existing Map** box and browse to the C:\CIA\Exercises\exercise_1 folder and open the EX1d.mxd file that is located there. Your map should have nothing in it. Position the ArcMap and ArcCatalog application windows so that you can see both of them on your desktop clearly. Parts of one may be behind the other, but this is okay. Make the ArcCatalog application window active (in front).

20. Now we want to add data from ArcCatalog to ArcMap. Find the C:\CIA\Geodatabase\Glendale_Data.Mdb\Features feature class, click on it with your left mouse button, and hold the mouse button down while dragging the entire feature class group from ArcCatalog to the ArcMap Display window or table of contents.

21. You can drag and drop any data from ArcCatalog to ArcMap, and you can also drag and drop entire feature class groups from ArcCatalog to ArcMap. This can be used for quickly creating a map in ArcMap. Just open a new ArcMap project and drag an entire feature class or group from ArcCatalog to the ArcMap display, and all of the data will now be part of the project. Use caution if you are going to pick a feature class group that has numerous types and formats of data because it may take a very long time to draw and render all of the data you dragged over.

22. You can also drag over layer files that are already set to display the data in the format you desire. Find the C:\CIA\Geodatabase\Layers folder. Go to ArcMap and turn off all themes that are currently in the project from the last drag operation. While turning these off, you should notice that they are all just some general color, and all features are represented as one graphic symbol for each record in the themes.

23. Go back to ArcCatalog and try to drag and drop the entire C:\CIA\Geodatabase\Layers folder onto the ArcMap project. It will not allow you to drop regular folders into ArcMap. Instead, open the folder, select all of the feature class names that end in .lyr, and drag and drop them onto the ArcMap project window. You should now see all of the layers displayed in their beautifully standardized symbols, colors, and styles. If these do not draw immediately, turn on each layer as needed.

24. Practice dragging and dropping data from ArcCatalog until you have a good grasp of how to do it.
25. Save your project, if desired, in your personal folder as MyEX1D.mxd, and close ArcMap and ArcCatalog.

Exercise 2: Graphics

Data Needed for This Exercise
- City boundary polygon.
- City street centerline.
- Beat polygon (if desired).
- Aerial photo of the area in question (useful but not required).

Lesson Objectives
- Open a new ArcMap document.
- Add data to the map display from shapefiles.
- Use the available GIS data to draw several graphics shapes to represent a citizen protest event and specific locations as directed.
- Label some of the graphics as needed to print a useful map.

Task Description
- The executive assistant chief (EAC) has entered your office and explains that he is going to a meeting within an hour and needs a map of the area between N 43rd Av and Bethany Home Rd to N 59th Av. He explains that a citizen group has requested a meeting with him to discuss a protest march they are going to be doing in one week. The plan is for them to collect at the southwest corner of N 43rd Av and W Bethany Home Rd and march along Bethany Home Rd to the Bonsall Park South (southeast corner of 59th and Bethany Home Rd) where they will have several speeches and a lunch for the protesters. The EAC wants a map showing the protest route and the total number of officers he will need for each intersection the protest will pass that may need traffic control (major intersections only; one officer on each side of Bethany Home Rd). He also wants the Special Events Unit to be stationed in at least the four corners of the park. The Special Events Unit will work the internal locations within the park, and he would like a marker within the park along the east side of the park, south of Bethany Home Rd to depict where the bandstand and special events squad will be mostly assigned. He would also like to have a line with an arrow on it showing the general path of the protest march from beginning to end within the park. He will be back in

20 minutes for one printed copy of the map, in color, and it only needs to be 8.5″ × 11″ in size, but he may need a much larger version at some later time.
1. Start the exercise by opening ArcMap from the Start menu.
2. At the dialog, check the **A new empty map** box.
3. Uncheck **immediately add data** if it is checked, and click the **OK** button.
4. There are a few ways to add data to the map:
    a. You can open ArcCatalog and drag the layers from it and drop them on your map.
    b. If your project crashes and cannot be recovered, try again, or load the March.mxd project from the C:\CIA\Exercises\Other folder to view a finished project as needed.
    c. You can click the **Add Data** button (DVD Figure 2–1) and then navigate to the location where the required data is located. I will use the CIA GIS course directory on my PC and add the following shapefiles from the C:\CIA\GIS_Data\ directory:
        i. Boundaries
            1. GLENDALE_CITY_BOUNDARIES.lyr
            2. GLENDALE_BEATS.lyr
        ii. Streets_Transportation
            1. GLN_SURROUNDING_STREETS.lyr
        iii. AerialPhotos
            1. CENTRAL_COLOR.tif; If you receive a "missing spatial reference" error message, you will need to redefine the projection for this file. To do this, click on the **ArcToolbox** icon (it looks like a little red toolbox) and select **Data Management Tools**, then **Projections and Transformations**, and then **Define Projection**. At this point, a dialog box will appear, and you will need to select this file (Central_Color.tif) as the "input data file or feature class" by using the dropdown menu. For the "coordinate system" click on the icon to the right to get a Spatial Reference Properties box. Select **import** and browse to the CIA\GIS_Data\Boundaries\Glendale_City_Boundaries.shp file. Click **Apply** and then **OK**, and you will see a coordinate system "NAD_1983_StatePlane_Arizona_Central_FIPS_0202_FEET" appear in the text box under "coordinate system." Select **OK**. This tells ArcMap to use the

same coordinate system for the Central_Color.tif file as the Glendale_City_Boundaries.shp file. Remember, it is important to use the same projections for all files included in a map.

   d. For more information about adding data, see Chapters 3 and 4 in the ESRI book entitled *Getting to Know ArcGIS Desktop*, which is described in the introduction to these chapter exercises.

5. After adding these layers, turn off the street layer because for the moment you do not need them to show, and it takes time for the streets to keep drawing. You need to zoom in to the extents of the aerial photo by clicking the Zoom tool (DVD Figure 2–2), and then draw a box around the aerial photo with it (see DVD Figure 2–3).

6. Make the beats invisible so that you only see the lines around them and not the colors. You can accomplish this by right clicking in the **Beats** theme in the table of contents (TOC) and then choose **Properties** from the drop-down dialog. Then go to the **Symbology** tab and click on **Features** to the upper left. Pick **Single Symbol**, and then find the word "Symbol" in the middle of the dialog and click on the small box that will take you to the symbol palette. When you are there, choose the **Hollow**, or no color, option and change the outline to a size of 3. Click **OK**. You may need to go back and change the outline color to black if it is another color. See DVD Figure 2–4 to compare your map to mine.

7. For more information about working with symbology in ArcMap, review Chapter 5 in the ESRI book entitled *Getting to Know ArcGIS Desktop*, which is described in the introduction to these chapter exercises.

8. The assignment we have been given is to quickly create a map showing these items requested by the chief, so dumb graphics are all we need to make this map quickly and efficiently. The first thing we need to do is find the starting intersection and the Bonsall Park area on the map.

9. To accomplish this, let's turn off the aerial photo, turn on the street centerline file, and use it to find the intersection of N 43rd Av and W Bethany Home Rd.

10. Right click on the **GLN_SURROUNDING_STREETS** layer and go to the **Properties** dialog. Click on the **Display** tab. At the very top is a check box that says "Show Map Tips (uses primary display field)." Check this box and then click on the **Fields** tab. In the top

of this dialog, there is a field called "Primary Display Field:" that we need to change to read *ANNAME*. Use the drop-down list to find and choose ANNAME from the list and click **Apply**. To close the Properties dialog menu, click the **OK** button.

11. Now, choose the Select tool (black arrow) and move it over the map. Notice the yellow map tips that now appear as you hover your mouse cursor over the street segments.

12. Save your map as March.mxd in your student folder.

13. Use these map tips to help you locate the N 43rd Av and N 59th Av and W Bethany Home Rd intersections. Zoom in to these areas so that only the path of the protest march shows in the map display.

14. Now let's load the city parks layer to find Bonsall Park South. Click on the **Add Data** button again. Find the GIS_Data\Other folder and the GLN_CITY_PARKS.lyr file. Load this into ArcGIS and move it around until it is visible and not hidden by any of the other layers (generally below the street centerline layer). You should have noticed that the order the themes are placed in the TOC makes a difference depending on if you are using point, line, or polygon features. Typically, aerial photos are all the way on the bottom, polygons are above that, lines above that, and points are on top. Because every theme draws from the bottom up, this keeps the information from covering other data.

15. Repeat the map tips process (step 13) with the Parks theme and name the primary display "Name."

16. We want to identify the Bonsall South Park because this is where the protest march will end with a barbeque and gathering. When you have found Bonsall Park, notice that there are two Bonsall parks. We want the one on the south side of Bethany Home Rd.

17. We need to draw the general path of the march. Go to the menu bar and find the menu item that says View → Toolbars → Draw and make sure there is a check next to the Draw toolbar item in the pop-up dialog. The Draw toolbar will typically be docked at the very bottom of the ArcMap window (see DVD Figure 2–5).

18. We need to click on the drop-down arrow near the white square (fifth from left on the draw toolbar) and find the Graphic Line Segment tool. It is in the middle of the eight tools and looks like a three-part straight line segment. Click on this tool and then click once at the intersection of 43rd Av and Bethany Home Rd to anchor the line. Now move your mouse west to 59th Av along

Bethany Home Rd and click again at N 59th Av to anchor it there. Now move it south and slightly east, and end it about the middle of Bonsall Park South. Double clicking will end the drawing of the line segment and leave your march path behind as a graphic with handles around it.

19. Right click on the graphic and choose **Properties** from the dialog box.
20. Choose the **Symbol** tab and make the line 6 points in size and bright red.
21. By experimenting with different symbol types you can change it to be dotted, dashed, or have arrow symbols in the line.
22. Now we need to add in the traffic enforcement units that will be present to block traffic at all of the main intersections along the path. The chief wants one officer stationed north of the march and one officer stationed south of the march along the entire route. To accomplish this, we will use the Draw tool once again, but this time we will add a point, which is the last tool in the same toolbox we got the Line Drawing tool from.
23. Add points along the intersections as needed, and change the symbology for each one to be a blue triangle. You will have to click on the Point tool after each click on the map because the tool deactivates itself after you have added each point.
24. To select multiple graphic points, select one, and then hold the Ctrl *or* Shift key down while clicking the others. Then right click any one of the points and choose **Properties** from the dialog box. Go to **Symbols**, then **Change Symbol**, and choose a triangle and make it about size 22 and blue.
25. Now we need to indicate where the Special Events Unit will be, and we need some text on the map to indicate what the bright red path is and what the blue triangles represent.
26. We will use another tool on the Draw toolbar for accomplishing this labeling. This is of course, the Label tool. It is the tool next (right) to the Draw Graphics tool that we used to add the line and the triangles. There are several label types we can add, but for our purposes we will choose the callout (center-left tool). Hover your mouse over the tools for tips on what they are.
27. Click on the Label tool, then put your mouse in the center of Bonsall Park South, and click and hold the left mouse button while dragging down and right. Let go of your mouse, and a box will pop up with the word text in it. Type *Protest March Rally End*

Point Special Enforcement Unit Responsibility and press **Enter** to stop adding text. The text will now appear in a yellow box and have a tail pointing back to the park. To wrap the Special Events Unit text to a new line, right click on the text, choose **Properties** from the pop-up menu, and go to the text. Place your cursor after "Point" and press **Enter** to move it down a line. Then click the **OK** button to finish.

28. Create two new text boxes that say *Blue triangles represent traffic enforcement units at each intersection along the protest march path*, and *Path of March*.

29. Now we want to label a few of the streets to get an idea of where this area is within Glendale. Another Label tool that looks like a sales tag (the upper right tool in the Text toolbox) allows us to add labels from the primary attribute field of a layer.

30. First, we need to make sure that the primary field name for our centerline file is set to ANNAME. Right click on the **GLN_SURROUNDING_STREETS** theme, choose **Properties** from the pop-up menu, and go to the **Fields** tab. Make sure the Primary Display Field shows ANNAME. If it does not, change it, then click **Apply** and **OK**.

31. Now click on the Label tool, and a new dialog menu should pop up (See DVD Figure 2–6).

32. Choose "Place label at position clicked", and choose a style on this pop-up menu.

33. Scroll to the bottom of the styles, and choose the banner rounded or banner style.

34. Now move this menu out of the way, and very carefully click on one of the street segments south of Bethany Home Rd on N 43rd Av. This is kind of tricky, but when it works it is great. You may get beats or some other layer labeled instead of the street segment, but with time and patience you can get it to work. I find that turning off all other themes except the one I am trying to label helps, and turning off the map tips when they are present can also help.

35. Label N 59th Av, N 55th Av and Bethany Home Rd at a minimum.

36. Don't forget to return to the black arrow (Graphic Select tool) when you are done using any other tool.

37. Save your project (March.mxd) in your student folder.

38. See DVD Figure 2–7 to compare your map to mine.

39. By changing the size of text, color, etc. using various symbols (right click → Properties → Change Symbol) you can create a wide variety of symbols and make the map more interesting and hopefully clearer.
40. Now we need to send this map to a printer. We could just send this map display to a printer as is, but this is not always a great way to do things because we want to make our maps look professional. This means we need to include such things as north arrows, titles, disclaimers, scale bars, legends, and anything else we feel would enhance the understanding of the map. An easy way to do this with only a few issues is to use a premade map template.
41. *Save your project.*
42. At the bottom of the map display is the Layout View button, which looks like a piece of paper left of the bottom scroll bar. Click this.
43. Now you have a map in portrait format (probably) that doesn't show much except for the map. Because this area is more landscape in form, we want to choose a template that is landscape and has some of the key items we want.
44. In the Layout toolbar, the far right tool is the Change Layout tool, which allows us to use a premade template for our map (see DVD Figure 2–8).
45. Click this tool, and in the dialog menu that comes up, choose **General tab → LetterLandscape.mxt**.
46. Click the **Finish** button at the bottom of the dialog menu.
47. Using the mouse, grab the map and resize it using the little blue handles that appear in each corner when you click on the map.
48. Double click the text sections and type in text as shown in DVD Figure 2–9.
49. Move the legend, north arrow, and scale bar to the left so it is not hidden by the map. To compare my final map to yours, see DVD Figure 2–9.
50. If for some reason ArcMap locks up when you choose a template, simply close everything, reopen ArcMap, and do steps 45 through 52 again. ArcMap has been known to do this a lot in version 9.1, but it may be improved or reduced in version 9.2.
51. *Save your project again. Remember to save often!*
52. Close ArcMap and any other applications you currently have open.

References

Boba, R. (2001). *Introductory guide to crime analysis and mapping.* Washington, DC: COPS, USDOJ.

Boba, R. (2005). *Crime analysis and crime mapping.* Thousand Oaks, CA: Sage Publications.

Brantingham, P., & Brantingham, P. (1981). *Environmental criminology.* Beverly Hills, CA: Sage Publications.

Brantingham, P., & Brantingham, P. (1993). Environment, routine, and situation: Toward a pattern theory of crime. In R. Clarke & M. Felson (Eds.), *Routine activity and rational choice: Advances in criminological theory* (Vol. 5). New Brunswick, NJ: Transaction Publishers.

Brantingham, P. J., Dyreson, D. A., & Brantingham, P. L. (1976). Crime seen through a cone of resolution. *American Behavioral Scientist, 20*(2), 261–273.

Bruce, C. (2004). Fundamentals of crime analysis. In C. W. Bruce, S. R. Hick, & J. P. Cooper (Eds.), *Exploring crime analysis: Reading on essential skills.* Overland Park, KS: IACA Press.

Bureau of Labor Statistics. (2007). *Women in the labor force: A databook (2007 edition).* Retrieved June 2008, from http://www.bls.gov/cps/wlf-databook2007.htm

Chamard, S. (2006). The history of crime mapping and its use by American police departments. *Alaska Justice Forum, 23*(3), 1, 4–8. Retrieved January 2008, from http://justice.uaa.alaska.edu/FORUM/23/3fall2006/forum233.fall2006.pdf

Clarke, R., & Felson, M. (Eds.). (1993). *Routine activity and rational choice. Advances in criminological theory* (Vol. 5). New Brunswick, NJ: Transaction Publishers.

Clarke, R. V., & Eck, J. (2005). *Crime analysis for problem solvers in 60 small steps.* US Department of Justice, Office of Community Oriented Policing Services.

Cohen, L. E., & Felson, M. (1979). Social change and crime rate trends: A routine activities approach. *American Sociological Review, 44*, 588–608.

Crowe, T. D. (2000). *Crime prevention through environmental design: Applications of architectural design and space management concepts* (2nd ed.). Boston: Butterworth-Henemann.

Dangermond, J. (2007). Taking the "geographic approach." *Arcwatch.* Retrieved February 2009, from http://www.esri.com/news/arcwatch/0907/feature.html

Gottlieb, S., Arenberg, S., & Singh, R. (1994). *Crime analysis: From first report to final arrest*. Montclair, CA: Alpha Publishing.

Harries, K. (1999). *Crime mapping: Principle and practice*. Washington, DC: US Department of Justice Programs. Retrieved March 2008, from http://www.ncirs.gov/htm/nij/mapping/pdf.html

International Association of Crime Analysts. (2005). *Frequently Asked Questions*. Retrieved August 2008, from http://www.iaca.net/FAQ.asp

Levine, N. (2003). CrimeStat II. *Crime Mapping News, 5*(2), 2–4. Retrieved January 2008, from http://www.policefoundation.org/

Mamalian, C. A., LaVigne, N. G., Staff of the Crime Mapping Research Center. (1999). The use of computerized crime mapping by law enforcement: Survey results. *NIJ Research Preview*. Retrieved January 2008, from http://www.ncjrs.gov?pdffiles1/fs000237.pdf

Paynich, R., Cooke, P., & Mathews, C. (2007). *Developing standards and curriculum for GIS in law enforcement*. Presented at 9th NIJ Crime Mapping Conference, Pittsburgh, PA. Retrieved December 14, 2008, from http://www.ojp.usdoj.gov/nij/maps/pittsburgh2007/papers/Paynich.pdf

Rosenbaum, D. P., Lurigio, A. J., & Davis, R. C. (1998). *The prevention of crime: Social and situational strategies*. Belmont, CA: West/Wadsworth.

Rossmo, D. K. (2000). *Geographic profiling*. Boca Raton, FL: CRC.

Rubenstein, J. M. (2003). *An introduction to human geography* (7th ed.). Upper Saddle River, NJ: Pearson Education.

Sacco, V. F., & Kennedy, L. W. (2002). *The criminal event: Perspectives in space and time* (2nd ed.). Belmont, CA: Wadsworth/Thomson Learning.

Shaw, C., & McKay, H. (1942). *Juvenile delinquency and urban areas* (5th ed.). Chicago: University of Chicago Press.

Weisburd, D., & McEwen, T. (1988). Crime mapping: Crime prevention. In R. V. Clarke (Ed.), *Crime prevention studies*. Monsey, NY: Willow Tree Press.

Social Disorganization and Social Efficacy

CHAPTER 2

▶ ▶ **LEARNING OBJECTIVES**

As discussed in Chapter 1, crime mapping is not a new practice of crime analysts. The literature is rich with theoretical justifications of ecological influences upon crime. Chapters 2 and 3 examine many of these theories. This chapter discusses several of these early theoretical approaches beginning with the Chicago School and ending with a discussion of Wilson and Kelling's more recent "Broken Windows" approach to crime prevention. After studying this chapter, you should be able to:

- Understand a brief history of criminology and how theories focused on social-level variables developed and fit within the larger discipline of criminology.
- Exhibit a solid understanding of the Chicago School and its contribution to the study of crime and crime mapping.
- Explain Burgess and Park's concentric zone theory.
- Explain how Shaw and McKay's Social Disorganization expanded upon the concentric zone theory by adding social-level variables to explain how people interact with their environment.
- Discuss the key elements to social efficacy.
- Identify the key elements to Broken Windows and critically analyze this approach through the lens of social and physical elements found in the theories discussed in this chapter.

▶ ▶ **KEY TERMS**

Broken Windows Theory
Cartographic School of Criminology
Chicago School of Criminology
Classical School of Criminology
Collective Efficacy
Crime Fuse
Determinism
Differential Association Theory
Ecological Fallacy
Gentrification
Heterogeneous
Positivist School of Criminology
Social Disorganization Theory
Utilitarianism

■ **Introduction**

At this point, you may be wondering why a textbook on crime mapping has several chapters dedicated to criminological theory. Theory is very important to the creation of crime maps. Theories provide useful suggestions about the types of variables and sources of data that should be included in maps of crime. Some examples might include measures of socioeconomic status and other social variables, population density, and land use types. As will be emphasized throughout this text, it is crucial that crime maps and analyses convey the whole picture.

Maps created without providing the accompanying environmental and social context are limited in their usefulness. Block (1998) emphasizes this point by stating that "The successful analysis of spatial patterns of crime requires that mapping technology be guided by theory that can link place to crime, can unravel the spatial characteristics of different types of crime, and can provide explanations and suggest prevention strategies for the high vulnerability of some neighborhoods or demographic groups" (p. 37).

Eck (1998) furthers this notion by suggesting that "everything displayed on a map should be of theoretical importance" (p. 381). In addition, "the choice of the features that describe the relevant context depends on the theory being examined. To interpret the dots we need a theory so we can display the relevant context and leave out the irrelevant context" (p. 383).

So let us begin with a *very* brief history of criminology. Theories of crime causation have come a long way since Cesare Beccaria published his classic essay "On Crimes and Punishments" (1764) and Jeremy Bentham put forth his *Introduction to the Principles of Morals and Legislation.* (1789). Prior to Beccaria and Bentham, crime was primarily viewed as a product of evil and demonic possession or inferior bloodlines. Punishments for committing crimes during the Middle Ages were barbaric and cruel and included various types of torture and execution. By the mid-18th century, however, the prevailing philosophy, rooted in **utilitarianism**, maintained that human behavior is, by and large, the result of reason and logic. Beccaria believed that people act on their own free will, and thus criminals make a conscious decision to commit crime. He advocated for humane forms of punishment and for punishment that "fit the crime." Bentham, also believing in free will and rational thought, argued that punishment itself was harmful and that if it was to outweigh the harm it produced, it must achieve four main objectives:

- The primary goal is the prevention of crime.
- When punishment does not prevent crime, it must persuade the offender to choose to commit a less serious offense.
- Punishment must persuade future offenders to use the least amount of force in committing a crime.
- It must prevent crime at a cost-effective rate.

Termed the **Classical School of Criminology**, Beccaria, Bentham, and others believed that crime was a product of rational thought; the major premise was that punishment should be swift, certain, and severe. If this goal was achieved and publicly exhibited to the general population, rational criminals, weighing the costs and benefits of their actions, would choose to commit either a lesser crime or no crime at all. Thus, criminals, under the classical view, were not inherently evil but were people with poor decision-making skills and weak morals. The classical perspective heavily influenced the development of the criminal justice system in America and is still very evident today. Modern initiatives designed to control crime, including mandatory and enhanced sentencing policies and three strikes laws, popular in the 1990s, are based on the rational offender model.

The **Positivist School of Criminology** gained prominence during the late 19th century and spurred new methods for studying crime. The scientific method heavily influenced the way positivist philosophers and researchers approached the study of human behavior and their immediate environment. Positivism holds two primary elements. First, human behavior is influenced by biology, psychology, and to some extent the larger environment (social positivists). Second, the best criminologists can learn about human behavior and solve social problems is to embrace the scientific method. Theorists from the positivist perspective by and large have a deterministic view of the world (**determinism**), arguing that human behavior is largely predetermined by psychological, biological, and environmental factors. A major assumption of positivist criminologists was that a criminal could be "cured" of some known or unknown biological and/or psychological defect. Crime control strategies stemming from this school of thought included sterilization (crime is genetic and thus can be propagated), lobotomies (to eliminate the "infected" part of the brain that causes criminal behavior), and various drug and chemical treatments to "fix" criminals. The Positivist School still retains influence today with modern research continuing to focus on biological and psychological factors and their relationship to criminal behavior. Contemporary positivists, however, would support less invasive methods than their earlier counterparts, including genetic counseling, drug therapies, and community-based programs that could help offenders overcome psychological, biological, and environmental risks of crime. The most influential remnant from the Positivist School is the application of the scientific method to the study of crime (which is key to crime analysis).

Both the early classicalists and the positivists were primarily focused on the individual and to some extent ignored many of the environmental factors that impact human behavior. Another group of criminologists, sociological criminologists, emerged in the late-18th through mid-19th centuries. These criminologists began to pave the way for modern theorists and analysts who study crime in a social context. Recall from Chapter 1 that Quetelet and Guerry are credited as being two of the first persons to use spatial analysis in their research and are thought to be the founders of the **Cartographic School of Criminology** (Levin & Lindesmith, 1971). Guerry, using several data sources, determined that crime in France was unevenly distributed across people and places. For example, property crimes were more prevalent in the northern regions of France (and occurred more often in the winter) while crimes against people were more prevalent in the southern regions (and occurred more frequently during the summer). Guerry also concluded that age, education, poverty, and population density were important in understanding the distribution of crime.

Quetelet also looked at how crime was distributed across France (Quetelet, 1842/1973). Using social and spatial variables in his analysis, he determined that crime was more heavily concentrated in the summer, in the southern regions, and among **heterogeneous** populations that, as a group, were uneducated and poor.

What Quetelet and Guerry discovered is that crime and other social problems are distributed unevenly across space and time, the fundamental justification for crime mapping. This is nothing new, even to the beginning criminal justice student and crime analyst. If crime was randomly distributed, there would be little need for crime mapping. So why study theory at all? The answer to this question is that criminological theory is instrumental in identifying the individual, social, and ecological factors that allow us to better predict where and when crime will be clustered. Theory can also help identify which variables we should target in crime control strategies that will be most beneficial in yielding the greatest reduction in crime. That said, the following sections examine the key theoretical arguments from major sociological and ecological theories of crime causation. In addition to helping us understand theories of crime causation, an examination of theory also helps us identify important variables and data sources to use in crime analysis and crime mapping. We will begin with the Chicago School of Criminology.

■ The Chicago School of Criminology

Scholars from the **Chicago School of Criminology** began the arduous task of developing hypotheses and constructing theoretical models about how crime was related to, and caused by, various social and environmental factors. The Chicago School of Criminology, named because the prominent theorists of this school were scholars at the University of Chicago, includes the work of several prominent sociologists, including Robert Park and Ernest Burgess. (For a more detailed description of this history, visit http://sociology.uchicago.edu/department/history.shtml.) Park (1915) was interested in myriad social problems that existed in cities. He observed that cities were made up of individual divisions that were based on race and ethnicity, socio-economic status and occupation, and the physical characteristics of structural components of the city, such as whether an area was used for business or residential purposes. Using concepts from the discipline of plant ecology, Park argued that cities grew from the inside out through a process of invasion, dominance, and succession. Burgess (1925), continuing Park's plant analogy, also studied the growth of cities. He argued that cities grow outward from the center in concentric circles starting with the inner loop (the business district in the center of a city), then the zone in transition (the zone characterized by high rates of crime and other social problems), and then several zones that contain the homes of people who commute varying distances into the city to work (modern day suburbs). He further argued that people who live in the loop immediately surrounding the business district (zone in transition) would experience the highest levels of social disorganization, including the uppermost crime and victimization rates of the city. Modern critics of the concentric zone theory make the very good point that this model does not explain all cities. Many cities in the West, for example, did not grow in concentric circles. In addition, modern cities that once fit this model have changed through the processes of **gentrification**. Areas in the zone in transition once characterized by high rates of delinquency and disorder are now desirable high-rent areas that are affordable only by the wealthy. However, what can be taken from the concentric zone model that is still applicable today is that crime and disorder is not randomly distributed throughout a city and that the areas most plagued by these problems will also have high levels of poverty.

Two other Chicago scholars, Clifford Shaw and Henry McKay (1942), furthered Park's and Burgess's works by studying juvenile delinquency and social structure variables within the zone in transition. Using the term social disorganization, Shaw and McKay sought to study the processes responsible for creating higher rates of juvenile delinquency in this area. They determined that three primary dynamics exist in socially disorganized neighborhoods; these include high rates of residential turnover, a heterogeneous population, and high levels of poverty. Essentially, they argued that the constant turnover of people moving in and out of the neighborhood combined with self-segregation of ethnicities inhibited the neighborhood cohesion necessary to solve common problems associated with high levels of poverty. That said, poverty is not the *cause* of crime, but it is correlated with other factors, such as high residential mobility and heterogeneity. When these factors are concentrated in a localized area, the likelihood of a high crime rate noticeably increases. Shaw and McKay also observed that this is true for *all* populations that live in socially disorganized neighborhoods. Thus, crime was not the product of a certain racial group of people but of any group living in high poverty areas with high diversity and turnover.

Edwin Sutherland (1947) furthered Shaw and McKay's work by suggesting that high crime areas were not socially disorganized, just organized differently than other areas, and thus the cultural values surrounding crime were different. That said, many juveniles who lived in these disorganized neighborhoods learned values and techniques *favorable* to committing crime via their peer associations. Termed **Differential Association Theory**, Sutherland focused on the social learning processes that aided in the cultural transmission of criminal values and argued that some neighborhoods are not disorganized but rather organized around different values. He argued that criminal behavior is learned amongst close-knit groups and that both the techniques needed to commit crime and the values favorable to committing crime are learned. Differential association theory is not meant to replace, but to supplement Social Disorganization Theory, and both theoretical approaches are still being tested today by modern theorists. Modern theories of crime attempt to integrate both the characteristics of physical space and social relationships to explain why crime occurs more frequently in some areas than others.

■ Social Efficacy

Recent research that examined social disorganization has determined that crime is not necessarily the problem but rather a symptom of inadequate social networks (or lack of collective efficacy) that exist in these neighborhoods (Sampson, Morenoff, & Earls, 1999; Taylor, 2001; Sampson & Raudenbush, 2001; Reisig & Cancino, 2004). The lack of **collective efficacy** in a neighborhood (inhibited by neighborhood turnover, heterogeneity, and poverty) is a more difficult and complex problem for law enforcement to address. Programs designed to weed crime out of an area by police sweeps and neighborhood clean-up programs ultimately fail in the long run if they do not address the notion of collective, or social, efficacy. Sampson and Groves (1989) contend that the lack of supervision, lack of community involvement, and reduced friendship and other social networks are the primary cause of crime and disorder in a neighborhood. That is, residents in socially disorganized neighborhoods do not know or trust one another, and thus the supervision of the neighborhood people and property is limited. In addition, when residents are aware of a problem, they are less likely to get involved in any collaborative efforts to develop a solution. Taylor (2001) studied neighborhoods he observed as having conditions of collective efficacy. These neighborhoods are characterized as having residents who hold common values about what is right and wrong, strong ties based on physical proximity, informal social control, and high levels of community participation in neighborhood programs. These neighborhoods typically have little to no crime problems. Neighborhoods without collective efficacy are more susceptible to crime and disorder and typically exhibit a host of social and physical incivilities signaling to motivated offenders that this is a good place to set up shop. For example, **Figure 2–1** shows an abandoned house in Providence, Rhode Island.

Abandoned houses such as this can signal to potential offenders that the owners are either unable or unwilling to care for the property. Abandoned houses also emit signals about the lack of social efficacy that is common in poverty-stricken areas, as well as areas where neighbors mind their own business and areas where people can act as they wish with little community disapproval. In addition, abandoned buildings serve as free housing for the homeless and act as places of business for drug activity, prostitution, gang activity, etc. In **Figure 2–2**, target hardening measures, such as barred windows, signal to potential customers (and potential witnesses) that this is not a safe store to shop at, decreasing the store owner's potential profits.

■ Broken Windows Theory: Can It Be That Simple?

An understanding of the relationship between community variables, such as social and physical disorder, and crime is not new to law enforcement. Within the last several decades, strategies based on the philosophy of community-oriented policing and problem-oriented policing have become very popular. Many of these strategies are geared toward social and physical incivilities within neighborhoods. The popularity of these strategies is often attributed to Wilson and Kelling's "Broken Windows" piece, which was published in 1982. However, most practitioners can tell you that community-oriented policing strategies were being implemented long before this work was even conceptualized, especially in rural law enforcement agencies where budgets are smaller, manpower is sparse, and officers or deputies are highly integrated into the community. The **Broken Windows Theory** (or

Figure 2–1 Abandoned Property

Figure 2–2 Target Hardening

Broken Windows Approach) is based on an experiment conducted by Stanford psychologist Philip Zimbardo in 1969. In this now classic experiment, Zimbardo placed two unoccupied vehicles in two different neighborhoods: Palo Alto, California and Bronx, New York. Within just 10 minutes of its placement, the car in Bronx was vandalized, and within 24 hours, nothing of value remained inside the car. After it was stripped of its value, passersby continued to vandalize the car by ripping its upholstery and smashing its windows. The car in Palo Alto, however, was left alone for almost a week. It was only after Zimbardo smashed it with a sledgehammer that people began to vandalize the car. Within a few short hours, the car was completely destroyed.

What Zimbardo's experiment tells us is that when the "no one cares" cue is sent out, crime can occur in any neighborhood. Wilson and Kelling (1982) contend that the only reason the car in Bronx was vandalized so soon is because the social and physical structure of the neighborhood emitted the message to potential offenders that no one cared. In Palo Alto, it was the smashing of the car by the researcher that signaled no one cared about the car. This suggests that if the car in Palo Alto had been left intact by the researcher, it may have sat for a much longer period of time before it was vandalized. Community-oriented and problem-oriented police strategies seek to address neighborhood problems that are thought to be root causes of crime. For example, proponents of community policing maintain that a law enforcement concentration on "minor disorders" will, in turn, "lead to a reduction in serious crime" (Mastrofski, 1988, p. 48). That is (in broken windows lingo), minor disorders, when not addressed, often lead to more serious disorders and ultimately to community decay. Supporters of community policing claim that its application can help break this destructive cycle through the coproduction of order and greater community involvement.

However, can crime prevention be as simple as cleaning up the signs of neighborhood decay? The answer is complicated. An illustration of some of the complicated relationships between geographic and social variables can be found in popular culture and movies. For example, in the popular movie *Boyz n the Hood*, Furious (Laurence Fishburne) takes his son Tre (Cuba Gooding, Jr.) and Trey's friend, Ricky (Morris Chestnut), for a drive in inner city Los Angeles. He stops at a residential neighborhood that could easily be described as a socially disorganized, or "broken windows" neighborhood (visible signs of both social and physical incivilities). Loud rap music plays in the background, and the audience can also hear a woman yelling. As Tre and Ricky exit the car, they look around nervously, and Ricky expresses to

Furious his level of anxiety about stopping in this neighborhood. Furious directs the boys toward a billboard sign that says "Cash for your home." He explains to them the message that the sign displays, that of gentrification. He explains the process as one where "they" bring the property value of a certain area down, get all of the current residents out of the neighborhood, then raise the property value and sell it at a profit. At this point in the scene, a small crowd of people from the neighborhood gather around Furious and the boys and begin a dialog. An old man from the crowd exclaims that it isn't people from *outside* the neighborhood bringing down the property value, but it is those *within* the neighborhood who do so by selling drugs and "shooting each other."

Furious then asks why there are so many gun shops in this community. Furious contends that the reason that gun shops and liquor stores are so prevalent in this community is because "they want us to kill ourselves" making a very noticeable distinction between "us" and "them," which can be interpreted in several ways, an important one to this discussion being the "haves" versus the "have nots."

In this movie, Laurence Fishburne's character touches upon several important ideas relevant to our discussion in this book. First, communities are set within a larger environment that must be understood when analyzing crime. Macro-level forces, such as economics, play a part in crime trends over time and must be factored into analyses that examine general crime trends. Ignoring macro-level influences in analyses presents an incomplete picture. Second, individual decisions and behaviors (for example, selling your home to move to a better neighborhood or putting bars on your windows to protect your property) may be beneficial for the individuals but are detrimental to the neighborhood at large. Third, structural components of neighborhoods (density of liquor stores and gun shops, for example) are important in understanding the neighborhood's social problems and how they impact the physical space. It is important to remember that relationships between geographic and social variables are complicated and in many ways reciprocal in nature. Fourth, incorporating the notion of a **crime fuse** (a theoretical concept that argues a society *allows* problems such as crime to exist in certain areas and not others), such as Furious alluded to in the "us" versus "them" statement, may help explain the source of some of the frustration in a neighborhood that can manifest itself as crime (Barr & Pease, 1990). Last, cultural components of violence (often escalated in heterogeneous neighborhoods) are not quickly or easily fixed by the criminal justice system. They require more complicated approaches that demand community participation in both the development and implementation of crimereduction strategies. Skogan (2008) argues that "we cannot arrest our way out of crime problems" and that effective problem solving for neighborhoods suffering from crime and social disorder problems "calls for examining patterns of incidents to reveal their causes and to help plan how to deal with them proactively" (such is the task in crime analysis and mapping) (p. 198). Furthermore, solutions will most certainly require the involvement of organizations inside and outside the criminal justice system and will depend upon community support and participation for their success.

The Chicago School did much to further our understanding of the relationship between crime and space, but it is not without criticism. First is the **ecological fallacy**, which, in simple terms, means it is inaccurate to make assumptions about individuals based on aggregate-level data. That is, there are plenty of people who happen to have characteristics that put them at a higher risk to become criminals, including living in socially disorganized neighborhoods. However, they do not always turn to a life of crime. They live under the same conditions with the same environmental cues and the same opportunities to commit crime, yet they do not. Thus, while it is important to include macro level forces in analyses, it is inappropriate.

A second problem with the Chicago School is that its approaches do not explain all crime types and are often tested using street crimes within urban areas. This, of course, is a problem with many criminological theories. A third problem has to do with misidentifying the causal order of the relationship between variables. For example, research on **Social Disorganization Theory** suggests that crime and social disorder have a reciprocal relationship (Markowitz, Bellair, Liska, & Liu, 2001) and thus crime control efforts must focus on *both* issues to be successful. Thus, simply cleaning up a neighborhood and not promoting activities to build social efficacy will only yield temporary results in crime reduction, giving the "Band-Aid on a bullet hole" analogy a whole new meaning.

A fourth problem with research on social disorganization is multicollinearity. This term will be discussed in Chapter 8, but essentially, when variables in an equation exhibit multicollinearity, it means that the concepts they are measuring overlap. Think about the social-level variables we discussed in this chapter, such as income and education. We know that both of these variables are associated strongly with crime. However, income and education, in what they represent, overlap at least partially. That is, to some extent, education level can be explained by income, and income is a function of education level. Another example might be overcrowding and substandard housing. Overcrowded neighborhoods tend to contain a lot of substandard housing. Thus, these variables are not completely separated from one another. Multicollinearity is highly problematic in some of the analyses that are routinely run in crime analysis and theory testing. Analyses containing multicollinearity can produce inaccurate and misleading findings.

Finally, research on social disorganization suffers from imprecise measurements. The concept of social disorganization itself cannot be measured precisely, and researchers are forced to employ proxy variables that measure *indicators* of social disorganization, such as residential mobility. Gau and Pratt (2008) found that citizen perceptions of disorder and crime were highly correlated, suggesting that treating these concepts as separate constructs is inappropriate. Research on social efficacy and its measurement of friendship networks and neighborhood supervision is more promising in its operationalization of social disorganization; however, these can be difficult concepts to measure. This is an important notion to keep in mind for various types of crime analyses.

■ Online Resources

- History of the Chicago School of Criminology: http://sociology.uchicago.edu/department/history.shtml
- NCJRS Publication on Social Disorganization and Rural Communities: http://www.ncjrs.gov/html/ojjdp/193591/page1.html
- Center for Spatially Integrated Social Science, *Clifford R. Shaw and Henry D. McKay: The Social Disorganization Theory*: http://www.csiss.org/classics/content/66
- Social Disorganization and Social Efficacy: http://www.crimeeducation.homeoffice.gov.uk/toolkits/fa0106c.htm

■ Conclusion

Regardless of these criticisms, understanding criminological theory is crucial to crime analysis and crime mapping. However, it is important to be aware of limitations of theory, just as it is important to be aware of the limitations of your data and analysis techniques. Eck (1998) sums it up best by stating:

> If the police do not know much about the area then they cannot form a testable hypothesis. This is the case in most situations when the police look for fast-breaking crime patterns so they can focus enforcement activity on a troubled area—by means of saturation patrolling, decoy operations, surveillance, or other tactics. Though the crime maps and even cluster detection algorithms may be useful for these operations, these mapping techniques are probably more useful when officers already

have a good understanding of the area in question. This suggests that the police, like researchers, should pay as much attention to criminological theories as they do to the data they examine (dots) and the methods they apply (maps). (p. 402–403)

The next chapter continues the exploration of crime causation but narrows the discussion to the individual characteristics and the physical properties of space, which are related to crime and victimization. It is important to note that the theories discussed in Chapter 3 do not conflict with the theories discussed in this chapter but supplement them. In some cases, they simply address crime at a different level and in some cases integrate several theoretical approaches into a larger and more comprehensive paradigm.

■ Recommended Reading

Akers, R. L. (2000). *Criminological theories: Introduction, evaluation, and application.* Los Angeles: Roxbury.

Cullen, F. T., & Agnew, R. (Eds.). (1999). *Criminological theory, past to present: Essential readings.* Los Angeles: Roxbury.

Cullen, F. T., Wright, J. P., & Blevins, K. R. (Eds.). (2006). *Taking stock: The status of criminological theory.* New Brunswick, NJ: Transaction Publishers.

Kelling, G. L., & Coles, C. M. (1996). *Fixing broken windows: Restoring order and reducing crime in our communities.* New York: Martin Kessler Books.

Sampson, R. J., & Wilson, W. J. (1995). Toward a theory of race, crime, and urban inequality. In J. Hagan & R. D. Peterson (Eds.), *Crime and inequality.* Stanford, CA: Stanford University Press.

Taylor, R. B. (2001). *Breaking away from broken windows: Baltimore neighborhoods and the nationwide fight against crime, grime, fear, and decline.* Boulder, CO: Westview Press.

■ Questions for Review

1. Why is it important to consider theories of crime when creating maps of crime?
2. According to early Chicago School theorists, how is the structure of a city related to the spatial distribution of crime?
3. How has *gentrification* changed crime distributions in urban areas?
4. When examining crime in socially disorganized neighborhoods, why must the analyst include both social and physical variables in his or her analysis to get a more complete picture of crime?

5. What is *social* or *collective efficacy*, and how is it related to the prevalence of crime in a neighborhood?
6. What are some of the problems in researching and/or applying theoretical principles of Chicago School theories of crime?

■ Chapter Glossary

Broken Windows Theory Broken Windows Theory was introduced by James Q. Wilson and George L. Kelling in the early 1980s. Essentially, it argues that if police focus on social disorder and minor crimes in neighborhoods that contribute to social disorder (for example, public drinking and street prostitution), they are likely to make an impact in reducing or preventing more serious crimes.

Cartographic School of Criminology Under this school, early 19th century researchers in Europe used demographic information to explain the spatial distribution of crime. Variables measuring population density and socioeconomic status, among others, were used to explain uneven distributions of crime.

Chicago School of Criminology Theories and research on crime classified in this school were developed by urban sociologists (primarily in Chicago) who were interested in the relationship between environmental conditions and crime. They studied social and physical variables to understand the distribution of crime in cities.

Classical School of Criminology Theories of crime in this school rested on the notion of "free will." Classical theorists maintained that criminals were by and large rational and chose to commit crimes for personal gain. Thus, crime was best prevented by implementing criminal sanctions that were high enough to deter potential offenders from committing crimes.

Collective Efficacy Sometimes referred to as "social efficacy," this term refers to the level of social control (supervision of neighborhood children, maintaining public order, etc.) wielded by communities. Collective efficacy is high in cohesive communities with mutual trust and is low in communities that are not cohesive and that do not have mutual trust.

Crime Fuse A theoretical concept that argues a society *allows* problems such as crime to exist in certain areas and not others.

Determinism The philosophy of determinism argues that human behavior is largely predetermined by psychological, biological, and environmental factors. In essence, people are victims of their own circumstances.

Differential Association Theory   According to Sutherland (1947), criminal behavior is related to the extent offenders were exposed to antisocial attitudes and values. Sutherland outlined nine principles explaining how criminal behavior is learned through association.

Ecological Fallacy   The problem that occurs when making assumptions about individuals based on aggregate-level data.

Gentrification   In the process of gentrification, existing buildings in a neighborhood, typically in urban areas, are replaced or upgraded. As a result, property in these areas becomes more desirable and results in a shift in population demographics of both residents and business owners. Usually, during gentrification, persons classified as having lower socioeconomic status move out of the neighborhood, and persons with higher levels of wealth move into the neighborhood.

Heterogeneous   Heterogeneous is termed used to denote variation. In the context for which it is used in this text, it refers to populations that have high levels of diversity, primarily in culture and ethnicity.

Positivist School of Criminology   The positivists used scientific methods to examine human behavior. Positivists maintained that crime was a result of biological, psychological, and social factors.

Social Disorganization Theory   Under this theoretical approach, crime is more prevalent in neighborhoods that exhibit social and physical incivilities. In these neighborhoods levels of social control is low due to a breakdown in institutions such as the family and school.

Utilitarianism   Attributed to Jeremy Bentham, utilitarianism refers to "benefit maximization." Essentially, decisions are made (rationally) in favor of less pain and the greatest good for the greatest number of people. Thus people are motivated to avoid pain or to pursue pleasure.

■ Chapter Exercises

Exercise 3: Tables, Addresses, and Maps

Data Needed for This Exercise
- All layers to start are in EX3.mxd.

Lesson Objectives
- Open C:\CIA\Exercises\Exercise_3\EX3.mxd.
- Add data to the map display from shapefiles.
- Work with tables:
  - Summarize
  - Sort
  - Statistics

- View address files in ArcGIS.
- Make an address file to use in ArcGIS.
- Add fields.
- Calculate fields.
- Export selected data.

Task Description
- City Council wants to know if there are any sex offenders on probation who live within 1000 ft of a Glendale school. The Crime Analysis Unit supervisor has assigned this project to you, and she wants to know a few things:
  - Are there any sex offenders living within 1000 ft of any school, and if so, how many?
  - If any are found, what are the crime descriptions for which these offenders are on probation?
  - Provide a list of these offenders in Excel via e-mail.
  - Are there any sex offenders living within 1000 ft of any of the following schools? If so, how many?
    - Glendale Elementary
    - Glendale Landmark Elementary
    - Glen F. Burton Elementary
- In addition to this work, the supervisor needs you to add the names of the City Council members to their respective council districts in a polygon data file.
- She also needs you to create a new table of massage businesses in Glendale to be later geocoded and used for vice operations.
- We will need to divide this work up into three or four different projects. Let's start with the last item and create a new list of businesses that offer massage within the city of Glendale. Hopefully there will not be too many and we can get this done fairly quickly. We will want to capture the name of the business, the address, the phone number if available, and, at a later date, maybe a field that indicates if they are candidates for vice operations based on their location. The easiest way to begin this assignment is to use the good old Dex online yellow pages.

1. Open the C:\CIA\Exercises\Exercise_3\EX3.mxd.
2. Minimize ArcMap after it opens.
3. Open Internet Explorer and browse to the **www.dexknows.com** Web site.
4. In the What field, type in *Massage*, and in the Where field, type in *Glendale, AZ* and click **Search**.

5. We want to "Search for businesses with 'massage' in the business info."
6. You should notice a plethora of listings. We need to refine that search so that we see only those massage business inside Glendale.
7. Select "Narrow my Search" near the top of the Web site listing your results. A dialogue box will appear giving you several options to refine your search. We want to use the one that says "Refine by Location—City." Open the field and choose Glendale from the pop up. This just reduced the total number to just 2 pages instead of 18 or more.
8. If you are an Excel expert (if not, the DVD that came with this book includes a tutorial on Excel that you may find useful), you can cut and paste the listings directly from the Web site and then massage (pun intended) them, move them around, and clean them up to get your table. If not, you need to do it the old-fashioned way. Open Excel, create three new field names, and start typing, using the Web site as a reference.
9. There are several other Web sites to find massage businesses on the Internet. These other sources include:
    a. Google maps: http://maps.google.com/
    b. Yahoo! Maps: http://maps.yahoo.com/
    c. Switchboard: www.switchboard.com (this site give you the less than upstanding ones)
    d. Yellowpages.com: www.yellowpages.com
    e. AnyWho: www.anywho.com
    f. Addresses.com: www.addresses.com
10. None of these Web sites will let you download a table of the data, and the limitation is often that if the business does not advertise with that service it will not be listed, and the old-fashioned paper copy of the yellow pages or the white pages may have to be used. You can buy products that allow you to download lists of businesses based on a category (like www.hoovers.com, www.selectory.com, and many others). The prices range from $100 to $5000, and you get what you pay for. Another source of this kind of data is the business tax license department for your city. This office will often have a huge list of anyone that does business within the city. When you have gained access to this list, you will need to learn what the classification codes are for the business types, and in many cases the business address is where the City sends the bill for the annual license renewal, which may or may not be the actual address of the business. (The corporate headquarters are in Houston, Texas? That's not in Glendale, is it?)

11. For the purposes of this exercise, we'll just use the addresses and names from the Dex Web site and the paper copy of the yellow pages.
12. In a future lesson we will use this file you have diligently typed, so save and store it in the C:\CIA\Exercises\Exercise_3\ folder as Massageparlors.xls. The final table should look like a spreadsheet with columns denoting the name, phone, address, city, state, and zip code, and the massage parlors listed in rows (see DVD Figure 3–1).
13. I found 34 listings. How many did you find? If you do not have all of these, for later geocoding exercises you can use the ones I found that are located in C:\CIA\Exercises\Other\Massage.xls.
14. Let us move on to adding the names of the City Council members to an already developed polygon shapefile.
15. Maximize the EX3.mxd project and click on the **Add Data** button. Find the C:\CIA\Exercises\Exercise_3\CCDistNames.shp data and add it into the project.
16. Right click the theme in the TOC, and open the attribute table for this polygon shapefile.
17. You will notice that there are six polygons listed in this attribute table, and each one represents a city council district in Glendale. Because you are likely unfamiliar with which council district member represents each section, we need to browse the Glendale Web site to find this information.
18. Open Internet Explorer and navigate to http://www.glendaleaz.com/cityofficials/index.cfm.
19. In the attribute table you have the name of the city council district for each polygon in a field named c_district. Now either write down the councilperson's name for each council district, or leave the Web browser open as we add the councilperson's name to the table.
20. First, we need to add a new field to hold the name values we are going to enter. To accomplish this, click on the **Option** button at the bottom of the attribute table window in our ArcMap document (you will have to switch from the Web page to ArcGIS to see this). Find the **Add Field** item in the dialog menu, and click on it.
21. You should now have the Add Field dialog box in front of you. In the Name box type in *Councilperson*. For the Field Type box, pick **text**. The field length for now will be fine at 50. Click the **OK** button to add the field to our attribute table.
22. An error message will pop up indicating that our field name is too long and will be truncated to CouncilPer for us. Click **Yes** to accept this change, and notice that the truncated field names can be no longer than 10 characters for a shapefile.

23. Now highlight any one of the fields between C_District and CouncilPer and right click on the field name. In the dialog box, choose **Turn Off Field**. Do this for each field between C_District and CouncilPer to make the table easier to use.
24. Now resize your attribute table so you can partially see the map display behind it and the table at the same time. You may also need to resize a few fields to be able to see both of them clearly.
25. At the very far right side of the attribute table is a small box to the left of the FID field value. Click on this little box next to the CHOLLA districts FID number 0. This will select the polygon on the map and in the table.
26. A blue line will now be around the polygon that is the farthest north in Glendale, and the record in the table is also blue.
27. Right click on the **CouncilPer** field name, and choose **Field Calculator** from the dialog box.
28. Click **Yes** to indicate that you wish to continue outside of an edit session.
29. In the Field Calculator dialog box, find the lower white box with CouncilPer= at the top and type in "*Vice Mayor Manny Martinez*" (or whoever is listed as the Cholla District's councilperson on the Web site).
30. Because we are entering data for a text field, we need to make sure we include the quotation marks before and after the text string we enter.
31. Make sure that at the bottom of the dialog box, the check box that says Calculate Selected Records Only is checked.
32. Click **OK**, and you should now see the text you typed in the CouncilPer row for the Cholla District.
33. Notice that with text characters and spaces "Vice Mayor Manny Martinez" takes up about 25 spaces. If we had made our field length 25 or less, this name may not have fit and would have been truncated off the right hand side. When choosing text field lengths, find all of the values that will be entered, count the longest one's total characters including spaces, add two to five spaces to it, and set the field length to that value. Because we used 50 characters, we don't need to worry, but in real life you will want to keep your field lengths down to a minimum for the data you are going to enter, or ever anticipate entering, because this saves disk space and reduces the size of files.
34. Continue selecting council districts and calculating each of the councilperson's names for that council district as we did in steps 25–33.

35. When you are done, close the attribute table, and save your project as MyEX3b.mxd in your student folder (but don't close ArcMap). Before closing the attribute table, make sure to turn on all of the fields you turned off so that they are available for the rest of the exercise. To do this, click the **Options** button, and choose **Turn All Fields On** from the dialog menu.
36. To use this new information, will we create a concatenated label for each council district using the labeling features in ArcMap.
37. First, let's change the colors for each district so they each have a different color.
38. Right click the **CCDistNames** layer in the TOC. Click **Properties** and go to the **Symbology** tab in the Properties dialog box.
39. At the left in the Show section, choose **Categories** and **Unique Values**.
40. In the Value Field box, find and choose **C_District**.
41. In the color ramp section, choose the color ramp at the very top.
42. At the bottom of the Symbology dialog section, click on the button **Add All Values**.
43. Now all of the unique values in the c_District field have populated the symbology area. Uncheck the box to the left of <all other values> because we don't want anything in our legend except for the data that is in our shapefile, and we do not want to see null values, etc. (*Even though we have none in this shapefile, it is important to remember to do this for the many times that null values exist.*)
44. Click the **Apply** button and close the Symbology dialog.
45. Check out your fine work on the map and the TOC.
46. You may notice that one of your council districts still has light blue lines around it, indicating that it is still selected. To unselect everything, we can go up to the **Selection** choice in the menu bar, and then choose **Clear Selected Features** in the pop-up dialog. The blue lines around the polygon should now be gone.
47. Now let's create some labels for the council districts on our map.
48. Right click the **CCDistName** theme in the TOC again, and choose **Properties**.
49. Go to the tab named **Labels**.
50. Check the box that says **Label Features in This Layer**, and then make sure the Method box says "Label All the Features the Same Way."
51. In the Label Field box you should see Entity. We want to change this to a combination of the district's name and the councilperson's

name. To accomplish this, we need to click on the **Expression** button to the right of the Label Field box.

52. Highlight and delete [**ENTITY**] in the bottom white box. Go up to the top white box where all the fields are listed and double click on **C_DISTRICT** to bring it down to the bottom box. Now find the **CouncilPer** field we created and double click it. It will come down into the bottom box and right next to the C_District field. We need to enter an & symbol in between these two field names to avoid a syntax error. The & symbol, or ampersand, is a separator for field names in several programming languages (like SQL, Microsoft Access, etc.). Click in between the two fields and add an ampersand like the following: [*C_DISTRICT*] & [*CouncilPer*]. Click **OK** to view the result.

53. You may not see all of the labels and the council district's and councilperson's names run together. We want the text to be a bit smaller, and we want the councilperson's name to appear below the council district name.

54. Reopen the **Properties** → **Labels** dialog menu and change the font size to 6, and click on the **Label Styles** button to get font style options. Go to the bottom of the styles and choose **BANNER**. (*You may need to change the font size back to 6 again after completing this task.*) Click **Apply** to see the results.

55. Reopen the **Properties** → **Labels** dialog again if you closed it, and go back up to the **Expressions** button and click it again. Click on the **Help** button and read about formatting text in the Help dialog. You will see that you have several options and capabilities to edit the way your text appears in this expression area. We will use some of this information to help our labels look nicer. In the Expression dialog where we currently have [C_DISTRICT] & [CouncilPer] listed, change it to: "*CC DIST: "* & [*C_DISTRICT*] & *vbNewLine* & [*CouncilPer*].

56. Click **OK** and **Apply** to make the changes effective, and close the Properties dialog to see the result.

57. This may still not be very pretty, but you can see the versatility you have with labeling features automatically using the Expression tools and a little bit of Visual Basic for Applications (VBA) code. You can also change many factors about the symbols being used (the yellow box and its outline) by clicking on the **Symbols** button and making various changes.

58. Now we come to the hardest assignments we were given today:
    a. Identifying if there are any sex offenders living within 1000 ft of any school, and if so, how many.
    b. Providing the crime descriptions for those offenders on probation if any are found.
    c. Providing a list of these offenders in Excel via e-mail.
    d. Identifying if there are any sex offenders living within 1000 ft of any of the following schools, and if so, how many.
        i. Glendale Elementary
        ii. Glendale Landmark Elementary
        iii. Glen F. Burton Elementary
59. First we need to define what the supervisor means by "sex offenders." After interviewing her, we understand that she wants to know which adult probationers are on probation for indecent exposure, molesting, or sexual assault and who are reported to be living within 1000 ft of a school. She is not concerned with registered sex offenders from the Arizona Department of Public Safety Web site because the person who requested this information has already been to the site and acquired the needed information.
60. Now we need to see if we have probationer data with sex offender information. In EX3.mxd's TOC, find the theme called **Probationers**. Right click this theme, go to **Properties**, and go to the **Definition Query** tab. You should already see a formula in this section that says ""DESCR" LIKE '%INDECENT%' OR "DESCR" LIKE '%SEX%' OR "DESCR" LIKE '%MOLEST%'".
61. There is a field in the Probationers theme's attributes called DESCR, which has the generic name of the Arizona Revised Statute Offense the person is on probation for. We are simply using that field to define the data we will see on our map and limiting it to just those who have "indecent," "sex," or "molest" in the offense description field. Isolating the adult probationers down to just those related sex offenses has already been done for you.
62. Click on the **Symbology** tab and notice that the legend has already been created for you as well.
63. Turn on the Probationers theme.
64. Expand the legend to see the symbol styles that are being used, and using the legend and the map, get an idea of what kinds of sexual-related crimes our probationers are responsible for. Do you see any specific spatial patterns?

65. Now we need to see how the Area Schools layer compares with the adult sex offender probationers, so turn on the **Area Schools** layer in the TOC.
66. Now we need to know what 1000 ft around each of these schools looks like, and with some sort of polygon we can count or capture information about the probationers within. To do this, we will need to create a buffer of 1000 ft around each of the schools and find all the probationers within that range for the first question.
67. The easiest way to accomplish this task is with the Select by Location tool in the menu bar. Go up to the menu bar and choose **Selection** then click on **Set Selectable Layers** and make sure *Probationers* is checked. Click **OK**.
68. Now go back up to the **Selection** menu item, and this time choose **Select by Location**. In the Select by Location dialog menu, set the following fields as indicated:

    I want to: *Select features from*
    The following layer(s): *Probationers*
    That: *Are within a distance of*
    The features in this layer: *Area_Schools*

    Check the box next to "Apply a buffer to the features of Area_Schools in: 1000 Feet."
69. Click **Apply** and then **OK**.
70. You should notice that there are several blue dots shining brightly on your map.
71. To answer question number one, because we have bright blue dots (selected records), the answer is yes, there are sex offenders within 1000 ft of area schools, and the second answer is 9 out of 136. How did I get this answer? Think attribute table! Open the attribute table for the Probationers theme and look at the bottom of the Table view. Write this answer down here. _____.
72. Now we need to export the data for these nine offenders to an Excel spreadsheet. Click the **Options** button on the attribute table view, and choose **Export**. We only want to export the selected records. We also want to put it in a location where we can find it later to e-mail it, so click the **Browse** button to the right of the Output Table box, put it in your student folder, and name it SexPO.dbf.
73. We will export this as a DBF file because this kind of file is easily read in Excel, and ArcMap won't directly export the selected records to Excel. Click **OK** to export the nine records, and do not

add it to the map at this time. We will pick it up with Excel in a little while.

74. Question 2 will be answered when we make the DBF table we just exported into an Excel file and e-mail it to our boss.

75. For the last questions, we need to combine a few select routines to get to the answer we need. First we want to select the three schools using a Select by Attributes query on the Area_Schools theme, and then repeat our Select by Location query on the probationers again. To begin, let's first make sure no records are selected by choosing **Selection → Clear Selected Features**.

76. Now go to **Selection → Select by Attributes**.

77. Make sure Area_Schools is in the box at the top.

78. The method will be Create a New Selection.

79. Double click on the Business_N field to bring it down into the Select From "Gln_Schools Where:" section is located.

80. Click on the **Get Unique Values** button to see all the school names in this data layer.

81. Click on the **Equal Sign** button to bring the equals (=) sign down after the BUSINESS_N field name, so that we always have a field name, operator and value in our queries. Then double click on **Glendale Elem. School**, and click on the **Or** button. Repeat this process for the other two schools until your query section looks like this: "BUSINESS_N" = 'Glendale Elem. School' OR "BUSINESS_N" = 'Glendale Landmark Elem. School' OR "BUSINESS_N" = 'Glenn F Burton Elem. School'

82. Click the **Verify** button to see if there are any errors in your SQL code. If you've done it correctly, you should get a message indicating that the expression was successfully verified.

83. Now click **Apply** and then **Close**, and you should see three little bright blue dots on your map, which are the three schools.

84. Now go up to **Selection → Select by Location**, and if you did not close ArcMap since you last ran our 1000 ft radius for probationers, all your settings should be the same as shown in step 68.

85. Make sure the check box next to Use Selected Features (three features selected) is checked.

86. Click **Apply** and **OK**.

87. Open the attribute table for Probationers and see if you caught anyone.

88. The answer is_____.

89. You actually need to go out about 2500 ft to find five probationers living near one of these three schools.
90. Choose **Selection → Clear Selected Features**, save your project in your student folder, and then close ArcMap completely.
91. Open Excel.
92. Click on **File → Open** and browse to the location where you placed your SexPO.dbf table. You will need to change the file type to DBF 4 to be able to see the file in the Browse window. Open your DBF file of the nine probationers, and save it as an Excel file called SEX_PO.xls.
93. This concludes this massive exercise. When you are familiar with each of these functions and processes and you logically put the steps together in the right order, you can accomplish anything in ArcMap. Each of these steps would take between 2 and 10 minutes for a working analyst.

Exercise 4a: Address Locators and Find Addresses

Data Needed for This Exercise
- Street centerline file.
- Sonorita high-resolution aerial photo.

Lesson Objectives
- Learn what an address locator is and how to create one. Use the Find function in ArcMap to find addresses quickly and easily.

Task Description
- You are working in the City's emergency operation center (EOC) as the GIS analyst on call for a chemical spill. When you arrive, the commander directs you to find the address of the spill on the map and display the map with an aerial photo and streets on the screen while resources are directed to the area.

1. You will need to create a new ArcMap project from scratch for this project.
2. Open ArcMap through the Start menu and begin adding layers to a new ArcMap project.
3. Use the **Add** button to add in the GLN_Surrounding_Streets data from the C:\CIA\GIS_Data\Streets_Transportation folder.
4. Also find and add the aerial photo for Sonorita in high resolution (C:\CIA\GIS_Data\AerialPhotos). This file may or may not be missing its spatial reference (you will receive an error message if it is missing this information). To define the projection for this file,

click on the **ArcToolbox** icon (it looks like a little red toolbox) and select **Data Management Tools**, then **Projections and Transformations**, and then **Define Projection**. At this point, a dialog box will appear, and you will need to select the file Sonorita_higRez .tif as the Input Data File or Feature Class by using the drop-down menu. For the Coordinate System, click on the icon to the right to get a Spatial Reference Properties box. Select **Import** and browse to the CIA\GIS_Data\Boundaries\Glendale_City_Boundaries.shp file. Click **Apply** and then **OK**, and you will see a coordinate system NAD_1983_StatePlane_Arizona_Central_FIPS_0202_FEET appear in the text box under Coordinate System. Select **OK**.

5. After loading these two sets of data, save your project as EX4a .mxd in your student folder.

6. The address where the chemical spill has occurred is 6023 W Ocotillo Rd.

7. We are going to use someone else's address locator that has already been created for this exercise, but we will create our own locator at a later time. Use the **Browse** button to find the address locator U.S. Address Finder Tele Atlas. (If you are unable to find this locator, skip to the directions for how to create your own (starting at step 14) and then return to this step.)

8. Click on the **Find** button in the Standard toolbar (it looks like black binoculars).

9. In the Find dialog box, click the **Addresses** tab. You should see a menu as shown in DVD Figure 4–1. Fill it out as follows:

    Choose an address locator: *U.S. Address Finder Tele Atlas*
    Country: *United States*
    Street or Intersection: *6023 W Ocotillo Rd*
    City: *Glendale*
    State: *AZ*
    Zip code: *85301*

    When you click the **Find** button you should get back the two addresses that show in the bottom of the Find dialog in the example. If you only get one result and it matches, use it.

10. Right click on the 6023 W Ocotillo Rd, Glendale, AZ 85301 match and choose **Add Graphic**. Close the dialog or move it to the side to see that a black dot is not on your map. Right click on the **Sonorita_higRez** aerial layer in the TOC, and choose **Zoom to Layer** from the dialog box.

11. You can also right click the address match in the Find dialog and choose **Zoom To**.

12. Now close the Find dialog and click on the black dot to get the handles around it. Right click on the black dot to change the symbol properties for it. Make it a red triangle about 18 points in size. Make sure you change your tool to the black arrow (Select tool) before clicking on the black dot or the Find menu will pop back up again!
13. We have completed the assignment for the EOC manager at this point.
14. While we are waiting for further orders, we are going to create our own address locator that we can use with this ArcMap project instead of the one that comes from the ArcMap software and the Internet. We do this because we know that our streets are much more accurate than these free streets, and we want accuracy to be part of everything we do.
15. The ArcToolbox (looks like a little red toolbox) tools can be used for a variety of things. Click on this icon and open the ArcToolbox.
16. Find the Geocoding Tools section and expand it. Find the Create Address Locator tool and double click it.
17. In the dialog box, set up the fields as follows (see DVD Figure 4–2):

    Address Locator Style: *US Streets with Zone*
    Reference Data: *C:\CIA\GIS_Data\Streets_Transportation\GLN_Streets_Plus.shp*
    Role: *Primary Table*

    You should also have the following:

Field Name	Alias Name
House from Left	L_FROM_ADD
House to Left	L_TO_ADD
House from Right	R_FROM_ADD
House to Right	R_TO_ADD
Prefix Direction	STREET_PRE
Prefix Type	<None>
Street Name	STREET_NAM
Street Type	STREET_TYP
Suffix Direction	STREET_SUF
Left Zone	L_CITY_NAM
Right Zone	R_CITY_NAM

18. When we set up an address locator, we need to determine what address styles we will try to find. The two most commonly used styles in law enforcement in Maricopa County are US Streets

and US Streets with Zone. US Streets just requires input of an address or intersection. US Streets with Zone requires the address or intersection and a zone (in our case we are using a field called L_City_Nam and R_City_Nam for the names of the cities). This can often be a zip code if your data is kept up-to-date well enough with the correct zip codes for each street segment and you have the zip code stored in your RMS/CAD system.

19. In this case we are also using the Gln_Streets_Plus shapefile as our reference source or the data in the street centerline we wish to match against.
20. We also set the address ranges along the left and right sides of the street along with the street name prefixes, names, types, and suffixes.
21. When we click the **OK** button, ArcMap indexes all of these fields to optimize geocoding and address finding and then tells us that the creation effort was a success.
22. ArcMap also automatically saves the address locator in the same directory as the street centerline shapefile we used as the reference layer.
23. Now open ArcCatalog from inside ArcMap by clicking on the button that looks like a yellow filing cabinet.
24. In the directory tree, browse to the C:\CIA\GIS_Data\Streets_transportation folder and view the contents. You should now see three files listed:
    - GLN_STREETS_PLUS.shp
    - GLN_STREETS_PLUS_CreateAddre
    - GLN_SURROUNDING_STREETS.lyr
25. The middle one, GLN_STREETS_PLUS_CreateAddre, with the reddish symbol to the left is the address locator.
26. Double click this file to open a dialog about this address locator.
27. The first thing we will do is check the box indicating we want to save relative path names. This means that if we move all our data in the same directory structure to another drive, everything will work correctly. If we do not check this, then we would have to recreate the address locator if we move the data anywhere.
28. We also want to set the offset to 100 ft and make sure that we check the box next to Add the XY Coordinates so that the process will add them to the attribute table of our data for potential use in the future (see DVD Figure 4–3).

29. The reason we use 100 ft is because houses or businesses are not located right down the middle of the street. Most are shifted north, east, south, or west of the street by 50–250 ft. When we use 100 ft, we are simply telling ArcMap to avoid putting the point directly on the street but offset it north, east, south, or west 100 ft from the street centerline where the actual building might be more logically located. This is an arbitrary distance based on your own experience, but most analysts I know use 50, 75, or 100 ft. The 100 ft distance is based on the average of actual measurements of several homes and businesses in Phoenix. The average distance from the middle of the road to a residential home was about 55 ft, and it was about 175 ft for businesses. Combining them all together gave me an average of about 101 ft, or 100 ft for short. The real key is to pick a distance for the offset that fits your city the best and then stick with it and don't geocode 45 ft today, 65 ft tomorrow, and 250 ft the next day. Be consistent to assist with data quality and analysis efforts.
30. You should also be aware that an address along a street segment is found through a method of interpolation of the closest approximate point and not the actual address in some or most cases (see DVD Figure 4–4).
31. In this case we are exaggerating the addresses to some that are not even actual addresses, but we do get this kind of data all the time. The red dots may not even be near the front of the address if we used 5350 W Glendale Av, but it would be approximated based on the length of the street segment it matched and the address ranges on it. The address ranges on the Glendale segment for this example are probably: left side, 5101–5499; right side, 5102–5498.
32. Geocoding and the various nuances of address locator settings can be a 3-day class all by themselves. Practice makes perfect for this as you see how various geocoding settings affect the accuracy of your data and analysis needs.
33. Close ArcCatalog after you have made your changes to the address locator.
34. Return to ArcMap, and click on the **Find** button again. Go to the **Addresses** tab once more, and this time click the **Browse** button next to the Choose an Address locator box.
35. Find the address locator we have been editing, and add it to this ArcMap project.

36. You will see that because we chose US Streets with Zone, we are only being asked for the street address and the zone, and the fields for zip code and country have been replaced.
37. Enter *6023 W Ocotillo Rd* and *GLN* as the zone and find it as before.
38. Right click and add a new black dot to the map for this new location based on our street centerline file.
39. Close the Find dialog and see how the 100 ft offset has changed the location of the point. Zoom in really close to the two dots to see that the red triangle is in the middle of the intersection/street and the black dot from our new address locator is more near the center on the south side of Ocotillo Rd where we would expect to find 6023 W. This demonstrates that the offset setting changes where the point is placed on the map and is "offset" from the centerline to be closer to the actual address in the real world.
40. This concludes this exercise. Save your work and close ArcMap.

Exercise 4b: Adding Tables Part I

Data Needed for This Exercise
- Everything is included in the Exercise_4 folder.

Lesson Objectives
- Adding and working with tables in ArcMap.
- Knowing when you need to geocode, when you can add XY coordinate event tables, and when you just need a table to help look up or add information to your geographic data to make it better.

Task Description
- This exercise will help you to recognize that there are three classes of data in tables that you will need to be able to work with in ArcMap. The first is a table of addresses that we will need to geocode. The second is a table of addresses that have valid X and Y coordinates that we can add to our map using Add XY coordinates to create an event theme. The final type of tabular data is data we can use to enhance the information already within a GIS layer and use for additional analysis.
- The first practical exercise is to geocode those massage businesses we created in a previous exercise for the vice squad. They are going to do undercover spot checks of each business to see if they will solicit sex, and they need to know where each business is so they can divide up the workload for their six detectives as equally as possible.

- The second practical exercise involves putting a list of liquor licensed establishments on the map quickly by taking an export table from the database program that already has XY coordinates in the table and making the dots appear on the map. The chief will use this map for a meeting with liquor control.
- The final practical exercise involves adding tabular information to a current GIS layer that will enhance services and information about the data on the map. The exercise will involve adding modus operandi (MO) factors to several burglary events to help change the symbols based on adding multiple additional fields in the table to assist detectives in seeing problems in an area of the Foothills Patrol Division.

1. Open EX3.mxd and then save it in your student folder as EX4b.mxd.
2. The address locator you created in the previous exercise should already be added to the Find button. If not, browse until you find it. Check to make sure your address locator has been added, and if not, add it to the Find button and find 8450 W San Juan Av.
3. You should notice that you do not find an exact match and only find a close address, which matched at 63% of 8301–8399 W San Juan Rd. Add the graphic point to the map and then zoom into the area to see why it did not find the address at 100%. You should see that San Juan does not appear to extend to 8400 W, and this looks like open area.
4. If you were to add and turn on the Central.tif aerial photo (found in the GIS Data folder), you would see that it shows that address to be in the middle of a farm field, which includes the streets we know are there between 8301 and 8399. This is another issue you will often run into if your city grows a great deal over time. The street centerline will not always keep up with the actual city growth depending on who in the city is editing it. You will have addresses and data that cannot be displayed on the map at times. Although this does not pose significant problems most of the time, you need to be aware of it and take steps to avoid it by making sure you keep up with the street centerline and always have the latest copy and most up-to-date version.
5. This will be our first practical exercise in geocoding. First, minimize (but do not close) ArcMap.

6. Find the Massageparlors.xls document you created in exercise 3. If you cannot find your Excel spreadsheet of addresses, you can use the copy in the C:\CIA\Exercises\Other folder.
7. We are going to open this Excel file and export it as a DBF 4 table. Before we can do that we need to name the data range and make sure our field widths are as wide as they can be to avoid any data truncation during the conversion.
8. Open the Massageparlors.xls file.
9. Highlight the entire range of data, and then go to the menu bar and choose **Insert → Name → Define**. Type in *database* in the name field area, and then click the **OK** button. You have just named this range of data and called it "database."
10. Now resize all your field columns so that each one is slightly larger than it needs to be for the data in the rows.
11. Make column A (name) about 35–40 characters wide and the other columns at least 25 wide.
12. Now go to **File → Save As** and save the data as Massage.dbf. You will need to change the Save As Type box to say *DBF 4 (Dbase IV) (*.DBF)*.
13. Close Excel and go back to your EX4b.mxd in ArcMap, then click on the **Add** button.
14. After adding the table to ArcMap, view the attributes of the table by right clicking it in the TOC and choosing **Open**. (Remember to click on the **Source** tab at the bottom of the table of contents to view the table you just added.) You will see that all the addresses should be intact, but there were a few really long business names that were slightly truncated here and there. Because the addresses are complete, we are okay and can proceed.
15. Look at the City field name values. They may have been truncated to GLENDAL. This doesn't matter because we are going to calculate these all to say "GLN" anyway to match the L_City_name and R_City_Name values in our street centerline file. Make sure there are no records selected, and then right click the **City** field name. Choose **Field Calculator** and type in *GLN* in the white box in the field calculator. Click **OK** and **Apply**. Every row should now have GLN instead of GLENDAL in the rows.
16. After you have made sure that all your addresses are complete, close the attribute table. If any addresses are truncated, remove the DBF from ArcMap, reopen the Excel file, resize the address field wider, and save it as a DBF again. Keep doing this until your data in each field is not truncated or is acceptable.

17. Right click on the **Massage** table in the TOC, and this time choose **Geocode Addresses** from the pop-up menu.
18. A new dialog will open and ask you which address locator to use. Choose the only one in the dialog menu. If you do not see one there, you will have to click **Add** and add in the address locator we previously created, which is located in the C:\CIA\GIS_Data\Streets_Transportation folder.
19. Click **OK**.
20. In the Geocoding Addresses Dialog make the following settings:
    Street or Intersection: *Address*
    Zone: *City*
    Output shapefile or feature class: Call it *MassBiz.shp* and put it in your student folder
21. Click **OK** to begin automatic geocoding.
22. You should get about 30 addresses that geocoded at above 80%, 1 at 60–80%, and 3 unmatched. You should also find two matches that tied.
23. Click on the **Match Interactively** button.
24. Our first address is entered as 55th & Acoma if you used the sample data and not your own.
25. To make this address geocode better, we need to change it to say N 55th Av & W Acoma Dr. Change the address in the box and press **Enter**. You should now see a 100% match in the white box at the bottom. Double click on this match to interactively match this record.
26. The next one has a similar problem and can be handled the same way as N 71st Av & W Maryland Av.
27. The last error says 7200 W Bell Av. We need to change it to 7200 W Bell Rd and then match it.
28. It is a very good idea to always check the ties like this as well to make sure that an address of 18234 N 59th Av is not accidentally geocoding to 18234 N 55th Dr or 18234 N 55th St. These are all valid addresses, but they're not the ones we want, and thus ties can be tricky sometimes unless you verify them.
29. When you have finished matching interactively, click the **Done** button, and you should see your points appear like magic on your map. You would then logically create a layout of the points with other themes to help the detectives find them and probably give them a list of all of the massage businesses in Excel so they can assign them for follow-up.

30. Save and close your project.
31. Open Windows Explorer and navigate to where you saved exercise 4b. Find and double click EX4b.mxd and watch while it reopens for you. Save the exercise as EX4C.mxd in the same folder, and continue this second portion of our practical exercise.
32. Liquor control is going to assist our officers with some liquor enforcement, and we need to give the chief a list and a map of all of the liquor licensed establishments in Glendale. He would like to know which ones may be bars and taverns versus markets and grocery stores.
33. You contact the city business license office and obtain a TXT file of the liquor licensed businesses in Glendale. This file is located in C:\CIA\Exercises\Exercise_4\liquor.txt. We need to open this file and check out the file structure before we bring it into to ArcMap.
34. Click **Start** → **Programs** → **Accessories** → **WordPad** (this may also be called "Notepad").
35. When you open the data, you can see that this is a comma delimited file (commas are between each column value, and each record is in its own row). There are also text items surrounded by quotation marks. A DBF file and a comma delimited text file are data sources we can bring directly into ArcMap.
36. Add the Liquor.txt file to your ArcMap project just like you previously did with the DBF table.
37. Review the data attributes and make sure that everything looks good and the fields are not messed up.
38. Close the table and right click on the **Liquor** table in ArcMap.
39. Choose **Display XY Data**.
40. In the dialog box, make sure that the following fields are set as:
    a. XField: *X_Coord*
    b. YField: *Y_Coord*
    c. Coordinate System of Input Coordinates: Make the following changes:
        i. Click the **Edit** button
        ii. Click the **Select** button in the next dialog menu
        iii. Choose **Projected Coordinate Systems** in the Browse for Coordinate System dialog menu
        iv. Choose **State Plane**
        v. Choose **NAD 1983 (Feet)**
        vi. Choose **NAD 1983 StatePlane Arizona Central FIPS 0202 (Feet).prj**

vii. Click **Add**
viii. Click **Apply**
ix. Click **OK**
x. Click **OK** to generate the XY Coordinate data layer

41. You may be wondering, "Where are my 271 liquor establishments?" Click on the little world icon in the Standard toolbar and see that your liquor points are about 35–40 mi west of where they should be. This little tricky and sticky situation happened because we forgot to ask the database folks what projection their data is in when we got it from them. A projection is a set of mathematical formulas that attempt to take the 3-D surface of the Earth and project it onto a 2-D piece of paper. If you imagine the Earth as a big light bulb with a piece of paper wrapped around it, the places where the paper is farthest away from the Earth's surface would be distorted, and the places where it touches would be more accurate. This is what projections do. State plane coordinates use a single reference point on the Earth's surface. A very detailed explanation of projections in general can be found at: http://www.colorado.edu/geography/gcraft/notes/mapproj/mapproj_f.html.

42. There are many other places on the Internet where you can find out more about projections. The key factor is knowing what projection is used for the data when you receive it from other persons and, of course, what data projection you use every day. You can still share data, even if it is not in the same projection, because you can specify the correct projection when you import the data, and the software will know how to line it up.

43. Let's go back to the drawing board and remove the liquor XY event theme from our project, and instead of using NAD 1983 State Plane, we'll use NAD 1927 as follows:

a. XField: *X_Coord*
b. YField: *Y_Coord*
c. Coordinate System of Input Coordinates: Make the following changes:
   i. Click the **Edit** button
   ii. Click the **Select** button in the next dialog menu
   iii. Choose **Projected Coordinate Systems** in the Browse for Coordinate System dialog menu
   iv. Choose **State Plane**
   v. Choose **NAD 1927**
   vi. Choose **NAD 1927 StatePlane Arizona Central FIPS 0202.prj**

vii. Click **Add**
viii. Click **Apply**
ix. Click **OK**
x. Click **OK** to generate the XY Coordinate data layer

44. Our points are now where they are supposed to be. This is currently just a temporary GIS layer that is displaying the dots and acts much like a shapefile. Some things may be out of our reach with an event theme like this because it is not a true GIS layer but is mimicking one. We can now easily make this into a shapefile that has all the bells and whistles by right clicking the event theme layer and choosing **Data → Export Data** and saving it as LiquorBiZAll.shp in our student directory.

45. Say **Yes** to adding it to the ArcMap project, and then remove the event theme and table from the project because we no longer need this data. They are just duplicates of the shapefile information, and we need to keep our work space uncluttered.

46. Now we need to see if we can determine which businesses are bars and taverns and which are other types. A field called Sic_Code_D should help us with that item.

47. Create a new legend for the liquorBizAll.shp layer that uses the Sic_Code_D as the Categories → Value field. (*Hint: Go to Properties for this layer and click on the **Symbology** tab.*) Add all of the unique values to the legend and apply it.

48. We want to see only those liquor licensed businesses that are like bars or taverns or drinking places. To accomplish this, remove from the legend all of the values that are not one of these types.

49. Right click the **Liquor theme → Properties → Symbology** and go into the little window where the individual symbols are located for each Sic_Code_D value.

50. We are going to remove any field names we don't want. To do this we click on a field or field names and then click the **Remove** button.

51. I left just the Drinking Places, Eating and Drinking Places, Liquor Stores, and Membership Sports_Recreation Clubs values of Sic_Code_D in the legend. You can uncheck <all other values> to avoid showing the other types if you want, or you can leave it checked, but make that individual symbol a very small blue dot (size 2–3 points) and still show it on the map. (Make sure to uncheck the other layers with points displayed on the map so you can see your results more clearly.)

52. Based on just your visual observation of the bars and other establishments, where do you think the majority of them are located? _____.

53. This map and a list of the locations in Excel would provide adequate information to the chief for this new project with the liquor enforcement folks at the State level. Save the changes to your map and close ArcMap.

54. Our final little assignment is to make a pin map for burglaries in 2004 for the detectives. They want to have the pins show point of entry (POE), type of entry (TYPEENT), and premise type (residential, etc.).

55. Open EX4C.mxd in the C:\CIA\exercises\exercise_4 folder.

56. In addition to the MXD, open the Access database in C:\CIA\Exercises\Exercise_4\Burg_MO_Data.mdb.

57. There is only one table in the Access database called BURGWMO. Open this table and browse through the contents. You will see that it has fields called ObjectID, Rptnum, POE (point of entry), POEX (point of exit), Tool (tool used to commit burglary), TypeEnt (type of entry made with tool), Premise (residential, business, etc.), and Property (the type of property taken).

58. Make the ArcMap project active, and view the attributes of the Burglaries theme. (If this theme is not added to your map already, add it now. You will find it in the exercise 4 folder.) You will notice that in this attribute table there are similar ObjectID and Report number fields, but none of the MO information stored in the Access table is in the attributes of the points on your map.

59. We need to get the information out of the Access table for MO characteristics and into ArcMap, and we need to join it up with the attributes that are already known about these points. We will accomplish this through what is called an attribute join.

60. In Access, click on the **BURGWMO** table and then go to **File → Export → Save As**. Type *DBF 5 (*.DBF)* and save this exported DBF table of the MO information to your student folder as BURGMOTBL.

61. Close Access when this has been completed successfully.

62. Go to ArcMap and use the **Add** button to add this new DBF table to your project.

63. Click on the **Burglaries** theme in the TOC, and then right click it.

64. Find **Joins and Relates** in the pop-up dialog menu and click it.

65. Click **Join** and enter information as follows:
    > What do you want to join to this layer? *Join attributes from a table*
    > Choose the field in this layer that the join will be based on: *OBJECTID*
    > Choose the table to join to this layer: *BURGMOTBL* (your file)
    > Choose the field in this table to base the join on: *OBJECTID*
66. Then click **OK** to join the BURGMOTBL data to the end of the attribute table of the Burglaries theme.
67. If a message comes up asking if you want to create an index, answer **Yes** and continue.
68. Reopen the attribute table for Burglaries and view the contents. You should now see extra fields added at the far right that represent all of the MO information for each burglary from the table we found in Access.
69. Close the attribute table.
70. We need to change the legend from a single field, categories-type legend to a *multiple* field, categories legend.
71. Right click on the **Burglaries** theme in the TOC and choose **Properties**.
72. Go to the **Symbology** tab.
73. Under the Show Categories section on the left, you will see that Unique Values is currently chosen. Change this to Unique Values, **Many** fields.
74. You will now see three value fields instead of one to the right. In the top value field box, choose **Premise**; in the second, choose **POE**; and in the third, choose **TYPEENT**.
75. Now click on the **Add All Values** button.
76. Click **Apply** and **OK**, then view your accomplishment.
77. Expand the TOC legend for burglaries as needed to view your new multiple category legend that shows MO factors for each burglary.
78. You may wish to go back to the symbology menu and uncheck the <all other values> section and reapply the legend.
79. Change each symbol type and color until you think the symbols do a decent job of showing the various burglary types. See DVD Figure 4–5 to compare your map to mine.

80. We almost forgot that we only want to see those burglaries that occurred in 2004 in this map, so we will create a theme definition in which only those burglaries that occurred in 2004 will be visible.
81. Right click the **Burglaries** theme again and choose **Properties**. Go to the **Definition Query** tab.
82. Click on the **Query Builder** button and double click on the **Burglaries.MIDPOINT** field name in the left box. This places it in the white box at the bottom of the Query Builder menu. Without clicking on any other field names, click the **Get Unique Values** button.
83. Click on the >= sign button to add it to the query window at the bottom and after the MIDPOINT field name.
84. Double click on the value item date '2004-01-01' and add it to the query window. Your completed query string should now be: "Burglaries.MIDPOINT" >= date '2004-01-01'
85. Click **Verify** to make sure you built the query correctly. If this indicates it worked okay, click **OK** and then **Apply** and **OK** again to close all the menus
86. See DVD Figure 4–6 to compare your map to mine.
87. Do you see any patterns of criminal behavior in this data? Do you think the detectives could also use any charts or tables to go along with the map to make the data easier to digest?
88. Save a copy of this project in your student folder and close it.

Exercise 4c: Adding Tables Part II

Data Needed for This Exercise
- Everything is included in the student's copy of Exercise 4c.

Lesson Objectives
- Learn how to make ODBC/OLE DB connections to data in Excel and Access directly to ArcMap. ODBC stands for "open database connectivity" and is a method developed by programmers to share information across database and programming platforms. OLE DB is another method or standard like ODBC. OLE stands for "object linking and embedding." A more detailed explanation of ODBC and OLE DB can be found at http://en.wikipedia.org/wiki/OLE_DB or http://en.wikipedia.org/wiki/Open_Database_Connectivity. Suffice it to say that this is the way

ArcGIS and Microsoft Access communicate and pass data and information back and forth.

Task Description
- This exercise will take you through the process of adding data tables to ArcMap via an ODBC/OLE DB database connection. You will use the data that you have already developed or used in previous exercises in Excel or Access. Instead of creating DBFs or comma delimited exports from those programs, you will connect directly to those database sources and tables.

1. Open the EX4C.mxd project from your student folder that you saved in your last exercise. If you forgot where you put this file, or if it has been lost, use the EX4C.mxd file in the C:\CIA\Exercises\Exercise_4 folder instead.
2. The data we will connect to will come from the following two sources:

    C:\CIA\Exercises\Other\Massage.Parlors.xls
    C:\CIA\Exercises\Exercise_4\Burg_MO_Data.mdb

3. The Massage.xls spreadsheet already has the range of cells named "Addresses."
4. In ArcMap, click the **Add** button like you have done in the past to add data.
5. In the **Look In** box, browse to find the Database Connections (see DVD Figure 4–7).
6. Click on **Add OLE DB Connection**.
7. In the Data Link Properties dialog menu, choose **Microsoft OLE DB Provider for ODBC Drivers item**. (Note: If you are connecting to an Oracle Database 10g, you would use the Oracle Provider for OLE DB instead or else the link may not perform adequately.) We are doing this to create and open a path or connection to the data in Access so that it can be used in ArcGIS (see DVD Figure 4–8).
8. Click the **Next** button.
9. Under "1. Specify the source of the data section," choose **Use Connection String**, and click the **Build** button (see DVD Figure 4–9)
10. A new dialog menu called Select Data Source will come up (see DVD Figure 4–10). You will have the option to choose a file datasource (DSN) or a machine datasource. If you want the data to be available to just you when you log on to this machine, you will choose the tab labeled File Datasource. If you want the connection to be available to anyone who logs on to the machine you will use

the Machine Datasource tab instead. For our purposes, we are going to use the File Datasource tab.

11. Where the DSN Name box is shown, leave this blank and then click the **New** button.
12. In the Create New Data Source dialog menu that comes up, choose the item in the window that says **Microsoft Excel Driver (*.XLS)** (see DVD Figure 4–11).
13. Click **Next**.
14. In the next portion click the **Browse** button and call this new DSN connection *Massage* and then click **Save**.
15. Now click on the **Next** button again.
16. Click **Finish** in the next screen.
17. In the ODBC Microsoft Excel Setup dialog menu (see DVD Figure 4–12) we need to point to the Massage.xls file, which is located in the C:\CIA\Exercises\Other\ folder.
18. Click on **Select Workbook** and navigate to C:\CIA\Exercises\Other\ folder and click on **Massage.xls** and then click **OK**.
19. Click **OK** again in the Microsoft ODBC Excel Setup dialog to get back to the DSN name menu.
20. Click **OK** and **OK** in the next menus.
21. In the Data Link Properties menu (see DVD Figure 4–13), click on the **Test Connection** button, and you should get a message that says Test Connection Succeeded. If you get anything else, consult the instructor.
22. Now you will be returned to the Database Connections dialog menu, and a new item named OLE DB Connection.odc (or something similar, like OLE DB Connection(2).odc) will be added.
23. In the white box that lists all these choices, click once on the new OLE DB connection name, and then once again about a second later. You should now be able to edit the name of this connection, so rename it *Massage.odc*.
24. Click on any other item to save the name change, then double click on the **Massage.odc OLE DB** connection item.
25. You should now see a named range within this Excel spreadsheet called Addresses. Click once on **Addresses** to choose it.
26. This adds it to the name box below.
27. Click **Add** to add this data to the ArcMap project.
28. Note: *You cannot have the Excel spreadsheet or Access database you are trying to connect to already open in those software applications.*

Windows "locks" the data files when they are in use by their native programs. An error message will indicate that the files are locked by some other user, and the data addition will fail.

29. In addition, for some reason you generally get an SQL connection string error message if you use the default name of "database" that Excel often uses to name ranges in a spreadsheet. If you have a named range called "database" when you go in to define a file, you can leave it there with no issues, but make sure to use the file you named instead when importing through this process into ArcMap.
30. Now the data table for massage businesses is part of the TOC, and we can geocode the addresses as we did with the DBF file.
31. Open the Addresses table and see that no data is truncated with this method. Unlike the DBF table, however, we cannot edit the City field to GLN where we see GLENDAL because OLE DB connections can only be edited in their source files and not in ArcMap (right click the field name and see that field calculator is grayed out). To correct this issue, you would remove the Addresses table, reopen the original Excel spreadsheet, make the changes, save and close Excel, and then reconnect to the data in ArcMap.
32. The same procedure is used to create a DSN OLE DB connection to an Access table.
33. Click the **Add** button.
34. Add a new OLE DB data source.
35. Choose **Microsoft OLE DB Provider for ODBC Drivers** again.
36. Choose **Microsoft Access Driver (*.MDB)** instead of the Excel driver.
37. Save this new DSN as *BURGMO*.
38. In the ODBC Microsoft Access Setup dialog, browse to the CIA\Exercises\Exercise_4\Burg_MO_Data.mdb file and select it.
39. Rename the new ODC record to BURGMO.odc, and then double click it.
40. You should see only one table called BURGWMO (see DVD Figure 4–14), so add it.
41. The table is now in ArcMap and can be joined to the Burglaries theme as it was in the previous exercise.
42. You do not need to save this exercise, so just close it and exit ArcMap.
43. Practice this exercise often because it is easy to forget the correct steps. You will find that your data is cleaner and often has fewer problems with truncated data when you use this process.

References

Barr, R., & Pease, K. (1990). Crime placement, displacement, and deflection. In M. Tonry & N. Morris (Eds.), *Crime and justice* (Vol. 12). Chicago: University of Chicago Press.

Block, C. R. (1998). The geoarchive: An information foundation for community policing. In D. Weisburd & T. McEwen (Eds.), *Crime mapping and crime prevention: Crime prevention studies* (Vol. 8, pp. 27–81). Monsey, NY: Criminal Justice Press.

Burgess, E. (1925). The growth of a city. In R. Park, E. Burgess, & D. McKenzie (Eds.), *The city*. Chicago: University of Chicago Press.

Eck, J. E. (1998). What do those dots mean? Mapping theories with data. In D. Weisburd & T. McEwen (Eds.), *Crime mapping and crime prevention: Crime prevention studies* (Vol. 8, pp. 379–406). Monscy, NY: Criminal Justice Press.

Gau, J., & Pratt, T. (2008). Broken windows or window dressing? Citizens' inability to tell the difference between disorder and crime. *Criminology & Public Policy, 7*(2), 163–194.

Levin, Y., & Lindesmith, A. (1971). English ecology and criminology of the past century. *Journal of Criminology and Criminal Law, 27*, 801–816.

Markowitz, F., Bellair, P., Liska, A., & Liu, J. (2001). Extending social disorganization theory: Modeling the relationships between cohesion, disorder, and fear. *Criminology, 39*(2), 293–320.

Mastrofski, S. D. (1988). Community policing as reform: A cautionary tale. In J. R. Greene & S. D. Mastrofski (Eds.), *Community policing: Rhetoric or reality?* (pp. 47–68). New York: Praeger Publishers.

Park, R. (1915). The city: Suggestions for the investigation of human behavior in the urban environment. *American Journal of Sociology, 20*, 577–612.

Quetelet, M. A. (1973). A treatise on man. In *Comparative statistics in the nineteenth century* (pp. 25–75). Germany: Gregg International. (Reprinted from *A treatise on man*, by M. A. Quetelet, 1842, Edinburgh, Scotland: William and Robert Chambers)

Reisig, M., & Cancino, J. M. (2004). Incivilities in nonmetropolitan communities: The effects of structural constraints, social conditions, and crime. *Journal of Criminal Justice, 32*, 15–29.

Sampson, R., & Groves, W. B. (1989). Community structure and crime: Testing social disorganization theory. *American Journal of Sociology, 94*, 774–802.

Sampson, R., Morenoff, J., & Earls, F. (1999). Beyond social capital: Spatial dynamics of collective efficacy for children. *American Sociological Review, 64,* 633–660.

Sampson, R., & Raudenbush, S. W. (2001). *Disorder in urban neighborhoods: Does it lead to crime?* Washington, DC: National Institute of Justice.

Shaw, C. R., & McKay, H. (1942). *Juvenile delinquency and urban areas* (5th ed.). Chicago: University of Chicago Press.

Skogan, W. G. (2008). Broken windows: Why—and how—we should take them seriously. *Criminology & Public Policy, 7*(2), 195–201.

Sutherland, E. (1947). *Principles of criminology* (4th ed.). Philadelphia: J. B. Lippincott.

Taylor, R. (2001). The ecology of crime, fear, and delinquency: Social disorganization versus social efficacy. In R. Paternoster & R. Bachman (Eds.), *Explaining crime and criminals.* Los Angeles: Roxbury.

Wilson, J. Q., & Kelling, G. (1982, March). Broken windows: The police and neighborhood safety. *Atlantic Monthly,* 29–38.

Environmental Criminology

CHAPTER 3

▶ ▶ LEARNING OBJECTIVES

Chapter 3 continues the discussion on theories of crime causation but narrows the scope to theories and approaches that use both individual behaviors and physical characteristics of space to explain crime. As discussed in the previous chapter, the physical characteristics of an environment communicate social clues to the people who live, work, and play there. These clues provide potential offenders and victims with information about how they should behave to ensure their safety, pleasure, and the successful fruition of their interests. For offenders, these clues identify the method by which they might commit a crime successfully and avoid getting caught. For potential victims, these clues provide information about how to avoid being victimized. This chapter reviews several theoretical approaches that are used in developing environmental design aspects to prevent crime. These aspects include lighting, landscaping, natural surveillance and crime prevention boundaries, and building design, to name a few. After studying this chapter, you should be able to:

- Identify and explain core elements to CPTED approaches.
- Explain the basic tenants to rational choice perspectives.
- Explain the Crime Pattern Theory.
- Explain and discuss the Routine Activities Theory and its utility in understanding and analyzing crime.
- Discuss lifestyle exposure approaches and understand their utility in understanding repeat offenses and repeat victimization.
- List and describe the types of *crime displacement*, and discuss the relevant research that examines the existence (or lack) of crime displacement.
- Explain the concept of *diffusion of benefits* and its utility in understanding crime patterns.

▶ ▶ KEY TERMS

Activity Space
Awareness Space
CPTED
Crime Displacement
Defensible Space
Diffusion of Benefits
Event Dependency
Hedonism
Risk Heterogeneity
Victim Facilitation
Victim Precipitation
Victim Provocation
Virtual Repeats

■ Introduction

Chapter 2 discussed several theoretical frameworks that identify the social and physical incivilities that are important in understanding community-level crime rates. These theories emphasize the flaws in the physical environment and the gaps in social networks that are common to high-crime areas. These theories also identify characteristics of individuals who are typically found (residing or working) in socially disorganized neighborhoods. This chapter narrows our focus to the individual characteristics and the physical properties of space related to crime and victimization through the creation of increased opportunities for motivated offenders to commit crime. Every person has routines and rhythms of their daily lives that become part of who we are and what we do. Have you ever been so absorbed by a project at work that on some Saturday or Sunday your significant other has asked you to run an errand, and you found yourself on the way to work or even pulling into the place where you work when you had intended to go to the store? That drive back and forth to work has become an important part of your social makeup. On our way to our activities and employment, we may look around and see a new construction site and ask ourselves, "I wonder what they are building there?" People who are prone to crime and hunting for targets do the same thing, only their question might be, "I wonder if that compressor is locked up or if I could steal it later tonight?"

Using environmental design to control human behavior is not only popular in developing crime prevention efforts; it is also used in the private sector to increase profit margins. Casinos, for example, are strategically designed to keep people gambling for as long as possible. Big casinos incorporate several design aspects to keep customers happy, comfortable, and, most importantly, gambling. For example, carpets exhibit bright, elaborate, and fun designs, while pumped-in oxygen, mirrors, and bright lights create a stimulating environment that makes it difficult to be bored or sleepy. Alcoholic drinks are also served 24 hours a day, often free of charge, to gambling patrons by casino staff members who are typically dressed in "barely there" outfits. Event halls and auditoriums are located adjacent to gambling areas, requiring patrons to walk through those areas upon arrival and departure. Even arrival to and departure from the Las Vegas airport requires people to walk past slot machines!

One does not have to look further than a local supermarket to grasp the concept of environmental design and how it impacts human behavior. In another example outside of criminal justice, to pick up a gallon of milk at a grocery store, customers must walk to the far corner of the store, usually past impulse and convenience food items, to reach the dairy section. Knowing that people are likely to pick up milk on their way home from work (and are probably hungry and tired), the store places various snack foods and ready-to-eat items on the path to commonly needed items (such as milk) in an effort to increase sales.

Thus, the notion of physical design to control behavior, particularly criminal behavior, also is not new. Attention in both the academic and practical realm shifted to the physical environment and how it could be modified to control human behavior when crime increased exponentially in the late 1960s and early 1970s. Jane Jacobs (1961) was one of the first researchers to propose a relationship between crime and the urban city environment. In her book *The Death and Life of Great American Cities*, she made several observations about crime and the physical environment. She was also one of the first to suggest that the physical environment could be manipulated by improving the natural surveillance (the ability of the persons living in an area to be aware of what is going on in their portion of the neighborhood) of an area to reduce crime. Other researchers during this time period, including Elizabeth Wood (1961) and Schlomo Angel (1968), focused their research on the effects the physical environment could have on human behavior and how environments could be manipulated to achieve social objectives.

The point to environmental design is to guide, manipulate, and/or encourage people to behave in a desirable manner in a given situation. In neighborhood crime control strategies, this can mean a variety of things. First, the environment needs to be created in a way that encourages informal social control efforts by the people who work and reside in a neighborhood. An example of informal social control is the existence of a block watch group in a neighborhood. All of the neighbors actively participate in reporting, interceding, or watching their portion of the neighborhood for criminal or socially unacceptable behavior. They cannot literally arrest someone, in most cases, but they can give the impression to unauthorized persons in the neighborhood that everything they are doing is being watched and will be reported to the police. Second, the physical properties of a space should allow for maximum visibility so that residents can observe what is happening in their surroundings. Third, environmental clues emanating from the neighborhood should send the message to outsiders and potential offenders that committing crimes in this place would be risky and unprofitable. A very important point here is that the *interaction* of physical design and informal social control are what creates an environment that is resistant to crime. If the environment cannot control human behavior in the way in which it was intended, the design aspects examined in this chapter will likely have few affects on crime.

The CPTED Approach

Let us first examine crime prevention through environmental design (**CPTED**) approaches. CPTED suggests that the design of physical space is important in understanding criminal behavior. C. Ray Jeffrey (1971) and his contemporary Oscar Newman (1972) are credited with providing the foundations that outline how physical space should be designed to maximize its crime prevention potential.

Newman's Defensible Space

Through the Safe Streets Act of 1968, grant funding was made available to research new crime-prevention efforts. The focus of Newman's research on crime-prevention strategies in public housing projects emphasized how architectural design could play a role in reducing crime. Newman's **defensible space** model argues that physical space can be structured in a way that fosters and reinforces a social structure that defends itself. Newman, focusing his work primarily on housing projects, suggested that by improving natural surveillance and encouraging tenants to assume responsibility for the public areas within the housing project, boundaries might reduce crime due to an increase in the risk of observation. Newman identified four key elements of defensible space: territoriality, natural surveillance, image, and milieu.

Territoriality refers to the ability of legitimate users of an area or physical space to frequently use and to protect the space from nonlegitimate users. For example, an outdoor courtyard nestled within several apartment buildings might have walkways and benches for people to use at their leisure. However, if legitimate users (residents of the apartment buildings) do not use the space, others, perhaps nonlegitimate users, may use the space in ways that are harmful. *Natural surveillance* involves designing physical space in a way that allows legitimate users to observe the behaviors of friends and strangers. In theory, this allows residents to augment and bolster law enforcement by being the eyes and ears of police and by taking action against criminal and/or other socially undesirable activities. Action could include calling the police or intervening when appropriate. *Image* has to do with fostering a neighborhood environment that creates the appearance that the neighborhood is well cared for and is not isolated from the communities that surround it or the people who inhabit the area. Environmental design of this type might include landscape lighting, clean grounds and garbage receptacles, working fountains, and manicured lawns and bushes. A great example of this is the many fantastic squares of historic Savannah, Georgia. There, the squares serve as tourist attractions, historic landmarks, resting and conversational spots, and speed constraints. When one thinks of Savannah, the beauty and romance of her squares surely create the imagery. The socially disorganized neighborhoods that are discussed in the previous chapter serve as good examples of what the concept of the image is *not*. Last, *milieu* (the French word for "environment") involves placing an area within a larger community or physical space that contains territoriality, natural surveillance, and image, creating a defensible space that remains free of criminal activity. If a neighborhood with defensible space borders neighborhoods *without* defensible space, milieu is not achieved. One might think of this as the total social landscape of a neighborhood or a group of neighborhoods that border each other. In other words, if you drive through your own neighborhood and adjacent neighborhoods today, what would you find there? Would all of the homes be clean, well cared for, and project the image of the entire neighborhood, or would there be some homes with uncut grass, shabby paint, or trash in the front yard?

Newman's work focused on a comparison of public housing projects in New York City. In his study, he examined several variables, including access points, the ability of residents to observe, the size of the project, and the building structure itself. He concluded that defensible space could be accomplished in public housing projects by installing doors and windows in places that allow for increased observation of surrounding areas by the residents. In addition, he argued that installing better lighting and creating common areas that residents could both use and control would also bolster crime prevention because residents would be better able to see the behaviors of legitimate and nonlegitimate persons and could thus take action when necessary. Newman identified four types of zones:

- Public spaces: These are areas that are open to the general public and serve a variety of uses, such as a public street.
- Semipublic spaces: These are areas, such as an apartment lobby, that are limited in their use but are still open to everyone. They are used more often by residents and their friends or families.
- Semiprivate spaces: These include areas that are more restricted in use, such as an apartment hallway or stairwell, which are open to nonresidents but are most often used by residents and their friends or families.
- Private spaces: Most notably the apartments themselves, these are areas that are not open to the public and are restricted to the use of residents and their friends or families.

In comparing two adjacent housing projects (one with and one without defensible space characteristics), Newman found that the project without defensible space had both higher maintenance costs and higher crime. He extended his research to include analyses of more than 100 housing projects in New York City and reported that there was enough substantial evidence to conclude that physical design aspects have important consequences regarding crime and disorder. Newman was also opposed to high-rise projects and argued that public housing should include as much private space as possible. Newman theorized that this would move people to maintain more guardianship over the spaces in and around their residences.

Subsequent tests of defensible space in other cities mirrored Newman's results. An examination of residential burglary in Boston identified several physical site characteristics that were important in predicting which homes were more likely to be victimized (Reppetto, 1974). Wilson (1978), in his analysis, found that areas of London that lacked defensible space characteristics suffered more incidents of vandalism.

Newman added much to the study of crime and its relationship to the physical environment, but his work is not without criticism. He has been criticized for ignoring the social elements of tenants in housing projects and of those residents in surrounding areas. That is, the physical design elements, while important, may offer far less value in predicting crime rates than social level variables. Taylor et al. (1985) found that the relationship between crime and physical design features was spurious and that neighborhood social status was much more predictive of crime. Merry (1981) also touched on the fact that while environments could certainly be redesigned to better suit defensible space, residents must want to assume a role in the guardianship of the space. Furthermore, her results suggest that physical design strategies may not be entirely effective in spaces with heterogeneous (culturally, racially, or ethnically diverse) populations.

CPTED: Theory and Research

In *Crime Prevention Through Environmental Design*, C. Ray Jeffery (1971) examined design aspects that contribute to crime prevention and how they could be applied in nonresidential areas, such as schools, to control human behavior. By incorporating elements of behavioral learning theory, Jeffery argued that the removal of reinforcements for crime from schools would reduce incidents of crime. While Jeffery's original model has been revised, the basic concept of crime prevention through environmental design (CPTED) is that by changing the environment (stimulus) we can change the behavior of the offender (response). Crowe (2000) outlines the assessment of CPTED with what he calls the "Three-D approach." Essentially, the Three-D approach is based on the notion that human space is designed to fulfill three functions (Crowe, 2000, p. 39):

- All human space has some designated purpose.
- All human space has social, cultural, legal, or physical definitions that prescribe the desired and accepted behaviors.
- All human space is designed to support and control the desired behaviors.

Space, then, can be evaluated based on *designation*, *definition*, and *design*. TABLE 3-1 utilizes the Three-D approach to assess a well-known theme park as a useful example. Essentially, space that is designated with a specific goal in mind, that is well defined, and that is designed appropriately for the purposes it was intended is least likely to experience crime or other socially undesirable behaviors. From the perspective of a theme park, spaces that are designed to provide hours of family fun with well-maintained areas for visiting, playing, eating, and shopping encourage people to engage in the structured fun that is provided. Furthermore, signs, landscaping, lighting, and other physical design features tell park visitors which activities are appropriate and which are inappropriate.

Kushmuk and Whittemore (1981) argue that there is an indirect relationship between crime and the physical design of the environment and that these changes operate through four intermediate goals: improved access control, surveillance, activity support, and motivation reinforcement. Their model is similar to Newman's defensible space. Good access control exists when we have the ability to regulate who enters and exits an area or building. Surveillance is the ability for the legitimate users of a space to observe their surroundings. Activity support and motivation reinforcement have to do with creating a community atmosphere where people feel vested in their neighborhood and watch out for one another. Kaplan et al. (1978) propose that opportunity, target, risk, effort, and payoff (OTREP) explain variations in crime across people and places. This model assumes that offenders are rational and that if physical design changes limit opportunities by increasing the risk and effort and reducing the payoff, crime will decrease. Thus, physical design elements that increase the offender's risk of getting caught or increase the effort required to commit the crime will, in theory, reduce the likelihood of crime.

TABLE 3-1 The Three-D Approach: Anywhere Theme Park, USA

Designation	
What is the designated purpose of this space?	Family entertainment for people of all ages. This space also caters to people with disabilities.
What was it originally intended to be used for?	Family entertainment, same as above.
How well does the space support its intended use?	Very well. Park designed with a high degree of safety and security for the patrons. Park structures are properly gated, metal is coated with rust-resistant, thick paint, and structures are designed with few hard or sharp edges. Grounds are kept in neat and safe order, trash is properly contained, and parking lots are well lit and well attended by live personnel and CCTV. Fences and gates are locked and signed as needed, pathways are relatively free of tripping hazards, restroom facilities are well marked and placed in numerous locations. Ride lines are housed under roofs and are relatively out of the weather. Exit routes are well marked, and emergency staff is on duty all open hours.
Is there conflict?	No. Nothing in the park suggests the area is to be intended for anything other than family entertainment. There is no structure, pathway, or area designed for anything unrelated to family entertainment or the administration of the park.
Definition	
How is the space defined?	Spaces are defined with clear pathways and borders, signs, lighting, music, and themes. Also, a wide array of colors is used to mark specialized areas, such as first aid and restaurant areas.
Is it clear who owns it?	Yes. Theme park insignia and corporate logos are everywhere. Ownership is a sense of pride.
Where are its borders?	The park has fencing and heavy signage around its perimeter. Individual areas within the park use landscaping, walkways, and physical obstacles, such as a waterway or animal enclosures and fencing, to define borders. In some areas the space is designated by music and the design of buildings. Generally an omnipresent theme designates each area, such as that of the food court, small children's area, gaming area, and general ride area.
Are there social or cultural definitions that affect how that space is used?	Yes. The culture of this park is about innocent family fun. Themes of friendship, fun, and happiness abound. Characters are typically on display and act as ambassadors of kindness, fun, and imagination. Nowhere will one find a theme related to negativity or harmfulness. Even those characters who may be thought of as harmful in the real world (such as pirates) are friendly, full of spirit and adventure, and neglect to harm a soul. While stereotypes of sexism may be observed, traditional gender roles are often combined, resulting in male heroes who take care of infants or female warriors who also take care of their families. Elements of racism are stricken in exchange for themes of multiculturalism and diversity.

Definition (continued)	
Are the legal or administrative rules clearly set out and reinforced in policy?	Rules are clearly printed on park handouts, such as maps and signs. There is ample staff present to enforce the rules, and all seem keenly aware of various restrictions and park guidelines. Visitors are searched upon entering the park to prevent various items from entering the park, identification is required to purchase tickets, close monitoring is conducted via CCTV.
Are there signs?	Yes. For every occasion, at every turn.
Is there conflict or confusion between the designated purpose and definition?	No. Each area is specifically designed for its purpose. This not only creates a richer guest environment but facilitates the movement of people through the area, increasing its overall productivity. The numbers of people who use these areas are too great to allow confusion. This would create a bottleneck in the flow of foot traffic and general dissatisfaction among the guests.
Design	
How well does the physical design support the intended function?	Very well. The color, design, and landscaping set the mood of family fun. Colors and characters are age appropriate, music is thematic, and structure sizes are appropriate to the user. There are no structures that do not contribute to the theme of the area or the park. Walkways are wide and well lit, clean, and level. Exits are clearly marked, doorways are well marked and wide, and walkways are indirect, leading to a sense of exploration and bigness.
How well does the physical design support the definition of the desired or accepted behaviors?	Very well. There are signs everywhere indicating directions, rules, and suggestions for park visitors to follow. Rides, games, and sitting areas are clearly separated or integrated, depending on the design of the area. Facilities and distractions are properly restricted where movement and traffic flow is necessary.
Does the physical design conflict with or impede the productive use of space or the proper functioning of the intended human activity?	No. All facilities and structures contribute to the purposeful movement of traffic. Again, restrictions are kept to a minimum so as not to create a greater problem in traffic management.
Is there confusion or conflict in the manner in which the physical design is intended to control behavior?	No. All facilities are designed to control and direct large amounts of traffic. Confusion would hinder the operation and productivity of the park.

In their review of the literature, Rosenbaum et al. (1998) divide CPTED strategies into the categories of target hardening measures, access control strategies, surveillance enhancement efforts, and community-building measures. Target hardening measures appear to be fairly successful in increasing the effort and risk for potential offenders. Measures including deadbolt locks, solid-case doors, window restrictions, steering column locks on cars, and alarms have all produced some measurable decrease in specific crimes. Access control measures can also be successful but may be difficult to implement. Research suggests that buildings and areas with multiple access points (and limited surveillance) have more crime. The most common type of surveillance enhancement is to increase lighting. However, the jury is still out on whether or not simply improving lighting is effective in reducing crime. Research in this area has found conflicting results. (See Farrington and Welsh, 2002, for a summary of the literature.) One reason for these conflicting results may be how improved lighting is operationalized. Another issue is the choice of crime measurement. (Chapter 6 examines the strengths and weaknesses of different data sources.) Pease (1999) suggests that lighting may reduce crime if it increases outdoor activities, which then increases surveillance abilities. In addition, lighting increases visibility and thus increases the risk to the potential offender.

The use of closed-circuit television (CCTV) for surveillance has grown in popularity in recent years, especially in England. In an evaluation of 13 CCTV ventures, Gill and Spriggs (2005) found mixed results on its ability to thwart crime. In some cases, areas with CCTV sustained more crime than control areas, and in other cases CCTV-equipped areas sustained less crime than the control areas. An interesting report by Jane Black (2003) in *Business Week* suggests that British residents are fed up with the use of CCTV cameras (used primarily for catching speeders) and that:

> The destruction of these surveillance cameras—which cost between 30,000 to 50,000 British pounds each (between $50,000 and $80,000)—has become a near-weekly occurrence in the British Isles. Farmers in Somerset have been charged with using speed cams and closed-circuit TV cameras (CCTV) for target practice. In Cambridgeshire, vandals set one afire. Earlier this month, one creative hooligan knocked down a speed cam by attaching a rope from the back of his car to the camera's pole and driving away—a mini reenactment of the toppling of Saddam's statue in Baghdad.

Furthermore, "Ten years later, it is clear that CCTV has done little to clean up the streets. Study after study shows that CCTV simply displaces crime to areas where no cameras are present rather than preventing it" (Black, 2003).

Lastly, well intentioned efforts designed to build community cohesion have found varying levels of success and failure. This may be in part due to the operationalization of key variables. For example, what exactly *is* a "community"? In addition, a primary assumption of CPTED approaches are that changing the physical environment will change the social environment. What happens if courtyards are constructed, benches are placed, and lighting improves, but the legitimate users of an area still do not use the spaces?

■ The Environment, Opportunity, and Decision Making

Felson and Clarke (1998) outline 10 principles of opportunity—the cornerstone for all criminal behavior. Essentially, opportunities are important in understanding crime, and these opportunities are usually highly specific, concentrated in time in space, and dependent on everyday movements. In addition, opportunities for crime differ with social and technological changes, and some opportunities are more tempting than others. Furthermore, crime opportunities can be reduced, producing significant impacts on crime with little to no displacement. The theoretical approaches framing these 10 principles of opportunity include the Rational Choice, Routine Activities, and Crime Pattern Theories.

Rational Choice Theory

The roots of rational choice perspectives can be traced back to the Classical School of Criminology. Recall that the major premise of the Classical School of Criminology is that offenders choose to commit crime based on their perceptions of risk and reward. Thus, offender behavior is guided by **hedonism**, the principle of pleasure versus pain. Essentially, criminals make decisions based on what is best for them at the time, seeking pleasure and avoiding pain. Sounds simple, right?

Recent rational choice perspectives argue that offender decisions are based on the perceived effort and rewards in comparison to the consequences of committing crime, including the likelihood and severity of punishment (Cornish & Clarke, 1986, 2003). Modern rational choice theorists identify multiple factors in offender decision making, including time constraints, cognitive ability, and available information. These theorists argue that decision making must be examined from a crime-specific focus. In addition, theorists contend that criminality and crime are fundamentally distinct concepts and must be separated under analysis. For example, Cao (2004) states that:

> Crime is an event; criminality is a personal trait. Criminals do not commit crime all the time; noncriminals may on occasion violate the law. Criminal involvement refers to the processes through which individuals choose to become initially involved in particular forms of crime, to continue, and to desist. The decision processes at these three stages are influenced by different sets of factors and need to be separately modeled. Some high-risk people lacking opportunity may never commit crime; given enough provocation and/or opportunity, a low-risk, law-abiding person may commit crime. The offender is seen as choosing to commit an offence under particular conditions and circumstances. The decision making is not always fully rational, or even properly considered; instead the perspective emphasizes notions of the limited rationality. The offenders thus are variable in their motives, which may range from desires for money and sex to excitement and thrill seeking. Offenders' ability to analyze situations and to structure their choice, to switch between substitutable offenses may also vary, as may their specific skills to carry out a crime. (p. 33)

Borrowing from the field of public administration, we can apply Simon's (1976) notions of "bounded rationality" and "satisficing" to the study of criminality. Bounded rationality assumes that people approach decision making with imperfect and incomplete information and that because it is impossible to make completely rational decisions with this information and we are constrained by time and other obstacles, we stop searching for solutions when we come across one that is good enough; in other words, we are not satisfied with perfect decisions but are satisfied with imperfect ones we think will work.

Apply this to known crime trends (discussed in Chapter 7). For example, most crime is intraclass. Working on a rational offender model, a poor offender committing robbery against a poor victim does not seem to make much sense. Wouldn't a rational offender want to rob someone with more money? The short answer is yes, and given the opportunity to do so, he probably would. However, it may be riskier or take more effort to find someone with more money. The robber may have to travel farther to find a wealthier victim, which further increases the effort required. In addition, if the offender has to travel to an area he is not familiar with, his perceived risk of getting caught also increases. In the end, the meager payout from robbing a poor person, in an area the robber is somewhat close to and familiar with, might in fact be a better deal. It is even possible that an offender's drug addiction may cause him to feel a sense of urgency, and thus close targets with little reward (that can be hit more frequently) could be part of his decision-making process to have enough money to buy the drugs he needs. Thus, recent rational choice perspectives offer a better framework for understanding imperfect decisions in an imperfect and dynamic environment.

In crime mapping and analysis, offender decision making is an important variable and must be understood for analysts to make useful predictions about future criminal offending. For example, using GIS to analyze a series of robberies in and around Phoenix, Arizona, Catalano et al. (2001) included some important assumptions about offender decision making in their model. First, they separated earlier and later crimes in this series based upon the assumption that decisions about where to commit crimes changes as the level of professionalism increases. Second, they only included those potential targets within three miles of a freeway (the average plus one standard deviation distance of the past targets' locations from freeway access), assuming that future decisions about the attractiveness of a target would be similar to past targets. Furthermore, an average daily net (the total score of a robbery divided by the number of days between it and the next robbery) was utilized to make predictions about when the next robbery would occur. Thus, Catalano et al.'s model attempted to make predictions about criminal decision making as to *where* and *when* the robbers might strike again.

As you can see, an understanding of criminal decision making is important to the analysis and mapping of crime. The next section takes the notion of rational decision making to the next level by including a discussion of which factors are most important to understanding crimes of opportunity. Specifically, the highest likelihood for crime exists in certain situations where the perceived risk and effort of crime is low and the perceived payout is high.

Routine Activities Theory

In Routine Activities Theory, both offenders and victims play a role in the criminal event. In everyday life, people travel back and forth to work, school, play, and home. It is during these normal movements and activities that potential offenders and victims come in contact with one another. Thus, through the normal course of business, people may increase or decrease their risk of victimization based on their patterns. For example, a person who works in an office building Monday through Friday and returns home after work probably has a low chance of becoming a victim of robbery or assault. However, when that same person attends a crowded or rowdy bar or takes part in a volatile situation (for example, a sports victory celebration that is out of control), the risk of being assaulted or robbed increases. In two recent instances we have seen a pattern of excessive substance use (suitable target, motivated offender) combined with the lack of a capable guardian result in the deaths of two young, college-aged women in Aruba (the Natalee Holloway case) and New York City (the Imette St. Guillen case). In addition, patterns of movement and activity have changed dramatically in the last 30 years. For example, many households today require two incomes. Having both guardians removed from the home every day (and removed from neighbors' homes as well) significantly reduces the offender's risk of getting caught during a daytime burglary. The routine activities perspective argues that crime is most likely to occur when the presence of three criteria exist. These criteria include a suitable target, a motivated offender, and the absence of capable guardians (Cohen & Felson, 1979).

Operating on the premise that most crime results from an exploited opportunity, this perspective holds that offenders, rather than engaging in extensive planning, choose to commit crime simply upon meeting an opportunity to do so, and some opportunities are more tempting than others. For example, to examine the suitability of targets for crime, Felson and Clarke (1998) put forth the VIVA risk quotient. VIVA stands for value, inertia, visibility, and access. Essentially, visible targets that are deemed valuable by offenders and are portable and easy to get at are at highest risk. Another useful acronym, CRAVED, (concealable, removable, available, valuable, enjoyable, and disposable) is expanded to include the elements of concealability and disposability (Clarke, 1999). That is, offenders must also be able to hide the item(s) during the commission of a crime and quickly dispose of the stolen items in the black market.

Roman (2005) operationalized several concepts from Routine Activities Theory to examine the spatial and temporal influences on violent crime. In his study of youth crime in Prince George's County, Maryland, he developed a model of opportunity that incorporated variable clusters measuring (1) place-associated risk, (2) the potential level of guardianship (from Routine Activities Theory), (3) informal social control measures of guardianship (from Social Disorganization Theory), (4) the potential for motivated offenders, and (5) the influence of violent crimes from bordering neighborhoods. Temporal influences were measured by the time of day, day of the week, and the time of year. (All of these were grouped based on the school session, for example, commuting times to and from school, evenings, curfew period, and weekends.) The analysis also included a spatial lag of violent crimes (computed by using a weighted average of crime rates in neighboring locations). Roman's findings indicate strong support for Routine Activities Theory and argue for the inclusion of temporal influences in understanding violent crime.

> The study results provide strong support for routine activities theory, as well as support for theoretical models that explicitly account for time of day. Individuals are vulnerable to violence during times when the flow of youth is highly concentrated. At certain times of the day there will be places with high concentrations of youth and limited adult supervision. Youth hangouts, schools and busy retail establishments all influence levels of violence, but their impact on violence is mediated by the time of day. For instance, places with youth hangouts generate 40% more crime in the after-school period than during the weekend. In addition, the findings suggest that weekend routine activities may bring about very different types of opportunity for violent crime. (Roman, 2005, p. 306)

Thus, routine activities approaches hold some interesting opportunities for crime mapping and analysis and can help guide strategies for crime control and resource deployment. The inclusion of contextual variables that measure various aspects of routine activities can improve the prediction power of crime maps and analyses, increasing their usefulness.

Crime Pattern Theory

Crime Pattern Theory intersects Rational Choice Theory, Routine Activities Theory, and environmental factors to provide a comprehensive explanation of crime (Brantingham & Brantingham, 1981, 1984, 1993). In crime pattern theory, individuals have both activity spaces and awareness spaces. A person's **awareness space** is comprised of those areas with which he is familiar. Rossmo et al. state that this is "similar to the concept of a comfort zone" (Rossmo, Laverty, & Moore, 2005, p. 106). An individual's awareness space is typically (but not always) derived from the individual's activity space. An **activity space** is comprised of various *nodes* of activity or locations that represent where people live, work, and play. The routes people take to travel back and forth from nodes are called *paths*. These paths are important in calculating the journey-to-crime distances that are discussed in Chapter 4. For now, the importance of these paths to Crime Pattern Theory is that potential offenders tend to search for opportunities to commit crimes along the nodes and paths of their own activity and awareness spaces. In addition to nodes and paths, *edges* are those areas on the peripheral (both physical and perceptual) of an activity space. These edge areas are premier locations for criminal offending. This is because the level of diversity encountered here (in people from both sides of the edge and their activities) limits the surveillance capabilities of potential guardians. **Figure 3–1** illustrates a simplified model of how a person's activity space may look.

Of course, depending on how far one travels and how many different nodes and paths an individual has, activity spaces can be vast and complex. Private investigators, for example, are likely to have very large and very complex activity spaces because their job requires them to drive to multiple locations on any given day. Awareness space also consists of places that individuals have visited or are aware of, such as landmarks, relatives' homes, etc. We also have to consider that some people are very transient and may move around a lot during their lifetime, and thus their awareness space and activity space changes over time.

Crime Pattern Theory provides a framework for understanding both offender and victim behavior patterns. Individuals, including offenders, create cognitive maps of areas they are familiar with while traveling from one node of activity to another. Offenders use these maps to help them choose targets of crime. What we see here can often be that an offender searches directly around them for opportunities to commit crime that are within their own awareness space, or they may also travel to specific locations (bus stations, ATM locations, shopping

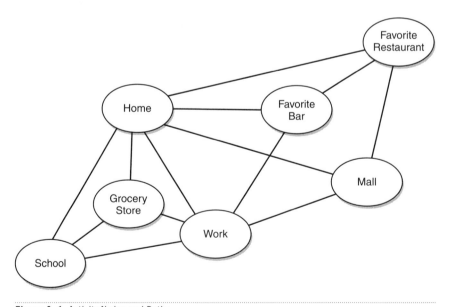

Figure 3–1 Activity Nodes and Paths

malls), often called "crime attractors," and wait for a victim. Through the processes of *recognition*, *prediction*, *evaluation*, and *action* (Smith & Patterson 1980), cognitive maps are created, targets are chosen, and crime is committed.

Whichever type of hunting one offender or another does may also relate to the obligatory time and discretionary time that they have available. An offender who works will most likely be obligated during his work hours and not be able to commit crime; however, when he gets off work, he can commit crime during this discretionary time. In any case, if the analyst can determine what type of thought processes the offender may be using, he or she can be better at predicting a new hit or determining where the offender may reside (see Chapter 4).

■ Lifestyle Exposure Approaches

Lifestyle exposure approaches to the study of victimization suggests that understanding offender behavior and decision making is only one component to crime and that victim behavior is also important, sometimes *more* important to understanding a criminal event. The notion of *shared responsibility* (while this term is not meant to place blame upon victims) implies that both victims and offenders contribute to a criminal event, sometimes in minor and sometimes in major ways. For example, a victim who leaves his car unlocked with the keys in the ignition (which was a common practice in cold weather climates during the winter season prior to the mass production of remote car starters) shares some responsibility in the theft of his automobile. A victim of robbery walking alone in a crime-ridden neighborhood at night shares some responsibility in his victimization. This is not to say that either of these victims *deserved* to be victimized. What we mean is that the decisions and behaviors of victims can create, in part, more attractive opportunities for motivated offenders. Recognizing this and taking active measures to reduce risks of victimization is the first step in preventing crime.

Karmen (2004) categorizes differential risks of victimization into several dimensions. First, the degree of *attractiveness* refers to the levels of risk and potential rewards for an offender. Victims who are likely to be carrying large amounts of cash (restaurant employees, for example) may be more attractive to potential robbers than college professors carrying bags full of books. However, the college professor, with her arms full, may be an easier target for sexual assault. Second, *proximity* refers to the geographic and social closeness an offender has with a potential victim. Targets that are easiest to reach geographically and socially are at highest risk. The *deviant place* factor (such as hot spots of crime), for example, allows greater access to victims for a variety of reasons (a large number of potential targets, lack of social efficacy, poor physical design characteristics). *Vulnerability*, a third dimension, has to do with how capable a potential victim or target is to resist being attacked. The college professor previously described would be more at risk as her arms were occupied by heavy bags of books (which could also be used as weapons if she has good upper body strength). Karmen also places victim behavior on a continuum as it relates to shared responsibility in the victimization. **Victim facilitation** refers to victims whose behavior was negligent in making themselves a more attractive or vulnerable target. **Victim precipitation** or **provocation** refers to situations where victims contributed *significantly* to the criminal event. A burglar getting assaulted by an awakened home owner would be an example of precipitation. In the case of provocation, the crime would not have occurred if it was not for the victim's behavior. An example here might be a pretty college student who dresses provocatively and then goes to a topless bar or strip club "just for the excitement." Although her intention may not be to be sexually assaulted, she places herself in a place and time, and amongst potential offenders of the crime deliberately, thereby increasing her risk of victimization.

Along these lines, lifestyle approaches suggest that the reason that victimization risks are higher for some and not others is because of the movements and activities associated with various factors. For example, younger persons in general (under age 25 years) are far more likely to be victims of violent crimes than are persons older than age 25 years. In addition, single persons are more likely to be victims of certain crimes (such as robbery) than married persons. These increased risks are not simply due to age or being married, but they are due to the changes in movement and activities that being older and being married brings. Young and single persons are more likely to be out and about in places or events (parties, concerts, sporting events) that may put them in a higher risk bracket for being victimized than older and married persons.

A person's profession may also affect his or her risk of certain types of victimization. Police officers, for example, held the highest rate of workplace violence compared to all other occupations studied during the years 1993 to 1999, accounting for 11% of all workplace violence (Denhardt, 2001). Those working in mental health occupations were second in their rates of workplace victimizations. College professors were victimized the least. Persons employed in the transportation (bus and taxi drivers, for example) or retail sales fields were more likely to be victims of robbery. Those employed in the transportation field held the highest rates of being victimized with a weapon. Persons working in retails sales had slightly higher risks of workplace violence overall than persons in transportation, teaching, and medical fields.

In addition to legitimate activities and movements, nonlegitimate activities can also place some persons at higher risks of victimization. For example, it's not hard to imagine why prostitutes, drug addicts, and gang members have very high rates of victimization compared to the rest of the population.

Repeat victimization (similar to repeat offending) can be explained in several ways. First, **risk heterogeneity** suggests that the characteristics of targets, or places that made them attractive to offenders in their first victimization, were also attractive to other offenders in subsequent victimizations (Gill & Pease, 1998). For example, an openly gay man practicing homosexual behavior in an atmosphere inhospitable to homosexuals may become a target for victimization of a hate crime. Continuing this behavior on subsequent evenings in the same atmosphere may make this person a target for future victimizations by the same types of offenders. **Event dependency** explains that an offender chooses to revictimize a person or place based on their successful past experiences. Places where offenders successfully committed crimes are attractive places for repeat performances. For example, it is not uncommon for a liquor store to be robbed several times by the same offender. Consider also the repeated victimization of individual citizens and shop owners by organized criminal enterprises offering protection from other thugs and troublemakers. Last, **virtual repeats** are when similar targets are chosen based on their similarities to past targets (Pease, 1998). For example, grocery or convenience stores with similar layouts may be hit in a robbery series.

Lifestyle approaches and explanations of repeat victimization help us understand why some people and places are victimized at greater frequency than others. They also provide valuable cues to crime analysts and mappers about which people and places are at highest risk of being victimized, allowing law enforcement officials to implement proactive measures aimed at reducing and preventing crime. Sometimes, however, in discouraging offenders from committing crimes against specific persons and places, we may encourage them to perpetrate their crimes elsewhere.

■ Displacement and Diffusion

Some research suggests that many crime intervention efforts do not reduce crime but simply displace it (Reppetto, 1976; Clarke & Weisburd, 1994; Barnes, 1995; Clarke, 1998). There are two general kinds of

crime displacement: benign and malign (Barr & Pease, 1990). *Benign* displacement is viewed as a success in that while criminals have not been prevented from committing crime altogether, they have at least been displaced to committing crimes that are less harmful or serious in nature. *Malign* displacement, on the other hand, is when criminals begin to commit crimes that are more serious and/or harmful due to the continued successful thwarting of their previous efforts. When evaluating the success of a crime control strategy, crime analysts must make maps and perform analyses to check for displacement, examining changes in different types of crime and changes in neighborhoods beyond the area targeted by the intervention(s) to ensure that crime was truly reduced and not simply displaced.

Types of Displacement

Reppetto (1976) identified five types of displacement. Spatial or *territorial* displacement occurs when crime is moved from one place to the next. This is often thought to be the most common type of displacement. However, research is mixed concerning the amount of territorial displacement that occurs. Santiago (1998) found that efforts to reduce auto thefts in Newark, New Jersey displaced the crime to adjacent areas. Canter (1998) also observed displacement of residential burglaries occurring in Baltimore. Chainey (2000) found that the use of CCTV was effective in reducing both street robbery and auto theft but that some displacement of the auto thefts occurred in surrounding areas. Displacement of street robbery, however, did not occur.

Temporal displacement occurs when crime is moved from one time to another but stays within the same area. This could mean that an offender or offenders move from operating on a weekend day to a week day or from early morning to late afternoon. This is important to understand, especially in a series analysis where patterns of offending may change to mitigate risks of getting caught.

Target displacement occurs when criminals choose another target due to target hardening and other strategies to reduce crime. Target hardening measures, such as house alarms or dogs, may displace a burglar to a different house in the same neighborhood without such protection measures. *Tactical* displacement is when criminals develop new methods to commit the same crime. For example, automobiles are much more difficult to steal today than they were 15 or 20 years ago. It is theorized that the rise in carjackings that occurred in the 1990s was due to the increased difficulty of stealing unattended vehicles that were armed with sophisticated alarm systems and other devices. While carjacking is technically a new crime, the offender's goal of stealing an automobile remains the same. The methods, however, have changed.

In carjacking, the offender takes (or attempts to take) a motor vehicle by force or the threat of force. This is a perfect example of malign displacement, where crime prevention efforts (target hardening of automobiles) created a change in tactics (and also in this example, a change of crime) that is more harmful and serious in nature than the original crime of auto theft.

Functional displacement occurs when the offender has difficulty committing one crime due to target hardening strategies and is forced to commit a different crime. The previous example of carjacking also works as an example of functional displacement because carjacking is a separate crime from auto theft. Another example would be a burglar switching to armed robbery due to the increased target hardening efforts, such as the installation of alarms by property owners.

The final type of displacement, offered by Barr and Pease (1990), is *perpetrator* displacement. Perpetrator displacement is most common in drug manufacturing and distribution crimes. For example, law enforcement efforts may cause a drug dealer to desist from further drug dealing (perhaps he was arrested and is in jail). However, in the drug market, these efforts create a vacant position for another drug dealer to step in and take his place. Thus, we have traded one drug dealer for another.

Diffusion of Benefits

Now that we have made you skeptical about whether or not crime prevention is ever truly achieved because of crime displacement, let us discuss another possible effect of crime prevention efforts (a much more positive one), the **diffusion of benefits**. The diffusion of benefits is defined as "the spread of the beneficial influence of an intervention beyond the places which are directly targeted, the individuals who are the subject of control, the crimes which are the focus of intervention or the time periods in which an intervention is bought" (Clarke & Weisburd, 1994, p. 169). Clark and Weisburd argue that two possible sources for diffusion exist: deterrence and discouragement. For example, imagine that your local police department has announced through the media that it will be targeting shoplifters at various local stores throughout the summer months. Potential offenders may be deterred from shoplifting at stores that are not identified in the announcement for fear that they are too close in location to the stores that will receive the enhanced enforcement. Likewise, it is possible that offenders will wait until late fall to return to the stores to shoplift, fearing that stepped-up enforcement may persist until sometime in September. This buffer of space or time provides the benefit of crime prevention efforts without actually engaging in them. Offenders are discouraged from shoplifting at the stores during these times because their perceived risk has been increased by the enhanced enforcement. What are often seen in typical police departments are enforcement activity areas, or what could be referred to as selective enforcement areas. These are small geographic sections of town that have been designated for extra patrols, increased offender apprehension, or other crime prevention programs. Depending on the effectiveness of the increased enforcement in this small geographic area, we may see beneficial diffusion results not only within the small area but surrounding it as well.

■ Online Resources

- Environmental Criminology Research Inc.: http://www.ecricanada.com
- Crime Prevention Through Environmental Design: http://www.cpted-watch.com
- Crime Prevention Through Environmental Design Training Conferences: www.cptedtraining.net
- The CPTED Page: www.thecptedpage.wsu.edu
- New South Wales Government CPTED: www.designcentreforcpted.org/Pages/EnvCrim.html
- CPTED Vancouver–Environmental Criminology: www.designcentreforcpted.org/Pages/EnvCrim.html

■ Conclusion

In conclusion, crime and victimization are not evenly distributed across people or places. Several approaches to the study of victimology suggest that understanding the targets, or victims of crime, that make them more attractive to offenders can provide valuable information to the examination and study of crime. Knowing which places or people are most likely to become victims of crime allows us to make better predictions about where and when crime is most likely to occur and better informs the mapmaking and crime analysis process—a critical part of preventing and reducing crime. The next chapter continues our discussion of the environment and decision making and examines the theory and research on serial offenders.

■ Recommended Reading

Dempkin, J. (Ed.) (2004). *Security planning and design: A guide for architects and building design professionals.* Hoboken, NJ: Wiley.

Fleissner, D., & Heinzelmann, F. (1996). *Crime prevention through environmental design and community policing.* National Institute of

Justice. Retrieved February 2009 from http://www.ncjrs.gov/pdffiles/crimepre.pdf

■ Questions for Review

1. How is environmental design used to reduce or prevent crime? What crime prevention strategies utilize environmental design elements?
2. What are the four key elements to Newman's *defensible space*? Explain each briefly.
3. How do public, semipublic, and semiprivate spaces differ? Give an example of each.
4. Explain how designation, definition, and design are used to asses space in crime prevention efforts.
5. What factors are included in offender decision-making models by modern day rational choice theorists?
6. According to Routine Activities Theory, crime is most likely to occur when three criteria exist. Briefly explain these three criteria and provide an example of a scenario where crime is most likely to occur.
7. How do VIVA and CRAVED explain offender decision making and opportunity? Based on these models, which items are most likely going to be stolen from a motor vehicle? Which items would be least likely to be stolen from a motor vehicle?
8. Explain the concepts *nodes*, *pathways*, and *edges* in Crime Pattern Theory. How does Crime Pattern Theory explain both offender and victim behavior? Draw a map of your own activity space.
9. According to Karmen, what are the differential risks of victimization?
10. In what ways do victims contribute to their own victimization?
11. Explain *risk heterogeneity*, *event dependency*, and *virtual repeats*. Provide an example of each.
12. What is the difference between malign and benign displacement?
13. Explain *diffusion of benefits*.

■ Chapter Glossary

Activity Space   In Crime Pattern Theory, activity space denotes the areas that offenders and victims are most familiar with on a daily basis. It contains nodes (home, work, school) and pathways (how people travel to and from their nodes on a daily basis), which define the borders (edges) of the space where people live, work, and play.

Awareness Space   In Crime Pattern Theory, typically a broader area than activity space. Awareness space includes any area that an offender is familiar with.

CPTED   Acronym for Crime Prevention Through Environmental Design. Attributed to Jefferey (1971), this approach examines the environment as a stimulus that potential offenders and victims respond to. Essentially, CPTED argues that we can reduce crime opportunities by changing the physical environment.

Crime Displacement   When offenders change their offending patterns in response to crime control strategies, crime displacement occurs. There are six types of crime displacement: territorial, functional, tactical, temporal, perpetrator, and target.

Defensible Space   Newman (1972) argued that areas with defensible space are less likely to suffer from high crime rates. Areas with defensible space could be characterized as having high levels of territoriality, natural surveillance, image, and milieu.

Diffusion of Benefits   The benefits of a crime control strategy extend beyond the borders where the strategy is employed.

Event Dependency   Situations where the same offender (usually) commits the same offense against the same target or victims.

Hedonism   The principle that people act to pursue pleasure and avoid pain.

Risk Heterogeneity   Something about the victim or target is attractive to different offenders and is thus more likely to be victimized multiple times.

Victim Facilitation   Victim facilitation is when victims, whether carelessly or unknowingly, make it easier for an offender to commit a crime against them.

Victim Precipitation   Victims who are hurt significantly in a crime in some way, usually violence related, contributed to the outbreak of the violence.

Victim Provocation   Victims who are actually responsible for their own victimization.

Virtual Repeats   When similar targets or locations are being victimized, usually by the same offender(s); for example, a robbery series of a chain of video stores.

■ Chapter Exercises

Exercise 5a: Selecting Features Part I

Data Needed for This Exercise
- All data is already contained in EX5.mxd.

Lesson Objectives
- In this exercise we will learn how to select features on our map interactively using the Select tool.

Task Description
- There are many ways to select data in ArcMap to get the result that we need when we need it. This exercise will cover the Interactive Select tool and the different ways in which data can be selected and then used when it has been selected.
- The Foothills patrol commander has requested a map showing only the Foothills patrol division, beats, grids, and streets. Because we do not have individual patrol division shapefiles, we will need to create them using the Select tool in ArcMap and then create new shapefiles for just that division.

1. Open EX5.mxd.
2. You should see a map of the patrol division boundaries for the entire city along with some main streets and grids showing.
3. We will first select only the Foothills division with the Interactive Select tool (select features) in the Standard toolbar.
4. Click at the bottom of the TOC on the tab that reads **Selection**.
5. Uncheck all the themes except for the Patrol Divisions theme.
6. Using the **Select** tool, click in the middle of the Foothills division (northernmost patrol division) and watch as light blue select lines show up around the polygon. (If you don't remember where the Foothills division is, turn on the label features by right clicking the Patrol Divisions theme to identify it.)
7. Now we want to make a brand new shapefile from this selected polygon.
8. At the bottom of the TOC, click on the **Display** tab to return to the regular legend.
9. Right click on the **Patrol Division** theme, and choose **Data → Export Data**
10. Make sure that the top box says **Selected Features**.
11. Save this file in your student directory as FHDivision.shp.
12. Answer **Yes** to adding it to the current map project, and then turn it off after you have verified that it is only one polygon of the Foothills division.
13. Go up to **Selection** in the menu bar and choose **Clear Selected Features**.

14. We need to create a new data frame to hold all of the new shapefiles we are going to create in this project. Creating a new data frame allows us to have more than one map section and store data to specific projects within the project.
15. Click on **Insert** in the menu bar and then **Data Frame**.
16. Drag and drop the FHDivision.shp theme to this new data frame and delete the one under the Citywide data frame.
17. When you created the new data frame, it automatically became the active frame.
18. Right click on the citywide data frame and choose **Activate**.
19. By repeating this process with each data frame, you can browse the data in each one as needed.
20. Right click the data frame called **New Data Frame** and go to **Properties**.
21. Under the **General** tab, rename this data frame to say *Foothills*.
22. Click **Apply** and then **OK**.
23. Save your project as EX5.mxd in your student folder.
24. Make the citywide data frame the active data frame again.
25. Now go to the **Selection** tab at the bottom of the TOC and make the Beat layer the only selectable theme.
26. Turn on the beats in the TOC so that they are visible.
27. Make sure the **Select Features** tool is the current tool on the **Standard toolbar**.
28. Go up to the upper left corner of the Foothills division beats and click and hold the left mouse button down while you drag to the lower right side of the beats.
29. Let go of the mouse, and now all of the beats in the Foothills division should be selected. An alternative way to do this is to click on the first beat, then hold the shift key down and select the other beats until each one has been selected.
30. If any beats in Gateway division are selected by accident, simply hold the shift key down and click on the beats to unselect them. Keep doing this as needed so that only the beats in the Foothills division are selected (they should have a blue line around them).
31. Export these beats as FHBEATS.shp in your student folder, and add it back to the Foothills data frame as we did with the Foothills division boundaries. Make sure to choose **Yes** to make a new layer.
32. Clear the selected features and make the Foothills data frame the active theme.

33. Right click on the FHDivisions.shp theme and choose **Properties**.
34. Go to the **Symbology** tab.
35. At the top right is a button labeled **Import**.
36. The Import Symbology dialog menu will appear, and we want to point to the Patrol_Divisions layer for our symbols (see DVD Figure 5–1). Click **OK** and then **OK** again, and accept **Regions** as the legend item value.
37. In the Symbology window, click on the **Gateway** value, and click on the **Remove** button to get rid of that legend item because we don't need it anymore.
38. Click **Apply** and **OK** to return to the map. Make sure that beats and the division are turned on in the Foothills data frame (see DVD Figure 5–2).
39. Do the same thing for the beats; however, use the Beats theme as the layer source for our legend items.
40. You will need to remove the legend symbology for beats 30–37 and then **Apply** the legend.
41. Repeat the entire process we went through for the patrol division with the Grid and Beats layers. Make sure you
    - make only the Grid layer theme selectable first
    - do your selections and exports
    - and then make only the Beats theme selectable
    - and repeat the process.
42. For the Main Street theme we could interactively select every single street, but because we still have all the grids selected in the citywide data frame, we can use these selected records to clip the city streets and create a new shapefile of those main streets that intersect the selected grids in the Foothills Patrol Division.
43. Click the little red toolbox to open the ArcToolbox functions.
44. Click on **Analysis** tools and then find Clip.
45. Double click **Clip** to open the Clip menu.
46. Complete the dialog menu as follows:
    Input Feature: *City_Main_Streets*
    Clip Feature: Grids
    Output Feature Class: Save as *FHMainSt.shp* in your student folder (see DVD Figure 5–3)
47. Click the **OK** button.
48. If the Clip dialog menu did not close after it completed, close it now.

49. The new theme has been added to the citywide data frame. Drag and drop it into the Foothills data frame, and turn it on after making that data frame active.
50. Close the ArcToolbox.
51. We are seeing every single street and not just the main streets, right? Why do you think this is?_____.
52. The legend for City_Main_Streets was controlling the street segments we could see, but all of the street segments were actually still in the attribute table because a legend only tells ArcMap what information to display and in what manner. It does not alter the underlying data.
53. As we did for the other themes, import the legend from the City_Main_Streets for our new theme.
54. Save your project and exit ArcMap.

Exercise 5b: Selecting Features Part II

Data Needed for This Exercise
- All data is already contained in EX5b.mxd.

Lesson Objectives
- In this exercise we will learn how to select features on our map based on attribute information within the tables behind our graphics on the map.

Task Description
- The **Select By Attributes** function allows us to select information from a theme's attribute table through a SQL query.
- In this exercise, the chief has requested that you display a map of the gang members in the following two gangs:
  - Homeboys
  - Southside Glendale Locos
- We will learn how to create a temporary layer file from our selection and use the Create New selection, add to selection, and the select from selection features to get what we need.
- A gang crimes detective has an aggravated assault case where the nickname of "Clown," or "Payaso" in Spanish, is used. He believes that the suspect belongs to the Crip Locos gang, but he is not positive. He also knows that the witness described the suspect as a Hispanic male (H/M), 25 years old, 5′10″, and 140 lbs.

1. Open EX5b.mxd.
2. You will see a citywide map with the Gang_Members theme at the top of the TOC. It is turned on, showing where all these imaginary miscreants of society reside.

3. Click on the **Selection** menu item and choose **Clear Selected Features** to begin. This makes sure that if any features are selected, they are all cleared before we begin. If this item is grayed out, then there are no selections in any of the themes.
4. Go to **Selection → Set Selectable Layers** and make sure that only the Gang_Members layer has a check mark next to it.
5. Go back to the Selection menu item and now choose **Select by Attributes**.
6. In the Select by Attributes dialog menu, enter the following formula to select just the Homeboys members.
    "GANG_NAME" = 'HOMEBOYS'
7. Make sure you click the **Verify** button before running the SQL code to make sure there are no errors in your code (see DVD Figure 5–4).
8. After you click **Apply** or **OK**, you will see that several points are now highlighted on the map. These are all the Homeboys gang members. We want to save this selection temporarily in this project by creating a layer file of these selected records. Right click the **Gang_Members** theme and choose **Selection → Create Layer from Selected Features**.
9. Turn off the Gang Member theme and see that the Gang Member selection layer only has the Homeboys gang members included.
10. Right click this new layer, go to **Properties → General** and rename this temporary layer to *Homeboys*.
11. Change the symbology for the Homeboys layer to a red triangle at 8 points in size.
12. Repeat this process for the Southside Glendale Locos gang members.
13. Remember to make sure only the Gang_Members theme is the only selectable layer and that you have cleared all previous selections before you begin.
14. Make sure to set the new theme name to "Southside Glendale Locos," and change the symbol to a blue triangle or other symbol as desired.
15. Now by turning on and off these two temporary layers, you can create two maps, one for each gang member list, for the chief.
16. The gang detective is looking for a subject nicknamed Payaso or Clown, and the witness described the suspect as being a Hispanic male, 25 years old, 5′10″, and 140 lbs. It is possible that he belongs to the Crip Locos gang, but the detective is not positive of this.

17. We can begin this search for our gang member in several ways. We could be very specific and find just those Crip Locos who meet the physical description and then find those with similar nicknames, or we could find everyone with the nickname and then those that meet the general description.
18. The general rule of thumb is to start your search with the item that will give you the most records and keep whittling down the data by adding additional search criteria until you've found what you want.
19. We will start with the simplest search that will give us the most records.
20. Go to **Selection → Select by Attributes** and enter the following formula in the SL query window:

    "RACE" = 'H' AND "SEX" = 'M'
21. Make sure you verify the query and then press **OK** to run it.
22. You will see that a whole lot of the points are highlighted, and very few are not selected.
23. Let's refine this search by adding some additional query criteria to the selected records.
24. Go to **Selection → Select by Attributes** again, and this time in the Method box change it from **Create a new selection** to **Select from current selection**. This will only query the records that are already selected.
25. Notice that the previous search criterion is still in the SQL query window. This will usually stay there unless you close this session of ArcMap or replace it with a new query string.
26. Replace the previous query string with the following:

    \"NICKNAME_A" = 'CLOWN' OR "NICKNAME_A" = 'PAYASO'
27. Click **OK**, and now you should see only three or so records highlighted in the map.
28. Open the attribute table for Gang_Members, and then click the **Selected** button at the bottom of the attribute table window.
29. Now you are seeing only those three selected records. Browse the data and see which subject might be the offender the detective is looking for.
30. None of these gang members belong to the Crips, but that was just a possible item anyway.
31. We can check to see if there are any Crips gang sets that may have a nickname like Payaso by adding some search criteria to this current query.

32. With the attribute table open, click on the **Options** button, choose **Select by Attributes**, and set the method box to **Add to Current Selection**.
33. Add a new SQL query that reads:

    ("NICKNAME_A" LIKE '%CLO%' OR "NICKNAME_A" LIKE '%PAY%') AND "GANG_NAME" LIKE '%CRIP%'

34. Make sure to click **Verify** to make sure your query contains no errors. If you receive an error message, check to make sure you have the correct spacing, number of characters, and that you have the parenthesis correct as shown. You should have added one record to the selected records and now have four records.
35. The newest addition happens to be a black male with the nickname of Clown and also belongs to the Crip Locos gang. You would normally provide this information to the detective by printing it out in an organized format.
36. Save and close the project in your student folder.

Exercise 5c: Selecting Features Part III

Data Needed for This Exercise
- All data is already contained in EX5c.mxd.

Lesson Objectives
- In this exercise we will learn how to select features on our map based on where they are located near other geographic areas or within polygons of our choosing.

Task Description
- This exercise will flex those select muscles and help you to manage your searches for data and information inside ArcMap.
- There will be three tasks we need to accomplish for this exercise that all involve a tactical prediction for a robbery series in which we have already determined the most likely hit location and point around which the suspect most likely lives. Using our ability to search by location and search by attributes together, we will find a robbery suspect from existing data and also figure out which store the offender will most likely hit next. This will enable us to provide good information to investigations that are based on our analysis.

1. Open EX5c.mxd.
2. You will see a project that is zoomed into an area of town where the video bandit robbery series has been committed. The standard deviation ellipse prediction and the mean center of the crimes are shown along with each of the robberies in the series.

3. We also see a theme for adult probationers, and Glendale_persons (which includes all persons listed in police reports in Glendale.)
4. We have two basic objectives for this crime series now that we have predicted the general area where the offender may live and the general area where the next crime in the series may occur.
   a. Find suspects from available data sources of persons that live near the mean center of the crime series.
   b. Find potential next targets in the series.
5. We will use the VideoSDE2 theme as a cookie cutter to select all of the businesses that fall within its boundaries. We will then use the Select by Attributes query we learned in exercise 5b to limit our results to just video stores.
6. Go to **Selection** in the menu bar and choose **Select by Location**. The Select by Location dialog menu will appear, and you will make the following settings (see DVD Figure 5–5):

    I want to: *Select features from*
    The following layer(s): Check *businesses* only
    That: *Are completely within*
    The features of this layer: *VideoSDE2*

7. Click **OK**.
8. Turn on the Businesses theme and see that only those businesses greater than one standard deviation and less than two standard deviation ellipses areas are now selected.
9. Open the attribute table for the businesses theme.
10. Click the **Options** button.
11. Choose **Select by Attributes** from the pop-up menu.
12. Choose **Select from Current Selection** as the method type.
13. Enter the following formula in the SQL query window:

    "BUSNAME" LIKE '%VID%'

14. Click the **Apply** button to run this refinement search.
15. Notice that the initial 200 and some businesses have been reduced to 18 records, all of which are video stores or have "video" in their names.
16. While the attribute table is still open, click on the **Options Button** and then **Export**.
17. Name the file PossibleTargets.dbf and save it to your student folder.
18. We want to make sure we are saving *only the selected records* (top box).

19. Minimize ArcMap, open Excel, and load the PossibleTargets.dbf file you just saved. We would generally trim this data and make it pretty for the detectives and either provide it to them in our bulletin (name and address of each business) or print it and hand it to them as needed so they would know which targets to stake out and wait for the robber. We will not do that now, but remember that you can for the future and probably should.

20. There is an extreme amount of versatility you have with combining the Select by Location and Select by Attributes searches. The only limitation is your logic, planning, and knowledge of the data you are working with. There are very few questions that GIS cannot answer if you plan out your project and searches to suit the analysis you are performing.

21. You could also right click the Business theme while these records are selected and then go to **Data → Export** and save as a new shapefile that we could use in our map for the bulletin. You could also use the **Selection → Create** layer from the selected records function we learned in a previous lesson to show just those records on the map.

22. Our next endeavor is to try to find a potential suspect from the basic suspect description we have for each robbery and the information on persons we have available to us. Turn off the Businesses theme and clear all selected features.

23. Turn off the VideoSeries and VideoSDE2 themes, leaving the Video MeanCtr theme on.

24. The offender has traveled about 2.2 miles between each hit (use the **Measure tool** to get the total distance and divide by five robberies).

25. We can also use the Measure tool to find the average distance from the mean center for each robbery. We find that:
    - Mean center to first robbery ≈ 1.29 miles
    - Mean center to second robbery ≈ 1.69 miles
    - Mean center to third robbery ≈ 2.58 miles
    - Mean center to fourth robbery ≈ 0.99 miles
    - Mean center to fifth robbery ≈ 0.86 miles

    The average distance from the mean center is approximately 1.48 miles.

26. This means that the offender may travel about that far from the mean center again about 50% of the time. It also means that there is a good chance that the offender in this series lives very near to the mean center of the crimes.

27. Turn on the Adult Probationers and the Glendale Persons themes to see how many people we need to get rid of in our search. It seems daunting, doesn't it? However, with some logical expressions and some thought, we can whittle this list down to a manageable size for the detectives in this case.
28. You can turn those themes off now to save the time it takes to draw them all.
29. Go to **Selection** → **Select by Location** again (see DVD Figure 5–6).
30. In this dialog menu, enter the following values:

    Select features from: *Adult_Probationers* and *Glendale_Persons*
    Are within a distance of: *VideoMeanCtr*
    Check the box next to Apply a Buffer to the Features in *VideoMeanCtr* and set the distance to *1.48 miles*
31. Click **OK** to run the search.
32. The search will take quite a long time to run because it will search several thousand records in these two themes.
33. Turn these two themes on and off to see that only a circle of records around the mean center were selected. You should also recognize that there are 588 adult probationers selected and over 28,000 persons selected in these two themes. Going to the detectives with this many possible suspects doesn't make much sense, does it?
34. The suspect description in our robbery series is: white male, 25–40 years old, 130–150 lbs, black hair, brown or hazel eyes, and one witness thinks he saw a tattoo on one hand with the word "Cry" in it. Another witness saw a tattoo on his left arm that looked like "Vanessa." In one robbery the victim told police that he thought someone had been waiting outside for the suspect and acting as a lookout, but he was not sure. In at least two of the other robberies, the victims felt that the offender left in a getaway car that was driven by another white male.
35. Now that we have found all the potential offenders that live within the potential anchor point locations, we can trim the choices down by using a **Select by Attributes** query on each theme.
36. Open the attribute table for the adult probationers and make sure you do not click anywhere on the rows of data. Doing so will unselect all of the records and only select the record you clicked on. *To avoid errors, keep your finger off of the left mouse button while browsing the data in the tables with records selected.*

37. Click on the **Selected** button to show only those selected records.
38. Click on **Options → Select by Attributes**.
39. Make sure the method is set to **Select from Current Selection**.
40. Enter the following SQL code in the window:
    [ETHNIC] = 'W' AND [GENDER] = 'M'
41. Verify the query, and then click **Apply** to run it.
42. This whittles your list down to 176 records.
43. Enter a new SQL code query as follows, but leave everything else the same on the Select by Attributes dialog menu:
    [HEIGHTFT] = 5
44. Apply this new search limiter.
45. We now have 116 records.
46. Enter a new SQL query, leaving the other settings alone, and changing the SQL to the following:
    [WEIGHT] >= 130 AND [WEIGHT] <= 150
47. Apply it and see that your total records reduce to 38.
48. This data has no Age field, and only some date of birth (DOB) fields have data entered. (The DOB field has values in a text field in the yyyymmdd format; e.g., "19721010," or the DOB is missing or null). Because we cannot use this text format and DOB data very easily to give us a possible age range, let's try searching for the "Cry" tattoo instead.
49. Because we don't know what other text may be in the tattoo, we need to use the Like operator as follows:
    [SCAR1] LIKE '*CRY*' OR [SCAR2] LIKE '*CRY*'
    Note: *The wildcard character changes depending on if we are working with a shapefile (%), a personal geodatabase file (*), a file geodatabase(%), or an SDE layer(%). For more information, query the ArcGIS Help for "wildcard."*
50. Apply this new search to reduce our search down to three records:
    - Michael Brown
    - Michael Contreras
    - Frank Crittenden

51. Only Frank's record shows that he has black hair, but they apparently each have similar tattoos that are worded slightly differently in the probation data.
52. The Descr field shows that the two Michaels are on probation for drug violations, and Frank is on probation for aggravated assault.
53. In real life we would export these records to a new Excel spreadsheet and add the information to our bulletin, but let's just move on to the Glendale Persons data and whittle it down the same way.
54. Turn off the Adult Probationers layer, turn on the Glendale persons layer, and open its attribute table. Work out how to make similar queries on this layer for a possible suspect. Remember, we only want those persons within 1.48 miles of our mean center. If this selection does not already appear, you will need to perform a **Select by Location** query using the directions in steps 29–31 above.
55. Possible query steps could include:
    - [MARKS] LIKE '*CRY*'
    - [RACE] = 'W' AND [SEX] = 'M'
    - [HEIGHT] >= 504 AND [HEIGHT] <= 508
    - [WEIGHT] >= 130 AND [WEIGHT] <= 150
56. You should wind up with about eight records or so, depending on what order you applied each of these search criteria.
57. Three of the eight records will have a tattoo description with "Crystal" instead of "Cry" in it, but five good records are returned.
58. We also may have forgotten that another tattoo might have begun with "Vanessa," so we can add records back into the selection by using the same **Select by Attributes** query and changing the method to **Add to Current Selection**.
59. We wind up with 129 records. (You should get 129 records if you did what I did. If your query was done in a different order, you may get more or less records.) However, looking at the map display shows that they are all over Glendale and not just within the 1.48 miles of the mean center.
60. Keep in mind that one search parameter overwrites the other search parameter and that generally we use the location search first and then attribute queries to whittle down the list. Doing it in the wrong order can cause you to go back to the beginning and start all over. We can continue to reduce this new list of 129 records by repeating the other queries for race, sex, height, and weight. Remember to change the method back to **Select from Current Selection** before you begin, and see how many records you can refine the list to.

61. You should be able to get back down to about 8–18 records, depending on your review of the records and information. By modifying the SQL query to [MARKS] LIKE '*cry *' and including a space after the "y" in "Cry," you can avoid getting back records for persons that have "Crystal" tattoos and information as well.
62. Save and close your project to your student folder as EX5c.mxd.

Exercise 5d: Selecting Features Part IV

Data Needed for This Exercise
- All data is already contained in EX5d.mxd.

Lesson Objectives
- In this exercise we will learn how to select features on our map based on where they are located within a graphic polygon we will draw and use to make our search.

Task Description
- This exercise will continue to flex those select muscles with a new tool called Select by Graphics.
- A patrol commander has come to you and indicated that he needs some quick numbers on calls for service for a focus area that they have been working for a few months. He indicates that the area is bounded by Camelback Rd on the south, Maryland Av on the north, 57th Av on the east, and 63rd Av on the west.
- You will need to draw a polygon that generally fits this area and then use it to select calls for service that intersect this graphic polygon. You will learn how to summarize data from a single field in an attribute table for selected records.

1. Open EX5d.mxd.
2. You should only see three themes in the TOC.
3. We will use the Select by Attributes search to find the streets that bound our focus area.
4. Choose **Selection → Select by Attributes**.
5. You will choose the **City_Main_Streets** as the layer to search on, and enter the following SQL query in the window (see DVD Figure 5–7).

    [ANNAME] = 'N 57TH AVE' OR
    [ANNAME] = 'N 63RD AVE' OR
    [ANNAME] = 'W CAMELBACK RD' OR
    [ANNAME] = 'W MARYLAND AVE'

6. Click **OK** to apply the search.

7. You should see something like the following in the map display (see DVD Figure 5–8) after applying this search query.
8. Zoom into the extents of the box that is formed by the selected streets.
9. Go to **View → Toolbars** and make sure that the **Draw toolbar** is turned on. If you are turning it on now, drag and drop it at the bottom of the project window (see DVD Figure 5–9).
10. The Draw toolbar has several functions, but for now we are just going to use the polygon Draw tool, which is the tool at the top center of the pop-up menu when you click the small down arrow next to the fifth item from the left in the toolbar (see DVD Figure 5–10).
11. Think of this tool as if you are stretching a rubber band over some anchor stakes at the corners of each point where the streets we want to follow change direction. Select the **New Polygon** tool and then click at the upper left corner of our selected streets. Make sure you have zoomed in as far as you can so that the placement of the boundaries you create will be semiaccurate.
12. Click at each place where the selected streets seem to change direction slightly, and follow the curve of the streets as best as you can. Clicking once at each location where you want to place an anchor *(secures the polygon's boundary line there).* When you are back up to the upper right corner of our focus area and in line with the first point we made to anchor our polygon, then double click to create the polygon.
13. You will end up with a polygon graphic that generally follows the selected streets and makes the box that we can use as a cookie cutter to select the calls for service that intersect this graphic (see DVD Figure 5–11). Notice in the image that there are small handles around the graphic. This means that the graphic is selected, and we want it to be selected for the next step. If you have accidentally unselected it, just find the **Standard toolbar**, make the black arrow your current tool, and then click along the edge of the graphic to get the handles back.
14. Turn on the Calls for Service theme in the TOC and then go to **Selection → Select by Graphics**, which should now be available because our graphic polygon is selected on our map display.
15. By simply selecting the **Select by Graphics** search option, the search goes out and selects all the calls for service that intersect it. This can take a while because there are several thousand records to search through.

16. Open the attribute table of the Calls for Service theme and click the **Selected** button again to show the selected records.
17. Notice that over 22,000 records are selected.
18. Scroll to the right of the table and find the field named Recvdate. Notice that it is in the format of yyyymmdd.
19. Right click on the field name once and choose **Sort Descending**. Do this again, but choose **Sort Ascending**. You should have observed that the data range for these records goes from 20010101 to 20041216 (January 1, 2001 through December 16, 2004).
20. The commander wants a summary table for just that data that was reported to police between July 1, 2004 and September 30, 2004, which were the dates of his action plan efforts. We therefore need to whittle this data down a bit with the **Select by Attributes query**.
21. Click on **Options → Select by Attributes**.
22. **Select from the current selection** and enter the SQL query of:
    [RECVDATE] >=20040701 AND [RECVDATE] <=20040930
23. Click **Apply**, and then close the **Select by Attributes** menu.
24. You should now have around 2997 records instead of 22,000.
25. Scroll to the left (remember not to click on any records inside the table) and find the field called Radioname. Right click this field name to bring up the pop-up menu (see DVD Figure 5–12).
26. Choose **Summarize** from the choices.
27. Save the summary table as CFSSUM.dbf in your student folder, and click **OK**.
28. Answer **Yes** to adding it to the current map project.
29. The TOC was automatically switched to the Source tab, and you will need to expand the folder name where your student files are located. When you do, you will find the CFSSUM.dbf table we created. Open it.
30. Right click on the **Count_RadioName** field and choose **Sort Descending**.
31. You now have a list and count of all the calls for service that occurred in this area between July and September 2004 for the commander.
32. In most cases we would then doctor it up to look pretty and professional in Excel before we give it to him.
33. Save your project as EX5d.mxd in your student folder and exit ArcMap.

Exercise 6: Buffers

Data Needed for This Exercise
- C:\CIA\GIS_Data\Boundaries\GLENDALE_CITY_BOUND-ARIES.lyr
- C:\CIA\GIS_Data\Streets_Transportation\
- GLN_SURROUNDING_STREETS.lyr
- C:\CIA\GIS_Data\Other\GLN_City_Parks.lyr
- C:\CIA\GIS_Data\Schools\Area_Schools.lyr
- C:\CIA\GIS_Data\Other\Glendale_Bike_Routes.lyr
- C:\CIA\GIS_Data\Offenses\Offense_data.shp

Lesson Objectives
- Learn how to use the various buffer tools in ArcMap.
- Customize the ArcMap interface to add buttons for useful tools.

Task Description
- When you arrived at the office this morning, you had four e-mails from different people within the police department asking for data from the crime analysis unit.
- Task 1: The motor sergeant wants to know how many DUI offenses/arrests were made "along" Bell Rd in 2003 and 2004.
- Task 2: The chief's secretary has forwarded an e-mail asking you to report on the number of auto theft cases "around" Bonsall Park in the past year (2004). The chief is not concerned with the park itself, although data from the park would be nice as well.
- Task 3: A school resource officer (SRO) at the Glendale High School is trying to do some problem solving and believes that kids walking home from school between 1400 hrs and 1600 hrs are committing vandalism on the homes "near" the school, and he wants to see how much of a problem it really is. He is not sure if they are doing it in the morning between 0700 and 0800 hours, but he believes the problem is worse in the afternoon. He needs you to advise him on "statistics" for the area.
- Task 4: A city council person has advised the chief that his neighbors have indicated that the bike paths "near" his home are attracting sexual deviants who are exposing themselves to people on the bike paths. He also feels like the number of robberies along the bike path has increased over the last 3 years. The council person lives at 5230 W Hearn and indicates it is the bike path north of his home that runs east and west and not the bike path that runs north and south.

1. Start this exercise by opening ArcMap by any method you prefer and add the six layers needed for this exercise. If you have any difficulties doing this step, EX6.mxd in the Exercise_6 folder has been created for you to use.
2. After adding these layers, make sure that you have all six layers in the project, and save your project as EX6.mxd in your student folder.
3. *Remember to save often during the course of your analysis to avoid losing any work.*

Task 1: DUI Offenses
1. We will begin with task 1. To fully understand how to complete this task, we have to know what data is available to us and where we can find it. Open the attribute table for the Offense data and browse through the fields to get an idea of what each field has in it and the format of the data. Generally speaking, if the field is left justified it will be text, if it is right justified it will be a number field. Notice also that these are offenses and not individual police reports. This means that multiple offenses could be listed for one report number in our data.
2. Close the attribute table, right click the **Offenses** theme, and go to **Properties**.
3. Click on the **Fields** tab to see the format of each field in the table, its length, etc.
4. Change the **Primary Display** field to *UCR Type.*
5. Scroll down through the list and find the field called UCRType. What type of field is it? _____. What is the field length set to? _____.
6. For any good analysis, the key is to know your data and what is available to be analyzed. Analysts who only know the rudimentary sections of their data will do rudimentary work. Analysts who become intimate with their data and know a lot about every field will be able to do more in-depth and valid research and analysis.
7. For the first task we need to know how many DUI arrests/offenses there were along Bell Rd. We therefore need to limit this data to show us only those DUI offenses that are in the table. Our RMS system does not store arrest information per se, so we cannot give the sergeant actual counts of arrests. It is okay to say "no" now and then.

8. Using Select by Attribute, search the Offense data where "UCR-TYPE" = 'DUI' and click the **OK** button to run the search.
9. Turn on the Offense layer if it is not already on, and notice all the blue points.
10. Right click the Offense layer and go to **Selection → Create Layer from Selected Features** from the pop-up menu.
11. Turn off the Offenses theme, and leave only the new Selected offenses layer on.
12. To find the DUI offenses "along" Bell Rd, we need to select the Bell Rd street segments within the City of Glendale. Bell Rd runs from 5100 to 8300 W, so we need to do a **Select by Attributes** query on the street file with the SQL string of *"STREET_NAM" = 'BELL' AND ("R_FROM_ADD" >=5100 AND "R_FROM_ADD" <=8300)*. Remember that the old math order of operations logic applies with every SQL query. In other words, where you place the and's and or's and the parentheses does make a difference on what gets calculated first, and thus, the final answer. Click **Verify**. If you receive an error message, double check that your characters and spacing are correct.
13. We could also interactively click on Bell Rd with the **Interactive Select tool**, but we would need to make sure we get every single street segment and do not miss one.
14. An alternative process would be to use a Select by Location search and find all the streets that intersect the city boundaries, and then use a **Select by Attributes** query to whittle the results down to just Bell Rd.
15. Choose one of these options to get Bell Rd within the city limits selected.
16. Open up the ArcToolbox (little red toolbox in the menu bar) for the next step.
17. Click on the **Analysis Tools** toolbox, and then **Proximity Analysis** to open it, and double click on the **Buffer** tool to open the dialog menu (see DVD Figure 6–1)
18. Pick the **Streets theme** as the layer we are going to create the buffer around, and set the buffer distance to *0.10* miles. Save the file as BufferofBellRd.shp in your student folder, and click **OK** to run the tool.
19. Close the ArcToolbox.

20. Find the new theme in the TOC and turn it on. If it is not in the TOC, click the **Add** button, find your new buffer, and add it to the project.
21. Zoom in on the new buffer by right clicking the new theme and choosing **Zoom to Layer** from the pop-up menu.
22. If we open the attribute table and view the new theme, we can see that it is really 24 separate buffers around the two different street segments that make up Bell Rd between 5100 and 8300 West. Open the attribute table for the Streets theme and see that there are 24 street segments selected as well (see DVD Figure 6–2)
23. Let's recreate this buffer, but this time we will make it one buffer for the entire section of Bell Rd.
24. We will turn on the ArcToolbox again, find **Analysis Tools → Buffer**, and enter the same information in the dialog menu, except in the field named **Dissolve Type (Optional)** we will pick **All**. This means we will dissolve the borders of the buffer parts wherever they touch each other. Name the output theme BufferofBellRd2.shp
25. Apply the Buffer tool and then turn off the first buffer and turn on the new one.
26. Now we wind up with a single buffer of all of Bell Rd that was selected (see DVD Figure 6–3)
27. *Save your project!*
28. This creates a permanent area (shapefile) of our research section, and if we ever need to redo this analysis, we can because this is a shapefile and not simply a graphic on the display that we have to delete to get it out of our way when we are not using it.
29. Now, use the **Select by Location** search to find all of the records in the Offense_Data Selection theme that intersect the BufferofBellRd2.shp.
30. We wind up with 213 offenses selected in the table.
31. The motor sergeant wants the total count for 2003 and 2004.
32. At this point it is wiser to export our 213 records as their very own shapefile than continue working with virtual copies of the data (remember that the layer file is only a temporary thing).
33. Save your project again!
34. Right click the **Offense_Data Selection** and **Export** the selected records to your student folder as DUIonBell.shp. Say **Yes** to add the exported data to the map as a layer.

35. Remove the Offense_Data Selection theme completely by right clicking it and choosing **Remove** from the pop-up menu.
36. You could also delete the buffers you created because they are shapefiles and you can load them back in at any time if you need them. Keeping the TOC uncluttered goes a long way in helping to keep yourself organized when there are multiple steps needed in any analysis effort.
37. Open the attribute table of the DUIonBell.shp file from the TOC and click on **Options** → **Add Field** (see DVD Figure 6–4).
38. Add a new text field or integer field called Year. I prefer using a text field, and it only needs to be four characters wide.
39. When the field has been added, right click the **Year** field name and choose **Field Calculator** from the pop-up menu.
40. If you had browsed the Offense data in detail, you would have seen that the Midpoint field was a Date-type field. We can pull the year out of this field using date parameter queries here in the field calculator. In Fields, select **Midpoint** by clicking once.
41. In the field calculator menu, find the check boxes that allow you to choose the type of field you are making a query for and check the "Date" choice.
42. Double click the **Datepart**() function to bring it into the SQL query window (see DVD Figure 6–5), and edit it as follows:

    DatePart ("yyyy", [MIDPOINT])
43. Click **OK** to run the SQL code.
44. You should now see the year in each of the records of this field.
45. Right click the **Year** field name and choose **Summarize**.
46. Save the new file as DUIYrSum.dbf in your student folder. Click **OK**.
47. Click **Yes** to add the results to the project. Remember to click the **Source** tab in the TOC to view your new table.
48. Open the new table, and you should see a new summary table showing the count of DUI offenses along Bell Rd for each year.
49. You should be able to see that the DUI offenses are increasing from year to year. You can advise the motor sergeant of this information and send him an Excel copy or cut and paste this information directly into an e-mail reply.
50. Save your project!

51. Clear any layers of selected items using the **Selection → Clear Selected Features** menu item, and prepare for the next buffer exercise.

Task 2: Buffering Polygons to Find Auto Thefts

1. In task 2 we are going to search for auto thefts "around" Bonsall Park. We are going to define "around" to mean within 0.25 miles of the park(s).
2. Turn on the City Parks theme, and use a **Select by Attributes** query to find Bonsall Park. Make sure you use an SQL query such as "NAME" LIKE 'Bonsall%' to get both parks that have "Bonsall" in the name.
3. **Zoom in** on the selected parks (right click the theme, then Selection, and then Choose Zoom to Selected Features).
4. You may want to click the **Zoom Out** button in the Standard toolbar a few times to get away from the park a little bit after you have zoomed in.
5. We are going to use a new buffer tool for this effort, which is called the Buffer Wizard.
6. To use this tool we need to customize the ArcMap project interface and add a button to our toolbar for this wizard. For some reason ESRI hid this tool, but it is very easy to use and is something I automatically put on my toolbar whenever I am working with a new copy of ArcMap on any PC.
7. To add new functions and tools, we need to go to **Tools** in the menu bar and then pick **Customize** from the pop-up menu (see DVD Figure 6–6)
8. Click the **Commands** tab, and then type the word *buffer* in the Show Commands Containing box.
9. Notice that the choices on the left are reduced based on what we typed.
10. On the left side we see Editor, Selection, and Tools. Click once on **Tools**. On the right side, you will now see an icon and the words Buffer Wizard.
11. Drag and drop this icon onto any toolbar. I generally drop it on the Main toolbar near where we launch ArcCatalog from.
12. Close the Customize dialog menu.
13. Now click the **Buffer Wizard** icon on the toolbar.

14. Set the theme to be buffered to the City Parks theme, and make sure the Selected Records item is checked (because we only want buffers around the Bonsall Parks and not all parks). Click **Next**.
15. Set the At a Specified Distance box to *0.25*, and change the Distance Units field to *Miles*. Click **Next**.
16. Dissolve Barriers Between Buffers should be set to Yes.
17. Click on the check box next to Only Outside the Polygon(s).
18. Save the new layer in your student folder as BonsallBuffer.shp.
19. Click **Finish**.
20. Zoom to the BonsallBuffer theme when it is added back into the TOC (right click and zoom to the layer).
21. Double click the **BonsallBuffer** theme, and go to **Properties → Symbology** and change the symbol to a hollow box with a line at 2 points and the color red.
22. Now go to **Selection → Clear Selected Features** to clear any selections we may have on any data.
23. Use a **Select by Location** search to find all offenses from the Offenses theme (the original theme, not the selected DUI theme that you made) that intersect the BonsallBuffer theme (see DVD Figure 6–7).
24. You should get around 3664 offenses selected in the Offense Data theme.
25. Now we need to trim this list down to just the auto thefts.
26. Do a **Select by Attributes** search with the following settings:
    Layer: *Offense_Data*
    Method: *Select from current selection*
    SQL Code: *"UCRTYPE" = 'AUTO THEFT'*
27. Click **OK** to apply the new selection, and browse your attribute table to see that you now only have 240 records selected. Show only the selected records, and verify that they are all auto thefts.
28. Turn the **Offense** theme on, and right click it. Choose **Data → Export** from the pop-up menu.
29. Export these selected auto thefts to your student folder as Bonsall AutoThefts.shp and add the new layer into your project.
30. Turn on the new layer and turn off the Offense_Data theme completely.
31. Now we need to whittle these 240 records down to just those that occurred in 2004.

32. Use a Select by Attributes query again, and use the following SQL code:

    "MIDPOINT" >= date '2004-01-01'

33. Change the method from Selection from Current Selection back to Create New Selection, then apply the query.
34. The answer for task 2 is 69 auto thefts occurred within 0.25 miles of the Bonsall Parks South and North. Save your results and the project.

    Task 3: Schools and Time Queries
1. This task will be about the same as the previous tasks, only this time we are going to summarize data from two different time frames and use a slightly larger buffer around Glendale High School.
2. First, find and select **Glendale High School**.
3. The easiest way to do this is to simply open the attribute table, sort the school names in ascending order, and scroll down through the list until you find Glendale High School. Then click on the empty box to the far left of the record to select it.
4. If we are going to do this analysis periodically, we will want to create a permanent buffer shapefile at a distance of 1 mile area around the school. If we know we are only going to do this once, a **Select by Location** search may be fastest and easiest.
5. I am going to use the **Buffer Wizard** tool to create my buffer at 1 mile.
6. Complete this step as you choose.
7. Select just the kinds of data you want to analyze from the source theme (Offense_Data in this case).
8. We want only vandalism offenses for this analysis.
9. I will create a theme definition for my Offense_Data theme where *"UCRTYPE" = 'VANDALISM'* first, and then use a **Select by Location** query to choose those vandalisms that intersect the buffer of Glendale High School.
10. I got 1886 offenses. If you got more than this, you may not have cleared the previously selected features from task 2. Clear all selected features and redo steps 1–9. Did you get 1886 this time?
11. Using a field already in the Offense_Data attributes, I created the offense data table shown in DVD Figure 6–8.
12. Does it look like the time frames of 0700 to 0800 hours or 1400 to 1600 hours are more problematic than other times?_____.

13. The offense data table represents all vandalisms, and the officer believes that the juveniles are involved in vandalism of residences near the school. What other searches and information might you have to pull to answer this question realistically? (*Hint: Look at your attribute table.*) _____

_____

_____

_____

_____

14. Which concern of the officer can you probably answer? _____

_____

15. Save your project, clear any selections, and remove any layers you do not need for the next task.

Task 4: Bike Paths
1. For this task you are on your own. Think out the steps needed to do the analysis. Generally these include:
   a. Turn off any unnecessary layers to avoid confusion.
   b. Find the theme that is the subject of the analysis *(the feature that applies to proximities such as around, near, close to, intersects, completely within, etc.)*. Remember the Find Address tool in the Standard toolbar that looks like binoculars. If an address locator does not show up, use ArcToolbox to create one as we did in exercise 4a.
   c. Select records from the Point, Line, or Polygon theme that is the subject of the analysis (the bike path).
   d. Create a buffer "around" those selected records at a logical distance *(use the Measure tool to see how far away incidents generally are from the selected features)*.
   e. Select the target theme records that meet the criteria needed in the final result *(robberies and indecent exposures)*.
   f. Create a temporary layer or a new shapefile of just those types of records.
   g. Use a **Select by Location** search to limit the target layer's records to just those within the buffer in step b.
   h. Export your final results as a DBF, bring them into Excel, and create a report.
2. Using a 0.5 mile buffer distance, you should get approximately 66 total offenses broken down as shown in DVD Figure 6–9.

3. Good luck, and if you have any problems, request assistance from the instructor.
4. *Save your project!*

References

Angel, S. (1968). *Discouraging crime through city planning.* Berkeley: Institute of Urban and Regional Development, University of California.

Barnes, G. C. (1995). Defining and optimizing displacement. In J. E. Eck and D. Weisburd, (Eds.), *Crime and place* (pp. 95–113). Monsy, NY: Criminal Justice Press.

Barr, R., & Pease, K. (1990). Crime placement, displacement, and deflection. In M. Tonry & N. Morris (Eds.), *Crime and justice* (Vol. 12). Chicago: University of Chicago Press.

Black, J. (2003, October 17). Smile, you're being watched. *Business Week*. Retrieved March 29, 2008, from http://www.msnbc.msn.com/id/3225985/

Brantingham, P. J., & Brantingham, P. L. (Eds.) (1981). *Environmental criminology.* Beverly Hills, CA: Sage.

Brantingham, P. J., & Brantingham, P. L. (1984). *Patterns in crime.* New York: Macmillan.

Brantingham, P. L., & Brantigham, P. J. (1993). Environment, routine and situation: Toward a pattern theory of crime. In R. V. Clarke & M. Felson (Eds.), *Routine activity and rational choice* (pp. 259–294). New Brunswick, NJ: Transaction.

Canter, P. (1998). Baltimore County's autodialer system. In N. Lavigne & J. Wartell (Eds.), *Crime mapping case studies: Successes in the field* (Vol. 1, pp. 81–91). Washington, DC: Police Executive Research Forum.

Catalano, P., Hill, B., & Long, B. (2001). Geographical analysis and serial crime investigation: A case study of armed robbery in Phoenix, Arizona. *Security Journal, 14*(3), 27–41.

Cao, L. (2004). *Major criminological theories: Concepts and measurements.* Belmont, CA: Wadsworth.

Chainey, S. (2000). Optimizing closed-circuit television use. In N. Lavigne & J. Wartell (Eds.), *Crime mapping case studies: Successes in the field* (Vol. 2, pp. 91–100). Washington, DC: Police Executive Research Forum.

Clarke, R. (1998). *The theory and practice of situational crime prevention.* Retrieved December 14, 2008, from http://www.e-doca.eu/docs/Situational_crime_prevention.pdf

Clarke, R. (1999). *Hot products: Understanding, anticipating and reducing the demand for stolen goods.* London: Home Office Police and Reducing Crime Unit.

Clarke, R. V., & Weisburd, D. (1994). Diffusion of crime control benefits: Observations on the reverse of displacement. In R. V. Clarke (Ed.), *Crime prevention studies* (Vol. 2, pp. 165–183). Monsey, NY: Criminal Justice Press.

Cohen, L. A., & Felson, M. (1979). Social change and crime rate trends: A routine activities approach. *American Sociological Review, 44,* 588–608.

Cornish, D. B., & Clarke, R. (Eds.). (1986). *The reasoning criminal: Rational choice perspectives on offending.* New York: Springer-Verlag.

Cornish, D. B., & Clarke, R. V. (2003). Opportunities, precipitators, and criminal decisions: A reply to Wortley's critique of situational crime prevention. In M. J. Smith & D. B. Cornish (Eds.), *Theory for practice in situational crime prevention* (pp. 41–96). Monsey, NY: Criminal Justice Press.

Crowe, T. D. (2000). *Crime prevention through environmental design: Applications of architectural design and space management concepts* (2nd ed.). Boston: Butterworth-Henemann.

Denhardt, D. (2001). *Violence in the workplace, 1993–1999* [Special report]. Washington, DC: US Department of Justice, Office of Justice Programs, Bureau of Justice Statistics.

Farrington, D. P., & Welsh, B. C. (2002). *Effects of improved street lighting on crime: A systematic review.* London: Home Office.

Felson, M., & Clarke, R. (1998). *Opportunity makes the thief: Practical theory for crime prevention.* London: Home Office Police and Reducing Crime Unit.

Gill, M., & Pease, K. (1998). Repeat robbers: Are they different? In M. Gill (Ed.), *Crime at work: Increasing the risk of offenders.* Leicester, UK: Perpetuity Press.

Gill, M., & Spriggs, A. (2005). *Assessing the impact of CCTV.* London: Home Office.

Jacobs, J. (1961). *The death and life of great American cities.* New York: Random House.

Jefferey, C. R. (1971). *Crime prevention through environmental design.* Beverly Hills, CA: Sage.

Kaplan, H., O'Kane, K., Lavrakas, P. J., & Hoover, S. (1978). *CPTED final report on commercial demonstration in Portland, Oregon* [Mimeo]. Arlington, VA: Westinghouse Electric.

Karmen, A. (2004). *Crime victims: An introduction to victimology* (5th ed.). Belmont, CA: Wadsworth.

Kushmuk, J., & Whittemore, S. (1981). *A reevaluation of crime prevention through environmental design in Portland, Oregon.* Arlington, VA: Westinghouse Electric.

Merry, S. E. (1981). *Urban danger: Life in a neighborhood of strangers.* Springfield, IL: Mombiosse.

Newman, O. (1972). *Defensible space: People and design in the violent city.* New York: Macmillan.

Pease, K. (1998). *Repeat victimization: Taking stock.* London: Home Office.

Pease, K. (1999). A review of street lighting evaluations: Crime reduction effects. In K. Painter & N. Tiley (Eds.), *Surveillance of public space: CCTV, street lighting and crime prevention.* Monsey, NY: Criminal Justice Press.

Reppetto, T. (1974). *Residential crime.* Cambridge, MA: Ballinger.

Reppetto, T. (1976). Crime prevention and displacement phenomenon. *Crime and Delinquency, 22,* 166–177.

Roman, C. G. (2005). Routine activities of youth and neighborhood violence: Spatial modeling of place, time, and crime. In F. Wang (Ed.), *Geographic information systems and crime analysis* (pp. 293–310). Hershey, PA: IDEA Group Publishing.

Rosenbaum, D. P., Lurigio, A. J., & Davis, R. C. (1998). *The prevention of crime: Social and situational strategies.* Belmont, CA: West/Wadsworth.

Rossmo, D. K., Laverty, I., & Moore, B. (2005). Geographic profiling for serial crime investigation. In F. Wang (Ed.), *Geographic information systems and crime analysis* (pp. 102–117). Hershey, PA: IDEA Group Publishing.

Santiago, J. J. (1998). The problem of auto theft in Newark. In N. Lavigne & J. Wartell (Eds.), *Crime mapping case studies: Successes in the field* (Vol. 1, pp. 53–59). Washington, DC: Police Executive Research Forum.

Simon, H. (1976). *Administrative behavior: A study of decision making processes in administrative organization* (3rd ed.). New York: Free Press.

Smith, C. J., & Patterson, G. E. (1980). Cognitive mapping and the subjective geography of crime. In D. E. Georges-Abeyie & K. D. Harries (Eds.), *Crime: A spatial perspective.* New York: Columbia University Press.

Taylor, R. B., Shumaker, S. A., & Gottfredson, S. D. (1985). Neighborhood links between physical features and local sentiments: Deterioration, fear of crime, and confidence. *Journal of Architectural and Planning Research, 2,* 261–275.

Wilson, S. (1978). Vandalism and defensible space on London housing estates. In R. V. Clarke (Ed.), *Tackling vandalism* (pp. 14–26). London: Her Majesty's Stationery Office.

Wood, E. (1961). *Housing design: A social theory.* New York: Citizens' Housing and Planning Counsel of New York.

Geography and Individual Decision Making: Victims and Offenders

CHAPTER 4

By Lorie Velarde

▶ ▶ **LEARNING OBJECTIVES**

Chapter 4 continues the discussion about how offenders interact with the environment to make decisions. Crime patterns are influenced by both the offender's knowledge of the environment as well as the spatial and temporal availability of targets. Essentially, offenders operate within the boundaries of their awareness space and are influenced by target availability, among other factors, when selecting locations to commit crimes. This chapter also reviews findings from studies about journey to crime, the distances offenders travel when committing crimes. Journey-to-crime research has identified patterns in the distances offenders travel. Just as offenders use clues about the environment to select targets, law enforcement uses knowledge about the criminal hunting process to find them. This chapter introduces the concept of geographic profiling, a police investigative methodology that uses the locations in a connected crime series to locate the responsible offender. This chapter concludes with a geographic profiling case example. After studying this chapter, you should be able to:

- Define and explain awareness space.
- Define and explain target backcloth.
- Understand the findings of several journey-to-crime studies.
- Define and explain geographic profiling and how it is used in a police investigation.
- List and describe the types of hunting methods used by offenders.
- Describe several different investigative strategies used with geographic profiling.

▶ ▶ **KEY TERMS**

Buffer Zone
Crime Series
Distance Decay
Expected Crime Pattern
Geographic Profiling
Hunting Methods
Investigative Strategies
Journey to Crime
Mental Maps
Search Base
Target Backcloth

■ Introduction

Chapter 3 discussed the theoretical frameworks for crime site selection based in environmental criminology, including Rational Choice Theory, Routine Activities Theory, and Crime Pattern Theory. Recall that Crime Pattern Theory, as discussed in the previous chapter, explains how offender interaction with the environment produces patterns of criminal activity. This chapter builds on this discussion of theory by discussing other concepts related to crime site selection, such as activity and awareness space, mental maps, and target backcloth. Studies examining the distance an offender travels from his home to a crime location are also discussed in the journey-to-crime section.

Researchers and law enforcement practitioners alike have been interested in the influence of the environment on where, when, and how crimes occur. By reviewing research conducted in the area of environmental criminology, the police have been able to use research empirical findings about offenders' interaction with their environment to help apprehend criminals. One promising methodology in this area is geographic profiling. The principles of this technique and a geographic profiling case example are discussed later in this chapter. But before we can discuss geographic profiling, we must discuss a little more theory.

■ Mental Maps and Awareness Space

Everyone has daily activities that they participate in. Such activities include going to and from work, school, shopping, or recreation areas. Over time, making repetitive journeys to various locations gives us knowledge of the routes and characteristics of these places. We form cognitive images of the places we frequent, the routes to these places, and the areas we pass through. These cognitive images are known as **mental maps** (Brantingham & Brantingham, 1984).

Mental maps are the interpretation of the environment by the individual and are the result of perception, coding, and interpretation of information. Mental maps not only reflect the spatial features of the environment but also the attitudes, experiences, and feelings of the individual about his or her environment. While mental maps vary from person to person because they are influenced by individual experiences, they also have much in common (Brantingham & Brantingham, 1984).

As people pass through a city, they begin to become familiar with areas. However, people do not know the entire city, either physically or temporally. There are areas that individuals know well and areas they have little or no knowledge of. The well-known areas, those places visited frequently, and the routes that connect these areas comprise our *activity space* (as discussed in Chapter 3) (Jakle, Brunn, & Roseman, 1976). If you wish to test this, sit down and draw a map of your world as you see it. Draw the major arterials and streets you travel on frequently. Also draw the places you go to shop, eat, work, attend

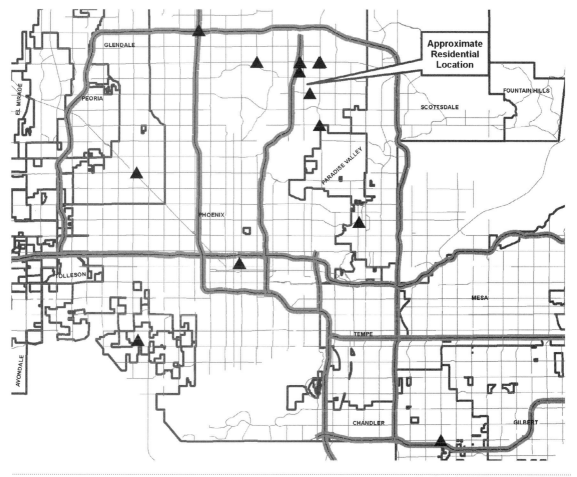

Figure 4–1 Sample Activity Space Map

school, meet friends, run errands, and other locations you may visit during a 2-week time period. When you are done creating your map, show it to someone who doesn't know where you live. It is likely that he or she will be able to roughly determine where you live based on where you spend most of your time (see **Figure 4–1**).

Criminals, just like noncriminals, have locations they visit frequently. An offender's activity space is comprised of his or her habitual geography, which includes locations he or she visits for both criminal and noncriminal reasons (Brantingham & Brantingham, 1981).

There are also locations that, while we do not regularly frequent them, we still know about them. Along with those places we visit as part of our day-to-day activities, these locations form our *awareness space*.

Awareness space is comprised of those locations for which we have some minimum level of knowledge. It is even possible to have awareness of places we have never visited (Clark, 1990). For example, an offender who regularly steals scrap metal may learn of a new target-rich location for this type of crime from a fellow thief. While the offender has not been to the new location, he is now aware that it exists and may eventually visit it to examine its potential crime opportunities.

When an offender decides to commit a crime, he searches through his awareness space for a location that provides an opportunity for crime. For some offenders, the search will be minimal, perhaps even opportunistic. For others, the search will be extensive before a suitable target is encountered (Brantingham & Brantingham, 1981).

An example of an opportunistic offense might be a juvenile who burglarizes an unlocked home while he was on his way home from school. On the other hand, an example of offenders who show more planning might be a group of adults who commit robberies of banks, or specifically, only of Wells Fargo banks that have limited security. Presumably, the robbers would spend more time searching and planning before committing their crime than an opportunistic offender.

Both types of searches at the aggregate level produce an **expected crime pattern**, areas in the city where crime is likely to occur. Some of these locations, through the overlap of multiple offenders, become hot spots of criminal activity. Target availability also affects the expected crime pattern (Brantingham & Brantingham, 1993). In the previous robbery example, if Wells Fargo banks in the city were few and far between, the target availability would affect the pattern of crimes in a series of these events.

■ Target Backcloth

The daily activities of an offender define the location and time that he is likely to commit a crime. The routine activities of potential victims also shape crime patterns. For example, it is from this distribution of victims or targets that offenders must choose. The distribution of targets in the environment is called **target backcloth** (Brantingham & Brantingham, 1993).

The understanding of target availability is important to the understanding of crime site selection because it is from this availability that the offender must make his criminal choices. Targets are typically assessed by criminals on the basis of suitability and risk. *Suitability* refers to the gain or profit of the target as seen by the offender. *Risk* refers to the probability of apprehension of the offender. Both criteria can differ significantly depending on the target selected (Felson & Clarke, 1998).

Target availability is comprised of both temporal and spatial opportunity structures. Because opportunity is different at different times of the day, criminal targets must be examined both geographically and temporally. Depending on the criminal offender's target preference, the availability of that target may vary significantly from neighborhood to neighborhood. Targets may also vary by time of day, day of week, and season of year (Rossmo, 2004).

Because the location of potential victims plays a key role in the determination of where a criminal event will occur, nonuniform or rare target opportunities can distort the spatial pattern of crimes. For example, some target opportunities require more searching by the offender to find them or may result in crime sites determined strictly based on the limited availability of that specific target. The locations of crimes against street prostitutes, for example, are mainly a function of the locations of red-light districts (Rossmo, 2000).

A uniform target backcloth provides the offender with many opportunities; the locations of his or her crimes are mainly a function of what he or she knows. A nonuniform target backcloth provides fewer opportunities; offenders are forced to the locations of the opportunities because their choices are limited and less can be determined about the spatial choices the offender has made because his choices have been a function of target availability (Rossmo, 2000).

Activity space, awareness space, and target backcloth all influence crime patterns. Law enforcement can use knowledge about how offenders interact with the environment to prioritize areas to search for them. This is discussed later in the chapter under geographic profiling.

■ Journey to Crime

Empirical research has examined the distances that offenders travel when committing crimes. The distances differ based on a variety of factors, including the type of crime, whether an area is urban or rural, and whether or not an area has a well-developed public transportation system. A **journey-to-crime** trip is the journey an offender travels when he commits a crime. While an offender may begin his actual journey to a crime from anywhere, researchers generally measure journey to crime as the distance between the offender's home and the location of his crime, regardless of the distance actually traveled on the day of the criminal event. Some crimes, such as domestic homicide, have a journey-to-crime distance of zero.

There have been many studies conducted on journey to crime. Some studies have analyzed crime trips based on the offense committed, while other studies have analyzed trips by factors such as demographics of the offender or day of week of the offense (Canter and Gregory, 1994).

Oftentimes, a journey-to-crime study will report its findings using measures such as the mean, median, or mode for the group studied (these commonly used measures of central tendency will be discussed in Chapter 8). While these describe the studied group in general, they may not apply to any one individual within the group. It is important to remember that crime trips will always vary by individual; some offenders make short journeys, while others make longer ones.

An example of a typical journey-to-crime study is found in **TABLE 4-1** from the City of Glendale, Arizona. Note that for the crime of burglary, offenders travel an average of 2.34 miles from home to commit a crime. Based on the table, it is apparent that the mean distances traveled by offenders vary somewhat by crime type, and the mean trips for auto theft, theft, and robbery are longer than trips for other crimes in the city.

By studying many journeys, researchers have found that the majority of crimes occur close to home. What "close" means is somewhat

TABLE 4-1 Average Distance to Crime in Glendale, Arizona

Crime Type	Number of Crimes	Mean Crime Trip (Miles)	Standard Deviation (Miles)	68% of Crime Trips Within (Miles)	Furthest Distance
Aggravated assault	8,526	1.18	2.86	4.04	6.91
Arson	62	1.32	2.68	4	6.67
Auto theft	1,058	2.88	3.55	6.43	9.97
Burglary	1,356	2.34	3.73	6.07	9.79
Curfew/loitering	164	1.59	1.85	3.44	5.28
Drug offenses	3,970	2.27	3.39	5.66	9.06
Murder	70	2.38	4.42	6.8	11.23
Other miscellaneous	18,237	2.42	3.9	6.32	10.22
Other sex offenses	404	1.61	2.66	4.27	6.92
Rape	85	1.35	2.79	4.14	6.94
Robbery	567	3.23	4.21	7.44	11.65
Runaway	5,958	0.19	1.29	1.48	2.77
Theft	5,139	3.18	4.28	7.46	11.75

Note: The date range was January 1, 2004 through July 24, 2004.
Source: Bryan Hill, Glendale, Arizona Police Department.

subjective, but the studies reviewed by Brantingham and Brantingham (1981) found that the majority of crimes occurred within 2 miles of the offender's residence. Research has also found that crimes against persons, such as homicide, assault, and rape, occur closer to home than property crimes, such as larceny and burglary (Brantingham & Brantingham, 1981).

Research has also found that crime trips show **distance decay** from the offender's residence. Distance decay refers to the decreasing probability of an offender traveling far from home to commit a crime. Generally speaking, the probability decays or drops off as distance from the offender's residence increases (Brantingham & Brantingham, 1984).

Most criminals, however, have a **buffer zone**. The buffer zone is an area around the offender's residence within which he commits fewer crimes. The buffer zone often exists for two main reasons: a perceived lack of anonymity and reduced number of targets. That is, while offenders often prefer to commit crimes in areas they are familiar with, they do not want to be in areas where they would be recognized (Rossmo, 2000). A burglar, for example, would prefer to steal from a stranger's home a few blocks away than risk being recognized while stealing from his next-door neighbor's home—even if he is more familiar with how to gain access into his neighbor's house. The stranger's house a few blocks away is still in an area familiar to the burglar, but the distance from his own home increases his anonymity and therefore lowers his risk of apprehension.

The number of potential targets increases with distance from a fixed point, such as the offender's residence. For example, if targets were randomly distributed throughout the environment, there would be twice as many targets available 1 mile away from the offender's residence than there would be one-half mile away. For this reason, offenders may not commit as many crimes close to home as they do a short distance away (Rossmo, 2000).

It is important to note that offenders do commit crimes within their buffer zone; however, there may be areas outside their buffer zone where they commit more crimes. The buffer zone represents an area of reduced activity, not an area of no criminal activity (Rossmo, 2000).

■ Geographic Profiling

Geographic profiling is a police investigation strategy that uses the locations in a connected crime series to narrow the search for likely suspects. A smaller search area can result in the apprehension of an offender faster and more efficiently, resulting in both fewer victims and a savings in expended resources. In larger investigations, this cost savings can be quite large. Geographic profiling has now been used with success in many different types of criminal investigation (Rossmo, 1995).

Geographic profiling is based on three theories that were discussed previously in Chapter 3: Rational Choice Theory, Routine Activities Theory, and Crime Pattern Theory. These theoretical approaches, while examining crime differently, share the perspective that crime setting and opportunity are very important determinants as to why crime occurs at one particular location and not at another (Rossmo, 2000).

Geographic profiling is a decision-making tool that can help law enforcement solve cases of serial crime, but it does not solve crimes by itself. It is only through a confession, eye witness testimony, and/or physical evidence that a crime can be solved and successfully prosecuted. The purpose of geographic profiling is to manage large amounts (sometime overwhelming amounts) of information and help focus limited resources on the most likely area of offender residence (Rossmo, 2000).

Geographic profiling is essentially a method that allows us to infer spatial characteristics of an offender from his crime patterns. For example, the offender's activity space is determined from the locations of the crime sites. Investigators, using this information, determine the target backcloth for the crimes, paths the offender is likely to be using, and other factors that may influence the crime pattern, such as offender hunting method (discussed later in this chapter), offender type, and land use in the area. From this information, sense can be made of the pattern from both subjective and objective perspectives (Rossmo, 1995).

Geographic profiling is typically applied to a **crime series**, a set of crimes believed to have been committed by the same offender. These crimes can be linked by physical evidence, offender description, or modus operandi. Essentially, characteristics about crimes in a series are similar in many ways, thus an inference can be made that they were committed by the same person(s). Identifying a crime series allows the crimes to be analyzed as a pattern rather than as isolated events. As each crime location gives an indicator of the offender's awareness space, more crime locations provide more information about the offender (Rossmo, 2006).

Hunting Methods

Criminals, especially predatory criminals, employ various **hunting methods** when selecting their victims. An offender's hunting method affects the spatial distribution of his crime sites, and any effort to predict the offender's residence must consider hunting style. There are criminals who set out specifically to find a victim, while others happen upon a victim while involved in other noncriminal activities (Rossmo, 2000). The various search and attack methods are discussed in the following paragraphs.

Offenders, when committing a crime, must first find a suitable target. Finding this target involves a search. Four types of search methods exist. The search methods are *hunter*, *poacher*, *troller*, and *trapper*. A *hunter* (also called "marauder" in the literature) is an offender who sets out from his home specifically to find a victim. A *poacher* (also called "commuter" in the literature) is an offender who also specifically sets out to find a victim, but his search is from a location other than his home, such as work or a recreation site. Poachers also can be offenders who commute to another city to commit their crimes. *Trollers* are offenders who are involved in their everyday activities when they opportunistically encounter a victim. *Trappers* are offenders who set up a situation in which victims come to them. This is usually done through an occupation, want ads, or the taking in of boarders (Rossmo, 2000).

There are three types of attack methods. These attack methods are *raptor*, *stalker*, and *ambusher*. A *raptor* is an offender who attacks the victim upon encounter. Most criminals use this attack method. A *stalker* is an offender who does not attack upon encounter but follows the victim away from the encounter point then attacks. An *ambusher* is an offender who attacks a victim that he has lured to a location he controls (Rossmo, 2000).

Investigative Strategies

An investigative strategy is any method used during the investigation of a crime that helps law enforcement identify the responsible offender. There are a variety of **investigative strategies** that can be used after a geographic profile has been prepared.

Investigative strategies that can be used with geographic profiling include prioritizing records from the various computerized police files, such as arrest files, crime report files, and jail booking files. These files often include the offender's address, physical description, and modus operandi. Other databases may also be used in conjunction with a profile, such as parole, probation, and sex offender records. A profile used with this information could help in the search for the offender by prioritizing persons based on their geography (Rossmo, 2000).

A geographic profile can also be used to prioritize areas for directed police patrols or area canvasses. This strategy can be especially effective if the offender is operating during a narrow time period. Area canvassing might also include providing specific information to local area residents and neighborhood watch groups. The police may also want to use the profile to direct community mailings and/or conduct a media campaign (Rossmo, Laverty, & Moore, 2005).

Geographic profiles can be useful in homicide and missing person cases. In a missing person case where the victim is suspected to be dead, a profile can help determine the likely location of the body. Geographic profiles also have utility in suspected homicide cases where the body is missing because they prioritize areas to be searched based on information about the probable suspect. Location information about the suspect is analyzed to determine where the body might be. Likewise, a profile can help direct the search for a criminal fugitive by determining probable hiding places based on new crimes, sightings, telephone calls, and areas that are familiar to the fugitive (Rossmo, Laverty, & Moore, 2005).

Computerized Geographic Profiling

Geographic profilers can make use of advances in technology to determine the most probable location of an offender's residence. Computerized geographic profiling systems assign scores to areas based on offender movement, hunting behavior, and journey-to-crime distances. The scores are used to create a probability surface (jeopardy surface) that is displayed over a map of the crimes. The profile map (geoprofile) is then used to prioritize locations in the search for the offender. There are three main computerized geographic profiling systems, which are discussed in the following paragraphs.

The Rigel geographic profiling system, developed by Environmental Criminology Research Inc. (ECRI) of Vancouver, British Columbia, Canada, uses the Criminal Geographic Targeting algorithm invented by Dr. Kim Rossmo in 1991 to produce a probability map. Rigel integrates with various mapping programs to display the geoprofile over city streets (Rossmo, 1995, 2000). Rigel is available for purchase by law enforcement agencies that have personnel trained it its use. Training in the use of Rigel is currently offered at several universities and police departments internationally as part of the Geographic Profiling Analysis training program, discussed later (ECRI, 2008).

The Dragnet geographic profiling system, created by Dr. David Canter of the University of Liverpool, England, uses a negative exponential mathematical function to produce a probability map. Dragnet does not integrate with mapping programs or GIS. Dragnet is available for purchase from the University of Liverpool's Centre for Investigative Psychology, or it can be downloaded for free as part of the Crime Analysis Unit Developer Kit from the Office of Justice Programs' National Institute of Justice Crime Mapping & Analysis Program (Crime Mapping & Analysis Program, 2007). Users are not required to obtain training before downloading and using the system, but training in the use of the system may be obtained by contacting Dr. Canter directly (Canter, 2008).

The CrimeStat spatial statistics program's journey-to-crime routine, created by Ned Levine & Associates of Houston, Texas, allows the use of any of five different mathematical functions to produce a geographic profile. Users also have the option of calibrating and using a model based on known offender behavior. The profile produced can be exported and overlaid on a GIS map of the city, not included with the system. This program is free to download via the Internet thanks to a series of National Institute of Justice (NIJ) grants, which funded the original development of this program and continues to fund program updates (Levine, 2007). Training in the use of this system is available from NIJ to a limited number of law enforcement students (Mapping and Analysis for Public Safety, 2008). The crime mapping exercises at the end of this chapter utilize CrimeStat.

All of the current computerized geographic profiling systems work by applying a mathematical function over the area of the crimes to predict a likely search base for the offender within the area bounded by the crimes. Because most crimes are committed by hunters or offenders who have specifically set out from their residence to find a victim, the profile usually outlines the offender's home area (Rossmo, 2000). However, not all crimes are committed by hunters (see the previous discussion of criminal search methods). Researchers have recently questioned whether the current systems have applicability in other types of cases, specifically cases for which the offender does not live within the area bounded by the crimes (van der Kemp, 2005).

Law enforcement practitioners in the field of geographic profiling have found that the area a geographic profile prioritizes is the offender's **search base**. A search base is the location from which the offender consistently begins his search for new targets. For most offenders, their search base begins at their home, and this is the location that a geographic profile prioritizes. For other offenders, such as poachers, their search base is somewhere else. This could be a freeway off-ramp, street intersection in a target-rich neighborhood, or supply location (Velarde, 2008).

Geographic profiling may be used to locate the offender's search base, a location of investigative value. By tailoring investigative strategies to the offender type, a geographic profile can be used to locate the offender, even if he does not live in the area (Velarde, 2008). One successfully geoprofiled case in which the search base for the offender was not his home is detailed later in the chapter.

Training in Geographic Profiling

While some of the computerized geographic profiling systems do not require any training before their use, training in geographic profiling methodology is essential, according to Kim Rossmo and Scot Filer, professional geographic profilers who authored "Analysis versus Guesswork: The Case for Professional Geographic Profiling." They warn that crime locations are just one part of the information used in the preparation of a geographic profile. The other information comes from factors such as offender type, site type, demographics, land use, location of arterial routes, hunting method, target backcloth, temporal patterns, and displacement. To ignore this information could reduce the accuracy of a profile, in some cases quite significantly. Rossmo and Filer stress the use of analysis based on training, experience, and research in the field rather than guesswork (Rossmo & Filer, 2005).

Formal training in geographic profiling is available through the Geographic Profiling Analysis (GPA) training program. This training program was originated to bring the methodology to local law enforcement agencies that are concerned with property crime (Weiss & Davis, 2004). The training, while open to anyone, is mainly taught to police detectives and analysts and involves classroom courses, work in a computer lab, and field evaluation/mentorship. The GPA curriculum consists of two 1-week courses. The first week provides an overview of environmental criminological theories, the geography of crime, crime linkage, and the geographic profiling methodology. The second week is taught in a computer lab and provides instruction in the use of geographic profiling software, developing a geographic profile map, casework exercises, and report preparation. As of 2004, approximately 400 people, representing 200 police agencies from nine countries, had been trained in geographic profiling analysis (Velarde, 2004).

■ Geographic Profiling Case Example

The Chair Burglary case is a good example of how law enforcement can use geographic profiling to help find the offender who is responsible for a crime series.

For several years, the City of Irvine, California, experienced a high number of residential burglaries, and it was suspected that a serial offender was active in the city. In years past, investigators and patrol officers from the Irvine Police Department had initiated surveillances, searched police databases of arrestees, and perused pawn shops, but they were unable to locate the offender or recover any of the stolen property, and the crimes continued. In early 2005, investigators decided to try another approach to the problem that involved searching for new evidence, crime forecasting, and geographic profiling (Rossmo & Velarde, 2008).

Upon reviewing the crimes, analysts decided they needed to identify which burglaries, if any, were linked. A preliminary analysis of all residential burglaries was conducted to determine if a crime series existed. This analysis produced a list of 42 burglaries believed to be linked by modus operandi and target characteristics (see **Figure 4–2**). The linked burglaries were then used to determine offender patterns, such as a target preference, temporal and spatial consistencies, and specific modus operandi.

The analysis concluded that the offender had a consistent pattern of behavior. The targeted residences were in upscale but older neighborhoods. He preferred corner lot homes, particularly those with a greenbelt or park to the rear. Entry to the targeted house was always made at the back through an open window or door, a pried door, or a smashed window. Many of the windows on the homes were metal framed with only one pane of glass (many newer home windows are dual paned and/or framed in vinyl). This assisted the offender in the

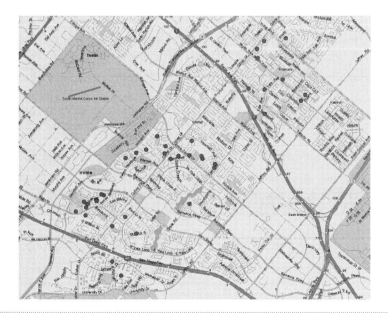

Figure 4–2 Irvine Chair Burglary Map
Source: Reprinted by permission from Microsoft Corporation

event that he had to break a window to gain entry. The offender consistently burglarized one neighborhood for about 6 months before moving to another neighborhood.

The offender often placed a lawn chair belonging to the home owner against the rear fence of the backyard. Based on footprints seen on the chair, it was believed the offender used the chair to make his escape over the fence after completing a burglary. The property stolen from the crimes was mostly cash and jewelry, items easily carried by the offender in pockets or a small bag. No fingerprints were recovered at any of the crime scenes. It became apparent to investigators that the offender was a professional burglar (Rossmo & Velarde, 2008).

Crime scene investigators from the Irvine Police Department had begun DNA swabbing at many residential burglaries (DNA swabbing involves running a cotton swab over surfaces that the offender may have touched). These swabs were retained for DNA analysis. In reviewing the evidence collected for the chair burglary series, it was found that DNA swabs were collected at 21 of the 42 crime scenes. In June 2005, these swabs were sent to the crime lab for analysis. The lab found that common male DNA, likely that of the offender, was present on samples from three of the crime scenes.

Now a suspect could be compared to crime scene DNA to see if there was a match. Investigators and analysts worked to develop a tactical action plan to gather suspects. Rather than attempting to find the Chair Burglar during the commission of a burglary, a very difficult task because of the offender's short exposure time, the plan instead targeted the burglar's search for new targets, a process that can take the offender several hours. It was decided that a temporal forecast would be used to predict when (day of week and time of day) the offender's next search would occur, and a geographic profile would be used to determine where that search was likely to be based (Rossmo & Velarde, 2008).

To complete the geographic profile, a crime analyst trained in the methodology analyzed the crimes for information about the offender's search process. The crime locations were then entered into the Rigel geographic profiling software, and a profile map was created (see **Figure 4–3**).

The analysis determined that the offender was likely a poacher, someone who did not live within the city limits but was traveling there because of an abundance of his preferred target. The profile therefore could not be used to find the offender's home, but it could be used to locate him as he began his search for a new target. Investigators used the profile for the strategic placement of undercover officers.

Geographic Profiling Case Example 163

Figure 4–3 Geographic Profile of Irvine Chair Burglary
Source: Reprinted by permission from Microsoft Corporation

In September 2005, undercover officers began to gather suspect information from the peak area of the geographic profile during the times and dates predicted by the temporal forecast. This information consisted primarily of the license plate numbers from vehicles seen driving in the area. The license plate numbers were researched for driver information through the California Department of Motor Vehicles (Rossmo and Velarde, 2008).

Investigators planned to continue this process for 5 weeks, but as luck would have it, the offender's vehicle, a rental car, was in the group of license plates gathered the very first night of surveillance. A DNA sample from the suspect was submitted to the lab and was a positive match for that found at the crime scenes. Raymond Lopez, a 47-year-old ex-convict from Los Angeles County, was arrested for the crimes. He later pled guilty and is now serving a 13-year state prison sentence (Rossmo & Velarde, 2008).

The Irvine Police Department received several awards for their work in solving this case, including the prestigious International Association of Chiefs of Police (IACP)/ChoicePoint Award for Excellence in Criminal Investigations (Rossmo & Velarde, 2008).

Online Resources

- Geographic profiling Wikipedia page: http://en.wikipedia.org/wiki/Geographic_profiling
- Texas State University – San Marcos page on Geographic profiling: http://www.txstate.edu/gii/geographicprofiling.html
- Mapping Crime: Principle and Practice – Chapter 6, Geographic Profiling: http://www.ncjrs.gov/html/nij/mapping/ch6_1.html
- National Institute of Justice page on geographic profiling: http://www.ojp.usdoj.gov/nij/maps/gp.htm
- Environmental Criminology Research, Inc.: http://www.ecricanada.com/
- Forensic Psychology Page on Geographic Profiling: http://www.all-about-forensic-psychology.com/geographic-profiling.html
- CrimeStat Chapter on Journey to Crime Estimation: http://www.icpsr.umich.edu/CRIMESTAT/files/CrimeStatChapter.10.pdf
- A Journey-to-Crime Meta-Analysis Paper: http://www.all-academic.com/meta/p_mla_apa_research_citation/0/3/2/3/4/p32344_index.html
- The Juvenile Journey to Crime Paper: http://www.allacademic.com/meta/p_mla_apa_research_citation/0/3/3/7/2/p33729_index.html
- Problem Oriented Policing Page on Journey to Crime: http://www.popcenter.org/learning/60Steps/index.cfm?stepNum=16
- National Institute of Justice Article on Predicting a Criminal's Journey to Crime: http://www.ojp.usdoj.gov/nij/journals/253/predicting.html
- Ned Levine paper on modelling the journey to crime using the Crime Travel Demand Model: http://www.cscs.ucl.ac.uk/events-1/cmc/programme/modelling-the-journey/

Conclusion

Offenders use their knowledge of the area, in the form of mental maps and awareness space, to find criminal opportunities or targets. Targets may not be evenly distributed through the environment; target availability varies by neighborhood. Crime patterns are driven by offender awareness space and the spatial availability of targets.

The commission of a crime often involves a crime trip for the offender. This crime trip, or journey to crime, typically starts at the offender's residence. Crime trips often show distance decay from the offender's residence; more trips occur to locations close to home, and fewer trips occur to locations far from home. Offenders typically have a buffer zone around their residence within which they are less likely to offend.

Geographic profiling is a law enforcement technique that helps find offenders based on the geography of their crimes. Based on Rational Choice, Routine Activities, and Crime Pattern Theories, geographic profiling uses knowledge about how criminals interact with their environment. The methodology takes into account offender awareness and activity space as well as information about target availability and criminal hunting methods. There are several different computerized geographic profiling systems available, and formal training in geographic profiling is important for profile accuracy. When properly applied in appropriate cases, geographic profiling can have significant investigative utility.

Recommended Reading

Rossmo, D. K. (2000). *Geographic profiling*. Boca Raton, FL: CRC Press.

Questions for Review

1. What are mental maps, and how are they related to offender decision making?
2. Explain the notion of *target backcloth* and how it explains when and where an offender may choose to commit a crime.
3. What is a *journey to crime*? What do we know about distances traveled by offenders?
4. How do the terms *distance decay* and *buffer zone* work together to explain the likelihood of an offender to commit crime within his neighborhood?
5. What computer programs are in existence to aid in geographic profiling?
6. What is a crime series? What factors can be used to link crimes in a series?
7. What are the four types of hunting methods offenders may use to commit a crime? Which method is most often used? Why?
8. What is an investigative strategy? Provide an example.

Chapter Glossary

Buffer Zone   An area around an offender's anchor point(s) where he or she is less likely to commit a crime due to risk of being identified.

Crime Series   A crime series is a set of crimes believed to have been committed by the same offender or offenders.

Distance Decay   Distance decay refers to the decreasing probability of an offender traveling far from home to commit a crime.

Expected Crime Pattern   Expected patterns of crime at the aggregate level indicate areas in the city where crime is likely to occur.

Geographic Profiling   Geographic profiling is a police investigation strategy that uses the locations in a connected crime series to narrow the search for likely suspects.

Hunting Methods   Hunting methods include the various search and attack methods an offender uses to commit crime. There are four types of search methods: *hunter*, *poacher*, *troller*, and *trapper*.

Investigative Strategies   An investigative strategy is any method used during the investigation of a crime that helps law enforcement identify the responsible offender.

Journey to Crime   A journey-to-crime trip is the journey an offender travels when he or she commits a crime. Typically, journey to crime is measured from an offender's home to the location of the crime. However, the offender could also be traveling from another anchor point, such as his workplace.

Mental Maps   Mental maps are cognitive images of the places we frequent, the routes to these places, and the areas we pass through.

Search Base   A search base is the location from which the offender consistently begins his or her search for new targets.

Target Backcloth   The distribution of targets in the environment is called target backcloth.

Chapter Exercises

Exercise 7: Journey to Crime Basics

Data Needed for This Exercise
- All data is contained in EX7.mxd.

Lesson Objectives
- Learn how to determine the best area where detectives can focus their efforts to find a potential suspect for a robbery series.

- Learn about average travel behavior for various crime types and criminals for the city of Glendale, Arizona.
- Use this information and the mean center to determine an adequate search region for possible offenders.

Task Description
- In this exercise, we will examine a rape series in North Glendale, Arizona. We are going to create a map that advises the detectives where the suspect most likely lives, works, or has some sort of anchor to within the crime series area. This search area will allow us to limit our search to just those offenders that match the description of our offender and prioritize any leads the detectives already have in the case (if any).

1. Open EX7.mxd
2. You should see rape cases plotted on the map. The mean center will also be displayed.
3. There are two general types of offenders, the hunter (marauder) and the poacher (commuter). The commuter lives outside the area and commutes into the area to search for targets and victims. This type of suspect is extremely difficult to identify through a journey-to-crime analysis or geographic profile. (While working for the Vancouver Police Department, Kim Rossmo gave the term "geographic profile" to the peak area from which an offender's anchor point is located in a crime series. His process and ECRI's software, Rigel, was developed to create these peak search areas in criminal cases. As discussed in the chapter text, analysts must attend, at minimum, a 2-week course in geographic profiling before using Rigel.
4. Depending on the area covered by an offender and the type of crime that was committed, we can use some good old common sense to give us a likely search area that we can use for predicting where the offender most likely has an anchor point or begins his journey to crime. In this series, the offender has limited his hunting to an area a little over 8 square miles (to our knowledge, but we could be missing cases).
5. Because of this, we can make an assumption that the offender may live, work, or have a parent or girlfriend that lives in the area, or he has some other location that anchors him to the general vicinity. We know from research that the offender's anchor point is often very near the mean center (mathematical center) or the center of

minimum distance (the point that is closest to all points in the series).
6. Turn on the **mean center** and the **center of minimum distance** in the TOC.
7. These two points are within 0.25 miles of each other, with the center of minimum distance (CMD) further south than the mean center (MC).
8. Several years ago, I pulled all the known offenders (someone was arrested) and took their reported home addresses and the locations of the offenses and calculated the distances traveled for each trip. Some initial information from a 2003 version of that study for all offender trips in Glendale is located in a memorandum located at C:\CIA\exercises\Exercise_7\distances_summary.doc.
9. Open this document now and review it, but do not close ArcMap.
10. At a later date, I performed this analysis again; however, I limited it to just those offenders who committed more than two crimes during their criminal history (we at least arrested them more than once for separate crimes). DVD Table 7–1 shows the results of this distance measurement analysis. All offenses that were reported at the same location as the offender's reported residence were excluded.
11. As we can see from the table, for rape offenses, the average distance traveled was 1.35 miles, and the median distance was 1.70 miles (indicating that there is some skewness to the data).
12. If we look at our crime series and the distance between the CMD or MC to each crime, we can see that if we use the MC or the CMD, almost all of the crimes in the series are within this 1.35 to 1.70 mile radius (except for the one on Northern Av, which was 2.6 miles away from the CMD).
13. We also can see that the MC and CMD are very close to each other, which is another indicator that this may be a good predictor of the offender's anchor point.
14. We can therefore search the person data right around the two center points for possible suspects.
15. The last victim stated that the offender had what she thought was a tattoo on one of his forearms of Taz, the cartoon character from the Saturday morning cartoons. She was not sure which arm but believed it was his right arm.
16. Based on a crime series description (you will perform this in a future exercise), we know that the suspect is a Black male,

17–20 years old, approximately 5′11″, 135 lbs, with black hair and brown eyes. We also believe he may have a Taz tattoo on one of his forearms.

17. Add the Adult Probationer and Glendale Persons layers from the Geodatabase folder.
18. We can see that our data shows well over 2,600 adult probationers and almost 170,000 persons arrested in Glendale. (Open the attribute tables to view the total records.)
19. Let's go to **Selection → Set Selectable Layers** and make sure that only the adult probationers and Glendale persons themes can be selected.
20. Then go to **Selection → Select by Location** and select the persons from those two themes that are within a distance of 1.35 miles of the CMD.
21. You may have to do each search individually because ArcGIS sometimes gets confused and may only select points in one of the layers if you check both at the same time.
22. Now we need to limit the search to offenders that meet our suspect's physical description.
23. Open the attribute table to the Adult Probationers theme.
24. Click the **Selected** button so we see only the selected records.
25. We have 584 adult probationers selected.
26. Click on the **Options** button and choose **Select by Attributes**.
27. Enter a query to limit the search results to just those Black male suspects as shown in DVD Figure 7–1.
28. Make sure the **Method** is selected from current selection before clicking **Apply** to run this selection routine.
29. We now have limited our list of over 500 to 55 persons.
30. Our suspect description also says the offender is between 17 and 20 years old, so let's do another attribute query, **Selecting from Current Selection**, where:

    [AGE] >= 17 AND [AGE] <= 20

31. We wind up with a total of six records. Browse the tattoo fields and see if any of them have a Taz tattoo.
32. Christopher T. Gerrie is a bit shorter than the description given; however, he does have a Taz tattoo on his left arm. He is a very good suspect, and he lives very close to the CMD point.
33. Let's repeat this analysis for the Glendale Persons theme.

34. In this layer, we still wind up with 406 possible candidates when we search the following criteria:
    [Race] = 'B' and [Sex] = 'M'
    [Age] >= 17 and [Age] <= 20
35. Approximately 406 candidates (even though many are duplicates) are way too many to be useful, so we need to further refine the search by adding the approximate height and weight to the attribute searches.
36. Create a new **Select from current selection** criteria as follows:
    [HEIGHT] >= 510 AND [HEIGHT] <= 600 (reduces our list to 81)
    [WEIGHT] >= 130 AND [WEIGHT] <= 140 (reduces our list to 10)
37. Reviewing the records, we see that Micheal Aspuro has a "Devil" tattoo on his back, and Demetric Sletton works as a fitness trainer, which could be interesting considering our suspect description and modus operandi (MO) of the offender.
38. We would probably want to also look at their criminal histories and photographs and then compile all of those results and give them to the investigators who are working the cases. We won't do this in our exercise because we don't have access to criminal history records and photos, but it would be the next step.
39. This is the basic process we would use with a journey-to-crime analysis.
40. CrimeStat III creates a journey-to-crime output when you input the XY coordinates of the crimes. (You will learn this in the next exercise.) Turn off the Persons themes and remove any selections you have made, then turn on the Journey to Crime Analysis Area theme. This was created by giving CrimeStat III the rape case shapefile as the primary file, setting the extents or minimum bounding rectangle (MBR) of the area to the area you see on the map display in CrimeStat, and then using the journey-to-crime routine and the negative exponential mathematical method. As you can see, this process also predicts that our offender very likely lives near the MC and CMD (red area).
41. Close this project.

Exercise 8: CrimeStat III: Journey to Crime Techniques

Data Needed for This Exercise
- This exercise starts by opening the C:\CIA\Exercises\EX_CrimeStat folder and opening the JTC.mxd file.

Lesson Objectives
- This lesson will help you to utilize the journey-to-crime routine in CrimeStat III and allow you to assist investigations by providing a most likely area where the offender in a series may live or have an anchor point from which he begins his journey-to-crime trip.

Task Description
- You have been working on a robbery series, and the detectives want to know where you think the offender has his home base or where he may live. They want to use this information to prioritize a few suspects they already have and decide where to place surveillance vehicles to search for a specific vehicle that has been seen leaving at least four of the robberies. You will help them accomplish this by knowing how and when to use the journey-to-crime (JTC) routine in CrimeStat III.

1. Open the JTC.mxd project located in the C:\CIA\Exercises\EX_Crimestat folder.
2. See DVD Figure 8–1 to view what the map should look like.
3. This robbery series has involved one Hispanic male suspect who has been hitting stop-and-go gas-type businesses. He generally walks up to the clerk, points a black sawed-off shotgun at the clerk's head, and demands all of the money from the cash register. He has also taken Budweiser beer and Newport cigarettes during all but one of the robberies. In the last robbery, he took the cash but was interrupted when several customers came into the store, and he left without taking the cigarettes or beer. In four of the crimes, witnesses have seen the suspect jump into a red four-door vehicle believed to be a Saturn. In the last crime, a witness got a partial plate of WE___9 as the vehicle fled the scene.
4. When doing a journey-to-crime analysis, there are really just a few things you need to look for:
   - Based on the crimes and the suspect's behavior, does it appear that the suspect is a hunter (lives near the center of the crime series) or a poacher (lives far away from the crime locations)? JTC analysis will not work when the offender is a poacher.
   - Do I have enough crimes in this series to do a JTC analysis? In general, CrimeStat III needs a logical minimum of five points on which to do an analysis. Although this is the minimum, it may not always be the best because the fewer the points you have, the less accurate your analysis will probably be. This does not mean you should not do the analysis, it just means you

should be somewhat cautious when relaying the accuracy of the prediction to the investigators.

5. One of the ways you can get a general idea if you are dealing with a commuter or a marauder is to simply look at the path between hits. In this map, this is represented by the green line with arrows. If we follow this path from hit 1 to hit 2 and so forth, we can see a possible pattern where the offender is skirting around the outside of a centralized location near crimes 2 and 11 in the crime series. Chances are the suspect will probably go south and east from the last hit in the series (10) if this pattern holds up and the offender continues to behave in the manner that he has in the past. When you see such a possible pattern that circles or crosses back and forth across the mean center, I have found that this can indicate the presence of a marauder. If the mean center calculation in CrimeStat III and the JTC analysis also highlight this same middle ground, then the chances are much higher, based on my practical experience.

6. On the other hand, if there does not appear to be this circling or crisscrossing pattern, a marauder is less likely, and the JTC analysis may be less useful.

7. We should also understand that although most JTC analysis routines are described as finding the home address for the offender, it is really predicting the location where the offender starts his or her journey to crime, and the actual location could be a residence, mother's residence, girlfriend's residence, work address, or perhaps a shopping center where the offender hunts for and follows his targets home (sexual assault example). This is often called an "anchor point."

8. The difficulty is that we probably do not know until after the offender is caught and interviewed what the anchor point represents to the suspect, but we make the educated assumption that it will be a home address. There is little research available that actually tests the validity of this assessment, but anecdotal success stories using the technique indicates a home address is often the anchor point.

9. Another indication that an offender may be a hunter and a good indication of how reliable your prediction may be comes from performing all of the methods of JTC analysis available in CrimeStat and comparing them. If they all highlight the same basic area,

then you can be fairly sure your prediction has a greater chance of being accurate. We will explore this concept and further explain it at the end of this lesson.

10. To do the actual JTC analysis we need a few things:
    - Do we have the minimum bounding rectangle (MBR) of the crime series points? This is also called a research area, study area, reference area, analysis region, and many other terms. It means that if we draw a rectangle around all the points so that all the points are contained in the rectangle, what would be the lower XY coordinate and upper-right XY coordinate of this rectangle?
    - Are the X and Y coordinates of the points in the attribute table?
    - Do we have known XY coordinates for known offender homes complete with XY coordinates of their crime locations? This can be used to create a calibrated model for JTC analysis based on offenders who live in your city.

11. We will draw a rectangle around the points in the crime series using the Draw tool in the Draw toolbar.

12. If it is not already turned on, go to the menu bar and click on **View → Toolbars →** and turn on the **Draw toolbar**.

13. Use the fifth item from the left of the Draw toolbar (New Rectangle) and draw a rectangle that completely covers all of the points. As a general rule, zoom out once or twice form the points and draw your MBR.

14. See DVD Figure 8–2 to view the rectangle that needs to be drawn, and notice how it completely covers the crime series points.

15. Right click on the rectangle and choose **Properties**.

16. Go to the Size and Position tab and click on the lower-left blue box.

17. Collect the X and Y coordinate values of this lower-left corner:
    X:_____ (approximately 599286)
    Y:_____ (approximately 919201)

18. Click on the upper-right blue box, and write down the X and Y coordinates here:
    X:_____ (approximately 632326)
    Y:_____ (approximately 943116)

19. We will need these coordinates later in CrimeStat III.

20. Now click on the **Area tab** and copy the total square miles of the rectangle (MBR) because we will need that information in CrimeStat III as well. *You can either write this value down, save it to an Excel or Word document, or remember it.*
21. Open the attribute table of the RobSeries theme, and make sure that there are fields that contain the X and Y coordinates of the points in the table. You will need to know what the field names are (in this case, "X" and "Y").
22. The last thing we need is the total street mileage under our rectangle.
23. Make sure the handles are still visible around the rectangle and that it is selected (use the black arrow Selection tool and click on the edge of the rectangle).
24. Make sure that the Streets layer is the only highlighted theme in the table of contents (TOC).
25. Go to the menu bar and choose **Selection → Set Selectable Layers**, and choose only the Streets layer.
26. Go to the menu bar and choose **Selection → Select by Graphics**, and in a few seconds you should see a bunch of light blue streets. Open the attribute table of the Streets theme and find a field called Length.
27. Right click on this field, and choose **Statistics** from the pop-up dialog menu.
28. Copy the Sum value _____ (approximately 2847760.469041).
29. Save your project in your student folder as JTC.mxd, and go to the menu bar and click on **Selection → Clear Selected Features**. You can also close the statistics results and attribute table at this time and minimize your ArcGIS project.
30. Open CrimeStat III from the following location:
    C:\CIA\Documents\Tutorials_Programs\CrimeStat_III\Crimestat.exe
31. After the start-up screen, click on the **Data Setup** tab and then the **Primary File** tab under it.
32. Click on the **Select Files** button and then browse in the pop-up window.
33. Find the C:\CIA\Exercises\EX_CrimeStat\RobSeries.dbf file and load it (see DVD Figure 8–3).

34. Check the **Projected (Euclidean)** box under Type of Coordinate System, check **Feet** under the Data Units, and leave the Time as **Days**.
35. Now click on the **Reference File** tab. Enter the lower-left and upper-right XY coordinates we got from the rectangle (MBR) we drew.
36. See DVD Figure 8–4 to view what your screen should look like.
37. Keep the Cell Specification setting to By Number of Columns set to 100. You can increase the smoothness of the final output by adjusting this up or down (50 to 250), but keep in mind that the more columns you have, the longer it takes for CrimeStat to produce a result.
38. Click on the **Measurement Parameters** tab.
39. Enter the square mileage of our MBR and the total street mileage we collected in ArcGIS. Check the box next to Indirect (Manhattan) as well.
40. Your entries should be similar to DVD Figure 8–5.
41. Now click on the **Spatial Description** tab at the top of the screen.
42. According to some research, the mean center or center of minimum distance also has a high accuracy for revealing the offender's home base where marauders are involved, so we will run several different Centro graphic analyses. This may also be an indicator that the offender is a marauder or commuter. If all of these Centro graphic measures indicate the same general location, it is more likely the offender is a marauder versus a commuter.
43. Check the following boxes and save the results in your student folder as indicated:
    - Mean Center and Standard Distance: Save as JTC1.shp
    - Median Center: Save as JTC1.shp
    - Center of Minimum Distance: Save as JTC1.shp
    - Directional Mean and Variance: Save as JTC1.shp

    CrimeStat will add some letters in front of JTC1 for all of these new shapefiles, so don't worry about them all being named the same at this point.
44. Now click on the **Spatial Modeling → Journey to Crime** tabs.
45. Check the box next to **Journey-to-Crime Estimation** (JTC).
46. Click on **Use a Mathematical Formula**.

47. Under Distribution, choose **Negative Exponential**. (*Hint: For a description of each of these match models and a more complete explanation of how they work in CrimeStat, read Chapter 8 in the CrimeStat III manual.*)
48. Leave the other settings for this model as they are.
49. Save the output as JTCNE.shp in your student folder.
50. Click the **Compute** button to run these routines.
51. After they run and you have reviewed the output screen as desired, close the output screen and return to the CrimeStat III **Settings** window. Go back to the **Spatial Description** tab and uncheck every item.
52. Go to the Journey-to-Crime Estimation tab again, and change the mathematical model from Negative Exponential to **Normal Distribution**, change the output name to JTCNorm.shp, and save it in your student folder.
53. Compute the new JTC routine.
54. Repeat this for each of the five mathematical models. Make sure to change the output file name for each run so it does not write over your results.
55. You will end up with five JTC results as follows:
    - JTCNE.shp (negative exponential)
    - JTCNORM.shp (normal)
    - JTCLogNorm.shp (log normal)
    - JTCLIN.shp (linear)
    - JTCTNE.shp (truncated negative exponential)
56. When we return to ArcGIS we will compare the results of each of these mathematical methods. Before we do this, however, let's create one last JTC estimation using a calibrated data file.
57. Check the box that says **Use Already Calibrated Distance Function**.
58. Now click the **Select File for Calibration** button at the top.
59. Click **Select Files** in the pop-up dialog and find C:\CIA\Exercises\Exercise_8\Calibration.dbf.
60. This file contains the X and Y coordinates of every known offender's home and the X and Y coordinates of the crime(s) committed by these offenders. This is done by querying your records management system for every known offender, pulling out their reported

address, and then geocoding them all. I usually also keep the report number as part of this geocoded data set of known offenders. You will likely geocode the same addresses in this file multiple times.

61. Get rid of any records that did not geocode or records that geocoded to the city jail, some default transient address you may use, etc. I also keep the subjects' dates of birth or ages, race, and sex for other analyses when I can.

62. Using this known and cleaned offenders file, create a new table of the locations where crimes occurred that these offenders were involved in, and geocode them. I generally also include the Uniform Crime Report (UCR) category or crime type involved.

63. Now combine the two tables based on the report number so that in one table you have at least the X and Y coordinate of each offender's residence and each X and Y coordinate of each crime location.

64. When dealing with calibration, you should consider the following:
    - If there are any residential and crime occurrence locations that are the same for an offender, you may wish to delete them. These would be crimes that occurred at the offender's home by the offender, and because no travel was actually done, these records should be deleted.
    - If you have any really odd low or high distances traveled *(i.e., the offender lives in Ohio and the crime occurred in Arizona)*, you should delete this record if you geocoded data such as this. Each situation is going to be a little bit different, but the key is to exclude those outliers that are likely going to make the analysis foggier and not clearer.
    - Consider creating multiple Occurred XY and ResidenceXY output versions for different types of crime or by race, sex, or age. If you know your offender is a teenager, you may only want to use travel distances made by known teenage offenders in your calibration.

65. It is always a good idea to have some concept of how far different offenders are willing to travel based on crime types. DVD Table 8–1 shows the average distances that offenders travel in Glendale based on UCR crime category.

66. As you can see, the average distances traveled by offenders based on different UCR categories varies quite a bit. Keeping a database

of 5–10 years of geocoded offender residence and crime location data by race, sex, age, and crime type can be useful to create different calibration files as needed based on the details of the crimes you are analyzing.

67. For this exercise, we are using the distances traveled by about 16,000 offenders in Glendale between 2000 and 2005 (approximately).
68. Now that we have selected the calibration.dbf table, we need to tell CrimeStat which are the home and occurrence XY coordinate fields.
69. Under Origin Coordinates, choose the **REZX** and **REZY** fields.
70. Under Destination Coordinates, choose the **OCCX** and **OCCY** fields (under the column heading).
71. Check the **Projected (Euclidean)** box again, and then click **OK**.
72. Click the **Select Output File** button, choose **Plain Text**, and save the output file to JTCCalib.txt in your student folder. (*Hint: We will use a .txt extension because I have found several bugs with using the .dbf output for this function, and the text version avoids the problems.*)
73. Click on the **Select Kernel Parameters** button and set the fields as shown in DVD Figure 8–6.
74. For further information on the various settings available here, read Chapter 8 of the CrimeStat III manual.
75. Click **OK** after completing the kernel parameters as shown.
76. Click the **Calibrate** button, and when it is done, view the graph of your distance decay function (see DVD Figure 8–7).
77. Click **OK**, and then click **OK** again to return to the JTC screen.
78. Click the **Save Results To** button and save the new JTC analysis we are going to do as JTCCalib.shp in your student folder.
79. In the Already Calibrated Distance Function section, click **Browse**.
80. Find the JTCCalib.txt file you created in step 72 and add it.
81. Make sure to change the file type to .txt instead of .dbf to find the file.
82. After you have loaded our calibrated file, compute the final JTC analysis.
83. Minimize CrimeStat, but do not close it. If anything went wrong with our analyses or settings, we can correct them and redo them in just a few minutes rather than having to redo everything from the beginning.

84. Open ArcGIS and find and add the following layers from your student folder:
    - DMJTC1.shp
    - GMJTC1.shp
    - HMJTC1.shp
    - MCJTC1.shp
    - MCMDJTC1.shp
    - MDNCNTRJTC1.shp
    - TMJTC1.shp
    - TMWTJTC1.shp
    - JTCJTCCALIB.shp
    - JTCJTCLIN.shp
    - JTCJTCLOGNORM.shp
    - JTCJTCNE.shp
    - JTCJTCNORM.shp
    - JCTJTCTNE.shp

85. After adding these layers, turn off the RobSeries Path and all polygon themes, leaving only the robbery series, Centro graphic point themes, and the streets on.

86. Observe where CrimeStat III has placed all these different ways to find a center of the crime series (see DVD Figure 8–8).

87. As we expected in the first few steps of the exercise, the centrographic measures seem to indicate that the center of this crime series is near crimes 2 and 11 in the series. With the exception of the directional mean and variance point, which is closer to hit 3 in the series, the other measures of centrality are within a half mile of each other. They are not in the same exact spot, but they are close. This is another indicator that the offender may live somewhere around hits 2 and 11 in the series and that he may be a marauder.

88. There is still some doubt, but this helps us increase our confidence in the final JTC estimation we will use.

89. Now turn off all of the point themes and turn on the JTCJTC-Calib theme.

90. Create a Quantities legend using seven natural break classes for this layer. (Right click on the theme and open **Layer Properties**. Click on the **Symbology** tab, then select **Quantities** and the graduated colors on the left side. The field called **Z** will be the Value field.)

91. Set the outline of the classes to No Outline, and you should wind up with something that resembles DVD Figure 8–9.
92. Turn on each of the polygon journey-to-crime estimations we created, and make the legend the same type for each of these themes (see DVD Figures 8–10 through 8–14).
93. We can see by observing these symbolized themes that the calibrated model, the negative exponential and truncated negative exponential models, and the linear model all basically highlight the same general area. This area is slightly larger or smaller in each instance, and the centrographic points we looked at earlier are also in about this same location. This is a strong indicator that the offender has some anchor point within this general vicinity. The log normal and normal runs are entirely different, which gives us some pause, but not enough to negate the indications that the offender is indeed a marauder and likely lives somewhere within the calibrated peak area.
94. We can delete all of the centrographic point layers and all but the calibrated model for the journey-to-crime estimation.
95. There are three basic reasons to create a journey-to-crime estimation or prediction.
    a. There are no suspects identified in the crime series, and you want to optimize where you searched geocoded police records on persons for someone who may fit the suspect description or own a red Saturn and lives in the area.
        i. This is a very time-consuming process and is generally not as successful as the other two reasons for creating a JTC model. Countless hours may go by while you search for possible candidates, and even with such a reduced area as the peak search area, you could be dealing with hundreds of possible suspects.
        ii. Police records may simply not have your suspect listed in them, so the search will be futile.
        iii. Turn on the layer called RedSaturnDrivers. Notice that only two people have been cited driving a red Saturn that live within the peak JTC area (dark red). One is an Indian male who is about the right physical description and age, but the other is a Hispanic female. Perhaps she has sons, brothers, or a husband who use her car often.

        iv. Generally, analysts will check several different databases, such as adult probationers, known robbery offenders from their own RMS system, recent parolees, and citation or contact information, and prioritize those persons that meet the general description, age range, vehicles they own or have driven, etc. The subjects that appear in multiple data sources within the highest JTC peak area would be looked at first, and so on.
    b. You can prioritize certain suspects or leads already known in the investigation. Let's assume this is a serial murder investigation, and you have 100 tips and leads indicating home addresses for possible suspects. You could geocode these addresses and place them over the JTC model, then work on those that are within the high peak area first. This could save some time and increase the chance that you will find the true offender much quicker.
    c. You can use the JTC model as a guide for surveillance and proactive lead and tip actions.
        i. As a guide to surveillance, you may have units in the area that are sitting on a predicted next hit location, and a new crime occurs at a place other than where they are located. The units can quickly drive to a point between the new crime location and the center of the JTC analysis high peak area and watch for a red Saturn to drive by, then they can catch the offender after the fact. In this case your predicted next hit location may not have been entirely accurate or was too vague to place units at specific businesses, so they need to wait for the next crime and then proceed to the JTC area to try to find the offender driving home from the robbery location.
        ii. You can use the JTC area as a place for canvassing, field interrogation, and other interviews. Maximize your agency resources by placing detectives and officers in the peak JTC areas and look for a red Saturn or gather information from the public, confidential informants, or other sources in the area to help identify possible suspects.

96. An example of the final output you would provide to investigators can be seen in DVD Figure 8–15.
97. Save and close your project. Close CrimeStat III, and have a nice day!

References

Brantingham, P. L., & Brantingham, P. J. (1981). Notes on the geometry on crime. In P. J. Brantingham & P. L. Brantingham (Eds.), *Environmental criminology* (pp. 27–54). Beverly Hills, CA: Sage.

Brantingham, P. J., & Brantingham, P. L. (1984). *Patterns in crime.* New York: Macmillan.

Brantingham, P. L., & Brantingham, P. J. (1993). Environment, routine and situation: Toward a pattern theory of crime. In R. V. Clarke & M. Felson (Eds.), *Routine activity and rational choice* (pp. 259–294). New Brunswick, NJ: Transaction.

Canter, D. (2008). *Offender profiling training.* Retrieved December 11, 2008, from http:// http://www.ia-ip.org/temp/index.php?page=gop

Canter, D. V., & Gregory, A. (1994). Identifying the residential location of rapists. *Journal of the Forensic Science Society, 34,* 169–175.

Clark, A. N. (1990). *The New Penguin dictionary of geography.* London: Penguin Books.

Crime Mapping & Analysis Program. (2007). *Crime analysis unit developer kit.* Retrieved July 10, 2008, from http://www.crimeanalysts.net/caudk.htm

Environmental Criminology Research Inc. (ECRI). (2008). *Geographic profiling analyst training calendar.* Retrieved July 10, 2008, from http://www.ecricanada.com/geopro/ref_training.html

Felson, M., & Clarke, R. V. (1998). *Opportunity makes the thief* (Crime Detection and Prevention Series, Paper 98. Police Research Group). London: Policing and Reducing Crime Unit, Home Office.

Jakle, J. A., Brunn, S., & Roseman, C. C. (1976). *Human spatial behavior: A social geography.* Prospect Heights, IL: Waveland Press.

Levine, N. (2007). *CrimeStat: A spatial statistics program for the analysis of crime incident locations (v. 3.0).* Washington, DC: National Institute of Justice.

Mapping and Analysis for Public Safety. (2008). *CrimeStatIII for crime analysts.* Retrieved December 11, 2008, from http://www.ojp.usdoj.gov/nij/maps/crimestat-training.htm

Rossmo, D. K. (1995). Place, space, and police investigations: Hunting serial violent criminals. In J. E. Eck & D. A. Weisburd (Eds.), *Crime and place: Crime prevention studies* (Vol. 4, pp. 217–235). Monsey, NY: Criminal Justice Press.

Rossmo, D. K. (2000). *Geographic profiling.* Boca Raton, FL: CRC Press.

Rossmo, D. K. (2004). Geographic profiling. In Q. C. Thurman & J. Zhao (Eds.), *Contemporary policing: Controversies, challenges, and solutions* (pp. 274–284). Los Angeles: Roxbury.

Rossmo, D. K. (2006). Geographic profiling in cold case investigations. In R. Walton (Ed.), *Cold case homicide: Practical investigative techniques* (pp. 537–560). Boca Raton, FL: CRC Press.

Rossmo, D. K., & Filer, S. (2005). Analysis versus guesswork: The case for professional geographic profiling. *Blue Line Magazine, 17*(7), 24–25.

Rossmo, D. K., Laverty, I., & Moore, B. (2005). Geographic profiling for serial crime investigation. In F. Wang (Ed.), *Geographic information systems and crime analysis* (pp. 102–117). Hershey, PA: Idea Group.

Rossmo, D. K., & Velarde, L. (2008). Geographic profiling analysis: Principles, methods, and applications. In S. Chainey & S. Tompson (Eds.), *Crime mapping case studies: Practice and research* (pp. 35–43). Chichester, UK: John Wiley & Sons.

van der Kemp, J. J. (2005, September). *When to use, when not to use, that's the question*. Paper presented at the Crime Mapping Research Conference, Savannah, GA.

Velarde, L. (2004, April). *Applying geographic profiling to property crimes: The geographic profiling analyst program*. Paper presented at the Crime Mapping Research Conference, Boston.

Velarde, L. (2008, July). *Operational geographic profiling of poachers*. Paper presented at the National Institute of Justice Conference, Washington, DC.

Weiss, J., & Davis, M. (2004, December). Geographic profiling finds serial criminals. *Law and Order*, 32, 34–38.

Research and Crime Data

SECTION II

CHAPTER 5	Research and Applications in Crime Mapping
CHAPTER 6	Crime Mapping and Analysis Data
CHAPTER 7	People and Places: Current Crime Trends

Research and Applications in Crime Mapping

CHAPTER 5

▶ ▶ LEARNING OBJECTIVES

Chapter 5 highlights the relevant research on crime mapping and its applications in the field. This chapter is meant to provide the student with a cursory examination of the existing literature on how characteristics of place and space impact crime and how GIS can assist criminal justice agencies in developing strategies that reduce crime, prevent victimization, and assist in the investigation and prosecution of offenders. After studying this chapter, you should be able to:

- Identify and discuss research in the field of crime mapping.
- Summarize and explain key research findings about crime at various places.
- Be able to identify different applications of GIS in criminal justice.

■ Introduction

The first chapters of this text introduced students to the basic knowledge needed to understand the theoretical justifications behind crime mapping and analysis. Chapter 1 introduced the basic concepts and terminology in crime mapping and provided a list of online resources where students can find publications, conferences, data, and assistance with crime mapping and analysis questions. Chapters 2, 3, and 4 examined key theoretical perspectives of why crime is unevenly distributed across people and places. These theories discussed the individual, social, and environmental factors thought to be associated with offending and victimization. Here, Chapter 5 provides a cursory review of the empirical research found in numerous publications geared toward both practitioners and academics. This literature discusses not only the use of GIS to examine spatial relationships of crime and the GIS applications utilized by various criminal justice agencies. In these studies, the use of crime mapping is based on the knowledge examined in the following chapters regarding statistics and data, and in the preceding chapters examining theory. Thus, students should understand that maps are not created out of thin air but are carefully constructed using the rules and assumptions provided by both statistics and theory.

■ Research

This chapter is organized into a number of sections and examines the research based on the non-mutually exclusive categories of research and application.

In GIS, the line between research and application is fuzzy at best. Many research studies are based on practical applications, and many applications are evaluated and then published in peer-reviewed journals. Thus, almost any given study could legitimately be discussed in a number of sections within this chapter. However, the studies examined in this first section focus primarily on spatial relationships of crime, and while the findings have practical utility, the primary research objective is to gain a better understanding of crime in a specific place or area. The following discussions examine the research and primarily analyze some places typically thought to be associated with different types of crime problems.

Public Housing

The common perception about public housing projects is that they are violent and dangerous places. However, only a few studies have empirically linked disorder and crime to public housing projects. Recall Newman's (1972) principles of defensible space and his research detailed in Chapter 3. Newman argued that there were key design flaws found in some public housing projects which facilitated victimization by reducing the visibility and the use of areas by legitimate users. More recently, Suresh and Vito (2007) examined hot spots of aggravated assaults in Louisville, Kentucky between 1989 and 1999. They found that a consistent clustering of assaults occurred around many of the public housing projects. Analysis revealed that several variables were important in understanding these hot spots. While income remained the most important variable, other variables of importance included the proportion of the population under age 18 years, the proportion of the male population that was unemployed, and the percentage of vacant housing. In addition, a distinct change in one of the historic hot spots of assaults in 1997 and 1998 (near the Park DuValle neighborhood) led the authors to investigate further. They discovered that:

> Park DuValle was one of the major historic hot spots until 1996. The disappearance of the hot spot at Park DuValle, in the years 1997 and 1998, suggested some type of intervention. Revitalization of two of the vulnerable low-income public housing projects; the removal of the adjacent dilapidated, unattended private apartments; and the relocation of the former residents of those two low-income public housing projects were the main reasons for this shift. (Suresh & Vito, 2007)

Other research suggests that certain crimes are more prevalent in public housing projects than they are in other places. For example, Holzman et al. (2005) examined public housing areas in three different cities. They found that while "clearly the image of public housing developments as violent, dangerous places finds support in this study," the risk of victimization for different crimes varied (p. 322). For example, aggravated assaults and robberies were higher in public housing areas, while burglaries, larcenies, and thefts tended to be lower in public housing areas in comparison to surrounding areas.

Holzman, Hyatt, and Dempster (2001) examined the victimization rates of aggravated assault of public housing residents in two cities. They found that black females who lived in public housing in both cities were at higher risk of being victims of aggravated assault than black or white women who did not live in public housing. In addition, the rate of victimization for black female residents of public housing varied due to "different agricultural design and geographic dispersion of the respective cities' public housing developments" (p. 662). In addition, the authors argue that one of the developments "offered less privacy and accessibility, thus discouraging would-be assailants" (p. 662).

Roncek's (2000) study on robbery in the Bronx found that after demographic, socioeconomic, and other variables were controlled for, the existence of public housing projects had no effect on robberies. Thus, the characteristics of *people* may be more important than characteristics of *places* for some types of crimes.

Fagan and Davies (2000) suggest that very little research exists on crime in public housing and that most is based on neighborhoods or areas where public housing projects are likely to exist. "The official eligibility criteria and social selection processes in public housing also contribute to a concentration of individuals with social deficits and below-average human capital" (p. 123). The authors point to three primary issues within the research on public housing, including:

- There are spatial and temporal differences in crime rates across public housing sites.
- Crime rates are not stagnant but are dynamic and have "crime careers."
- Various problems exist in calculating population denominators.

One reason Fagan and Davies suggest that people perceive crime to be so prevalent in public housing projects is that these areas have high "spatial and social concentrations of poor people" (p. 124). In addition, there is a general lack of informal social control in the socially disorganized neighborhoods where public housing areas are typically located. Furthermore, the physical structure of some public housing projects limits the "guardianship" within the area. Thus, Fagan and Davies suggest that crime could be both migrating into public housing and out of public housing, producing a two-way diffusion effect. Their study examined interpersonal violence in 82 public housing projects and the immediate surrounding areas in Bronx County, New York. Using data from the census, the New York City Housing Authority, and official crime reports for the study areas, they found "significant diversity in both social characteristics and the violent crime rates across public housing projects" (p. 127). They found that, in general, crime is greater within a 100-yard perimeter surrounding a housing project than it is within the boundaries of the project itself. Violent crime, while still high, begins to decrease between 100 and 200 yards of the project. Using a two-stage least-squares regression to identify the direction that crime travels, Fagan and Davies concluded that assaults appear to both come from public housing and travel to surrounding neighborhoods and travel into public housing from surrounding neighborhoods. Limitations of the regression analysis included the omission of potentially important variables, including residential mobility, social interaction between neighbors, processes of informal social control, and income inequality. In addition, random selection of public housing sites was not accomplished in this study, limiting the ability of these authors to generalize their findings.

Businesses

Wernicke (2000) employed GIS in a multipronged strategy aimed at reducing incidents of thefts at construction sites in Overland Park, Kansas. The first layer of the strategy was to alert the public that the police department would be cracking down on construction site burglary. The second layer was to work closely with builders and contractors to help them identify ways to better secure the sites. The third layer was to identify known offenders in the area who specialized in this type of burglary. Last, through the use of checkpoints and with detectives working weekend hours, the Overland Park Police Department noted a 26% reduction in stolen equipment.

In Chula Vista, California, an analysis of calls for service (CFS) data indicated a high concentration of calls for service from several of the motels on the west side of the city (Eisenberg & Schmerler, 2004). In 2003, 2091 calls for service were documented in Chula Vista motels, resulting in 129 arrests. The analysis indicated that five motels accounted for a disproportionate number of CFS. The typical reasons for disturbance CFS included: guests wouldn't leave and/or pay; guests/visitors were having loud parties; guests/visitors were arguing or fighting; and finally, people were loitering and would not leave the property. A temporal analysis revealed that CFS were highest at about noon and again at about 11:00 p.m. Using theory to guide the analysis, it was determined that several factors were related to the high number of CFS from motels. These included lower room prices (attracting a lower class clientele), local residents renting the rooms (indicating that persons had nowhere else to go or wanted a temporary place to party), a higher than normal percentage of guests on probation or parole compared to the population, and a long length of stay. A basic environmental assessment found that many motel rooms did not have chain locks, deadbolts, or peepholes, and very few signs were posted to display the rules of the motel. Several strategies were implemented, including strict screening of both guests and visitors, increased access control, employing security guards, clearly conveying the rules to guests, and posting signs around the motel property that displayed the rules. In addition, employee training, code enforcement, and limits on the length of stay were implemented in efforts to reduce the number of CFS from the motels (Eisenburg & Schmerler, 2004).

Schools

Reno (1998) utilized GIS to examine residential burglaries in Shreveport, Louisiana. The analysis identified that daytime burglaries were more prevalent than nighttime burglaries and that a main concentration of daytime burglaries was located near a high school. Based on this analysis, officers were deployed to target areas located near the school and performed investigative stops. They also coordinated their efforts with the high school, focusing on the problem of truancy. The strategy was successful in reducing the burglaries by 67% and virtually eliminating the truancy problem.

Hendersen and Lowell (2000) focused on campus crime at Temple University. Using GIS to map self-reports of assaults (instead of relying on official reports), the researchers identified four clusters. In these four areas, they added security kiosks, improved lighting, increased patrols, and provided escort services. Subsequent self-reports suggested a dramatic decrease in assaults (from 57 to 16).

Roncek (2000) analyzed the proximity of robberies to schools. His research at the block level in the Bronx concluded that proximity to schools accounted for a very small percentage of crime and that the type of crime associated with schools varies across different schools. In addition, ordinary least-squares regression analysis (explaining 36.4% of the variation in robbery) identified other variables, including the block population and the presence of a subway stop, as important in understanding the number of robberies. Furthermore, "higher concentrations of the three largest minority groups—African Americans, Asians, and Hispanics—are associated with more robberies, as is lower socioeconomic status" (p. 160). A subsequent negative-binomial regression found that:

- 29.5% more robberies occurred in blocks adjacent to public senior high schools.
- 30.8% more robberies occurred in blocks adjacent to public junior high schools.
- 11.4% more robberies occurred in blocks adjacent to public elementary schools.
- 21.3% more robberies occurred in blocks adjacent to private grammar schools.

The relationship between schools and crimes is complicated, and while it appears that the proximity to some types of schools in some areas is important in understanding some crimes, other variables, such as demographic and socioeconomic factors, may have more influence over the crime rates within these neighborhoods. Although some of these statistics sound like they may be conflicting or confusing, the real benefit comes with the knowledge that research is being done related to crime and place. Many of these studies can be duplicated by analysts within their own jurisdictions. Analysts can thereby add to the research and eventual publications that deal with these crime issues to benefit everyone involved in crime prevention and control.

Transit Stops

Block and Block (2000) examined street robbery in proximity to rapid transit stations in the Bronx and Chicago. They argue that characteristics of transit stops make them ideal places for motivated offenders.

> Transit stops provide cover for potential offenders. There are transitional breaks in transportation, where standing around is not suspicious activity. By definition, transit stops are easy to enter and exit. Potential targets usually live some distance from the transit stop, are not always familiar with the surrounding area, and are unlikely to have previously met potential offenders. (p. 138)

Block and Block's data included both attempted and completed street robberies in Chicago in 1993 and 1994 and New York City police data on street robberies that occurred between October 1995 and October 1996. Hot spot analyses indicated that 10 of the 13 hot spots in the Bronx either contained a transit station or were adjacent to one. Every transit station in the Chicago analysis was found in a hot spot (11 total hot spots, 10 hot spots that contain transit stations). The authors, citing Angel's theoretical model about different types of land use and crime, suggest there is a "critical intensity zone" (p. 146) of land use where there are enough targets to attract offenders while potential witnesses are insufficient in numbers to deter the offenders. Utilizing an intensity analysis of these hot spots, Block and Block determined that 39% of street robberies that occurred in Chicago's Northeast Side in 1993 and 1994 were within 1000 feet of a transit station, with a critical intensity zone located between one and two blocks away from the station. A second peak occurred approximately five blocks away on a commercial strip that parallels the transit line. In the Bronx, approximately 50% of street robberies occurred within 737 feet of a transit station, with the critical intensity zone located within 400 to 500 feet of a station. "In both cities, the number of robberies peaks where the density of targets is still high but civilian and police surveillance are declining" (p. 148). Furthermore, to understand overall crime patterns, it is important to understand the characteristics of areas in which transit stops are located. For example:

> The backcloth, the stations and shops and services that serve their transit riders, may attract predatory offenders. The blocks surrounding the EL stations in Chicago are sometimes a convenient market location for both drug buyers and sellers. These areas also have cash transaction businesses frequently patronized by transit travelers. For example, a small strip mall near the Howard station contains a liquor store, a laundromat, a video shop, a pawn shop, and an adult book store. (p. 149)

Hot Spots

The literature on hot spots is as prevalent and as varied as the definitions of hot spots. Generally, hot spots are viewed as small geographic areas that experience higher than average levels of crime for a consistent period of time. Taylor (1998) suggests that three primary forces are leading hot spot research: frustration, tools, and theory. Identifying hot spots is not as simple as counting the dots that appear on the map. "Such complexities suggest that it may not be easy for police to define crime hot spots precisely and allocate patrol resources accordingly. This raises general questions about the clarity of the concept of hot spots" (Taylor, 1998, p. 3). In addition, as we will discuss in Chapter 10, many analysis programs produce hot spots in the form of circles and ellipses, but real hot spots do not conform to exact geometric shapes.

In his review of the relevant literature, Taylor suggests that hot-spot-based strategies, in general, have produced minimal impacts on crime and that these impacts are typically only achieved for the short term.

Basing his suggestion on the literature, he also advised that "we know more about reducing disorder than about reducing crime" (p.9).

Other research on hot spots suggests that the identification of hot spots based on one's experience with an area (such as law enforcement) is not necessarily an easy task. For example, Ratcliffe and McCullough (1998) asked police officers in South Nottingham, England, to pick out hot spot areas of crime. When these areas were compared with actual hot spots of crime (identified through analysis), the researchers found that police perceptions about hot spots did not match actual hot spot locations. In another interesting study, Paulsen (2003) compared "media" hot spots with actual hot spots. Paulsen utilized GIS to create maps of homicides covered by the local media in Houston, Texas. Upon comparison with actual homicide hot spots, he found that the actual homicide hot spots tended to be closer to the center of the city than hot spots identified by media coverage.

Martin, Barnes, and Britt (1998) examined arson hot spots in Detroit, Michigan. Using GIS to map arsons, they developed a multi-pronged approach to reduce arson hot spots. The strategies included forming a citywide task force, identifying community resources, and mapping current arson hot spots and abandoned buildings that may become future targets. In addition, letters were sent to residents to alert them about the arson hot spots and explain the strategies that were being developed to address and reduce those threats. This strategy resulted in fewer fires and the elimination of several hot spots. Unfortunately, a number of hot spots still remained after implementing several of the new strategies. These "stubborn" hot spots were situated in areas of high crime and a large number of abandoned buildings, a low level of community organization, and high levels of poverty. The factors suggest that additional strategies may need to be developed to address these factors in the neighborhoods with remaining arson hot spots.

Lockwood (2007) made several interesting observations when he examined the relationship between violent crime and land use in Savannah, Georgia. First, assaults were related to the type of land use within an area, specifically with retail/commercial/office, renter, and public/institutional land uses. Second, both homicide and assault were strongly and positively correlated with social disadvantage. In addition, commercial areas with high robbery rates tended to be adjacent to residential block groups with high rates of both homicide and assault.

In examining the distribution of homicides in three cities, Adams (2001) found that although the physical environment was important in understanding homicide hot spots, the more important variable was economic deprivation. The author suggests that more research to examine the impact of rejuvenation projects aimed at both physical and social issues is necessary.

Drugs

GIS is also useful in researching drug-related crime and the spatial dispersion of drug activity. Romig and Feidler (2008) utilized GIS to better understand methamphetamine (meth) activity in the state of North Dakota. They created map layers displaying meth-related arrests, meth lab seizures, and various demographic characteristics. Using completed maps, they identified uneven distributions in abuse, arrests, and distributions of meth across North Dakota.

> When examining the spatial pattern of methamphetamine-related arrests, our research discovered that they are more likely to take place in counties that are urban, closer to containing a larger city, highly populated, higher per capita income, have a larger proportion of vacant houses, a higher percentage of the population in their late twenties, and a growing population. (p. 217)

For clandestine lab seizures, the trends are very different. Lab seizures were higher in "counties that are more rural, further away from larger cities, and have smaller populations." (p. 217) In their research, they discovered an anomalous county, Williams County, that suffered from both a high seizure rate *and* a high arrest rate. The authors also note that the distribution of rehabilitation programs throughout the state is limited and currently cannot support the numbers of addicted persons who need treatment.

■ Applications

Again, the line between research and application is by no means clear. This section examines the research conducted on applications of GIS in various capacities and is organized by the criminal justice function that it serves. While reading the following sections, it is important to keep in mind that the utilization of GIS varies from department to department. In some agencies, GIS is used extensively for multiple purposes. In other departments, GIS is employed sparingly for only a few types of analyses. A 2002 publication by the Arizona Criminal Justice Commission indicates that only 14.5% of law enforcement agencies in Arizona used any type of crime mapping at all. A notation in the executive summary of this publication indicates, "Furthermore, the average length of time that agencies report having had crime mapping capabilities is approximately 4.5 years. This is actually longer than was found nationally (3.3 years) and would strongly indicate that the potential of mapping in criminal justice systems is in its infancy" (p. 5).

Ratcliffe (2004) proposes an interesting "hotspot matrix" that looks promising as an approach to planning both long-term and short-term policing strategies. Using an example for a housing estate, the matrix is composed of spatial patterns (dispersed, clustered, and Hotpoint) and temporal patterns (diffused, focused, and acute); matrix cells provide possible police strategies, such as uniform vehicle patrol, foot patrols surveillance units, and CCTV, among many suggestions.

Investigation and Prosecution

There are endless applications of crime analysis and mapping for investigation purposes. The Lincoln Police Department in Nebraska, for example, used GIS to successfully stop a series of automobile break-ins. In addition, with the use of GIS, the LPD investigated and solved a sexual assault case of a young boy by linking previous incidents of indecent exposure in the area near the crime (Casady, 2003). In another case, detectives in Knoxville, Tennessee, used maps of known sex offenders who lived in the area of a rape incident to conduct a photo lineup for the victim. The perpetrator was quickly identified, apprehended, and subsequently confessed (Hubbs, 1998).

In Spokane, Washington, GIS was used in the investigation of two high-profile cases: the Brad Jackson case and the Robert Yates case (Leipnik, et al., 2003). Brad Jackson was a suspect in the suspicious disappearance of his 9-year old daughter, Valerie, on October 18, 1999. In the Jackson case, the Spokane Police Department used GPS tracking and GIS mapping capabilities to follow Jackson's whereabouts for a period of 18 days. The information about his travels and length of stay at various locations was collected and analyzed and led to the discovery of a temporary gravesite and a permanent gravesite where his daughter's body was found.

In the Yates case, police used GIS analysis of body dump sites and of supermarkets in efforts to prioritize areas of Spokane where a serial murderer resided. In the murders, several of the victims' heads were covered with plastic bags. The bags had store markings that could be traced to individual franchise stores. The stores where these plastic bags came from were mapped using GIS capabilities, and an area of probable activity was created. This helped investigators narrow their search by giving them a high probability area to start investigating. Although GIS was not instrumental in solving the case (blood evidence in a white Corvette once owned by Yates was the zinger), Leipnik et al. (2003) argue that:

> Although GIS and geographic profiling did not solve this case, in retrospect it could have narrowed down the focus of interest to a five square mile area out of 1800 square miles within the greater Spokane metropolitan area. The suspect's home residence turned out to have been located within this five square mile "high probability" area; in fact he lived less than a mile from the mean locations (centroids) of both the body dump and the grocery store clusters. (p. 179)

An interesting twist to the Yates case is that a GPS unit containing 72 waypoints was recovered from Yates's home. That information is still being analyzed to help determine whether or not those waypoints represent dump sites for yet unidentified victims (Leipnik, et al., 2003).

In a series of indecent exposure cases, GIS analysis identified information about spatial and temporal clustering of incidents that was disseminated to the public and informed the development of surveillance operations. The subject was ultimately arrested (Woodby & Johnson, 2000). Investigators in Seattle, Washington, used crime mapping to identify and make predictions about a robbery series. Based on the findings of the analysis, crime bulletins were created and distributed to the public via television and newspaper. The suspects were apprehended based on a tip provided by a confidential informant. Additional analysis was performed to review suspect activity and help discover additional crimes perpetrated by the suspects in which weapons were stolen and used in several robberies (Robbin, 2000).

In Edmonton, Alberta, Canada, Warden and Shaw (2000) examined clusters of break-and-enter crime. Spatial and temporal analysis identified patterns that were useful to surveillance, which led to the arrest of the individuals responsible for the break-ins.

Crime incidents are not the only points that are plotted during investigations. For example, the murder investigation of a young man in St. Petersburg, Florida, was facilitated by mapping the locations of cellular telephone transmitter sites. In this case, a young man's dismembered body was found in Tampa Bay. The victim had been involved in a bungled drug deal and was subsequently tortured and killed. The times and the locations of the suspects' cell phone use was plotted and mapped to corroborate eye witness testimony of the times and locations that the offenders and victim were seen together (Moland, 1998).

> In the end, the weight of the expert testimony provided by the two detectives, the AT&T cell phone technician and other witnesses was sufficient to convict the two suspects and sentence them to life in prison. This case shows the difficulty of explaining complex, technical but verifiable information in its purest form. The old expression "a picture is worth a thousand words" was proven during this trial. (p. 72)

In another similar example, several violent hijackings of persons over the course of a few days occurred in South Africa. Four persons were suspected in these incidents. Of these hijackings, only two victims (a father and son) were left alive. Two couples were shot and killed. Of these, one woman was raped before she was killed. DNA analysis of evidence implicated two suspects in the rape; they were shot and killed in a police shootout. Two suspects remained at large. During the investigation, it was learned that the woman's cell phone was missing, so phone records of calls made and received were acquired. Further investigation revealed that one phone number was called multiple times at about the same time she was last seen alive. The recipient of these calls happened to be one of the suspects who had been shot and killed in the police shootout. Subsequent investigation of cell phone records revealed that suspects were in constant contact between one another on various cellular phones and landlines. These phone records were mapped and used in court to corroborate eye witness testimony of the suspects' whereabouts (Schmitz, Cooper, Davidson, & Roussow, 2000).

Overall and Day (2008) employed a post hoc analysis to determine the prediction ability of a probability grid method (PGM) outlined by Hill (2003) on the Hammer Gang robbery series in the city of Durban, KwaZulu-Natal, South Africa, between October 2003 and January 2004. The analysis proved to be highly successful in that:

> The resultant probability map produced two very high-probability offence locations in close proximity to each other, one being the seventh and last offence in this series.... Both of the very high probability target locations were located within the first ring buffer and the calculated radius was in line with the offenders' directional path. Taking into account the number of potential targets, the PGM reduced the potential target locations from eleven to two. (Overall & Day, 2008, p. 60)

In another example of the potential prediction ability of crime mapping and analysis, Casady (2008b) discusses the arrest of a bank robber in Lincoln, Nebraska. In this case, GIS was used to prioritize potential sites for deploying stakeouts. In the analysis, spatial, temporal, and other characteristics of bank robberies in a series were analyzed to produce predictions about when and where the next bank would be hit:

> Crime mapping helped determine Erving's *modus operandi*, helped locate his potential victim banks, helped determine the best locations for stakeout and helped identify likely escape routes. The case is illustrative of the nexus of the art and science of policing: the creativity, initiative and intuition of experienced police officers informed by geographical crime analysis and traditional investigative analysis. (p. 68)

Proactive Policing

The term "proactive" policing typically means employing strategies to *prevent* crime and to address the root causes of crime rather than just responding to calls for service ("reactive" policing). Depending on the agency, this might mean randomly running license plates throughout an officer's shift or developing multileveled strategies to address social and physical incivilities associated with crime in a neighborhood. Whatever the case, proactive policing assumes that rather than simply responding to crimes that have already occurred, law enforcement can develop and implement tactics that prevent future crimes. The previously mentioned case in Overland Park, Kansas, would be a good example of creating a multifaceted strategy to reduce or prevent crime.

Casady (2008a) describes an automated system implemented by the Lincoln Police Department (Nebraska) in which maps and reports that alert officers to increases in crime and recently issued warrants, display crime trends for crime types such as burglaries, and monitor sex offender registrations are automated, providing officers with up-to-date information that is useful to their daily operations.

Event Planning

Mapping techniques are useful to police and other agencies in preparation of large events. For example, the Lincoln Police Department in Nebraska uses GIS to plan traffic and crowd control details for large events, including the University of Nebraska's Cornhusker football games (Casady, 2003). Crowd control, traffic management, and other related concerns are planned in advance to ensure public safety and minimize public inconvenience before, during, and after the games.

Community Policing

The Lincoln Police Department used GIS to develop a number of proactive approaches to crime and other community problems that are related to proximity to the college. First, "using GIS LPD analyzed the number of complaints received from the public concerning wild parties, and found a 27% reduction in complaints from neighborhoods located within one mile of the campus, as compared to the same period in 1997" (Casady, 2003, p. 115). Second, an analysis of burglaries identified trends in residential burglary in which offenders entered open garages. Officers armed with informational fliers went door to door and told homeowners to shut their garage doors (Casady, 2003). A similar strategy was employed in Overland Park, Kansas. An analysis that used GIS identified several neighborhoods that experienced high numbers of "garage shopping" in which residential burglars entered open garages and stole valuable tools, bikes, and other items. Officers targeted these neighborhoods and handed out brochures to residents alerting them of these thefts. This effort resulted in a reduction of these burglary incidents and facilitated approximately 1200 citizen contacts in the process (Wernicke, 1998).

The LPD has also used GIS for other proactive strategies, including reducing the number of police incidents at a specific apartment complex, decreasing the number of thefts of self-service gasoline, supporting community development in weak neighborhoods, and keeping the public informed about crime in their community and the LPD's commitment to reducing crime (Casady, 2003).

The Knoxville Police Department (Tennessee) used GIS to map the area's crimes and calls for service, including shots fired calls and auto burglary, and overlaid this information on top of parolee addresses. They identified a clustering of several different crimes and disturbance calls near the end of a cul-de-sac where a parolee (he was also a drug dealer) resided. The parolee was arrested after a search of his home, and it was later discovered that many of the drug dealer's friends were responsible for the crimes and disturbance calls that initiated the analysis (Hubbs, 2003). Hubbs argues that police can effectively use mapping capabilities to track known sex offenders. He cites several successful cases from the Knoxville Police Department, including one where they found an abducted little girl and arrested a serial rapist. The KPD has also used GIS to extensively map and analyze gun violence in Knoxville.

In Hartford, Connecticut, law enforcement believed that drug calls in the Blue Hills neighborhood were severely underreported. The officers who worked in this neighborhood brought this to the attention of the residents, which resulted in a 170% increase in the number of drug calls for service. These calls resulted in the subsequent arrest of 11 people (Rich, 1998).

Citizen satisfaction can also be affected by the use of GIS and crime analysis. For example, Canter (1998) examines the impact of Baltimore County's Autodialer System—a system that utilizes GIS to systematically alert citizens to crime problems occurring in the community and provide suggestions about how they can reduce their chances of being victimized. The system has been successful from a community-oriented policing perspective in that it provides a direct line of communication from police to citizens about their community. An evaluation of the system indicates that the citizens really like it.

Problem-Oriented Policing

Herman Goldstein argues that police should pay less attention to "incidents" and more attention to "problems" if they want to make any impact on crime (Goldstein, 1990). The identification of crime problems can be achieved through the use of multiple data sources. In addition, solutions to these crime problems should include both legal and nonlegal applications where appropriate.

> Problems are defined either as collections of incidents related in some way (if they occur at the same location, for example) or as underlying conditions that give rise to incidents, crimes, disorder, and other substantive community issues that people expect the police to handle. By focusing more on problems than on incidents, police can address causes rather than mere symptoms and consequently have a greater impact. (Cordner & Biebel, 2005, p. 155)

In problem-oriented policing, problems are addressed utilizing the SARA model (scan, analyze, respond, and assess), with the ultimate goal being to eliminate the problem rather than simply continually responding to it (Eck & Spelman, 1987). Under SARA, officers identify related crime problems in their jurisdiction, analyze the problem to identify trends or patterns that are useful to uncovering solutions, develop a response (or responses) to the problem, and assess the success or failure of the response(s). Problem-oriented policing, then, is a focused, proactive approach designed to attack the root causes of crime rather than just crime itself. Hill (2003) argues that police departments are very good at the SAR stages but fail in the assessment stage. Utilizing GIS can improve officers' ability to assess the effectiveness of their responses. In the crime analysis community, the acronym CCADE (collect, collate, analyze, disseminate, and evaluate) is also used. SARA and CCADE are very closely related; one looks at problem solving from the police officer side (SARA) and the other looks at it from the analyst side (CCADE; see **Figure 5–1**).

For example, the Springfield Police Department and Illinois State Police collaborated to reduce gun-related incidents and gang-related shootings (Bitner, Gardner, & Caldwell, 2000). An analysis indicated that gang violence was emerging in an area where recently relocated gang members from Peoria, Illinois, resided. Utilizing GIS, a summer strike force that targeted gang- and gun-related violence yielded substantial reductions in the rates of violence and forced gang members to relocate to another area. Otto et al. (2000) used GIS to identify gang hot zones in Akron, Ohio, by mapping graffiti that marked gang boundaries. Hot

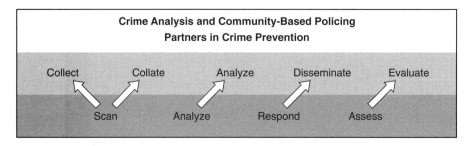

Figure 5–1 SARA Model

zones were those areas where gang boundaries overlapped and were thus more likely to experience intergang-related violence.

Using repeat address mapping (RAM), Gersh and Beardsley (2000) utilized GIS analysis to identify the top 10% of addresses accounting for drug calls for service to guide the development and implementation of a crackdown (Operation Clean) in the Langley Park area of Washington, DC. Mapping and analysis of the area after the crackdown indicated a substantial reduction in the number of hot spots and in the number of calls for service within the remaining hot spots. In the Lincoln Park area of San Diego, gang boundaries and overlapping areas were mapped along with personal crimes, property crimes, and calls for service to identify problem locations concerning gang-related crimes. The use of civil injunctions was used to prevent gang members from hanging out in specific areas, associating with other known gang members or affiliates, and engaging in various gang behaviors. This strategy netted Lincoln Park a crime rate reduction of 29.3 to 18.1 (per 1000 inhabitants; Polk, Hammond, Yoder, & Burke, 2000).

In the early 1990s, the Boston Gun Project and other strategies were implemented simultaneously to produce substantial reductions in gang and youth violence in the city of Boston. GIS was utilized to map gangs and gang territory based on the current gang intelligence of police officers, probation officers, and street workers. A strong consensus existed amongst the practitioners about the active gangs, their rivals, and their territories. Using network analysis, key gangs and their conflict network were identified.

The Boston Gun Project is a problem-solving exercise aimed at preventing youth violence in Boston by: convening an interagency working group; performing original research into Boston's youth violence problem and illicit gun markets; crafting a city-wide, interagency problem-solving strategy; implementing that strategy; and evaluating the strategy's impact. Key participants in the project have included gang officers from the Boston Police Department, probation officers whose jurisdictions incorporate those Boston neighborhoods at high risk for youth gun violence, and city-employed "streetworkers"—outreach specialists focused on preventing and mediating gang disputes and diverting youths from gangs. (Kennedy, Braga, & Piehl, 1998, p. 228)

Deemed the "Boston Miracle" by many (Winship, 2002), youth homicides dropped dramatically:

The '90's were a remarkable decade for Boston. Not only did Boston enjoy a period of nearly unprecedented economic prosperity, but it was also a time of historically low crime rates and unusually good police–community relations. Most dramatically, the number of homicides plummeted over the decade. Whereas there had been 152 homicides in 1990, in 1999 there were only 31. Much of [the] drop occurred for individuals twenty-four years of age and less. Whereas in 1990 there were 73 homicides in this age group, there were only 15 in 1999 and 2000. For the twenty-nine month period ending in January 1998, there were no teenage homicide victims. Thus, the so-called Boston Miracle. (Winship, 2002, p. 1)

Compstat

In Newark, New Jersey, the problem of auto theft was examined in semimonthly Compstat meetings. An analysis indicated that the time of day these auto thefts were occurring was a critical factor. The last known addresses for people who were arrested for receiving stolen property and the vehicle recovery locations were mapped. Strategies employed in the East district, based on the analysis, yielded a decrease from an average of 40 to 30 auto thefts per week (Santiago, 1998).

McGuire (2000) argues that the key to Compstat's success in New York City was the involvement of multiple agencies including the District Attorney, New York State Department of Parole, City Probation Department, and City Corrections Department. Other city agencies, such as Consumer Affairs, Social Services, and Environmental Protection, also attend the Compstat meetings. Compstat is attributed as being responsible for the 41% drop in crime that New York City experienced between 1993 and 1997, which included a 60% reduction in murder and nonnegligent manslaughter.

Other Miscellaneous Applications of GIS Analysis

Rieckenberg and Grube (1998) examined traffic fatalities in Cook County, Illinois. They collected and mapped information about the type of accident, location of accident, the condition of the driver and roadway conditions at the time of the accident, and the time and day of the accident to inform a concentrated enforcement effort in high-accident areas. Using a zero-tolerance approach, they achieved a 35% reduction in traffic fatalities in the first 6 months.

In Lowell, Massachusetts, mapping was used to pinpoint and chronicle evidence in a homicide trial (Cook, 1998). In Florida, information on the whereabouts of probationers and parolees being tracked through the use of GPS is stored and accessed using a Web-based interface system whereby participating agencies can query tracked offenders to determine if they were within 1000 feet and 30 minutes of a crime (Frost, 2005). In addition, the system signals zone violations of offenders leaving their "inclusion" and "exclusion" zones. The system then automates an e-mail to officers, including a hyperlink to the offender's track points.

Freisthier et al. (2007) employed mapping techniques to look at the effects of alcohol outlets on child protection services (CPS) referrals, substantiations, and foster care entries in the state of California. Using data at the zip code level, they found a positive relationship between the number of off-premise alcohol outlets and the percentage of black residents with higher rates of maltreatment.

> More specifically, the model derived estimates that an average decrease of one off-premise outlet per zip code would reduce total referrals to CPS in the 579 zip codes by 1,040 cases, substantiations by 180 cases, and foster care entries by 93 cases. Characteristics of adjacent zip codes also were related to maltreatment rates in local neighborhoods, indicating a spatial dynamic to this relationship. (p. 114)

■ Online Resources

- Great criminal justice-related articles and publications, including crime mapping studies: http://www.ojp.usdoj.gov/njj/maps
- National Criminal Justice Reference Service: http://www.ncjrs.gov
- Police Executive Research Forum: http://www.policeforum.org

■ Conclusion

As you can see, the possibilities for crime mapping in small-, medium-, and large-sized departments are endless. The applications range from the facilitation of simple presentations to the development of complicated prediction models. In addition, the enhanced supervision and monitoring of parolees and probationers is also aided by the use of GPS and GIS.

> The most common approach is to have a crime analyst use GIS to analyze the spatial and temporal factors associated with a series of crimes or to detect patterns, trends, and exceptions. In most police departments,

crime analysts view GIS as in important but non-essential tool. . . . The reason that many departments use GIS selectively is that it is a relatively new technology in policing. (Leipnik & Alpert, 2003, p. 5)

As the familiarity of GIS technology increases, crime mapping applications in both research and practice will continue to expand. Future chapters examine the key processes in some of the more commonly employed spatial analyses.

■ Recommended Reading

Asbell, B. (2003). National Guard Bureau counterdrug GIS programs: Supporting counterdrug law enforcement. In M. R. Leipnik & D. P. Albert (Eds.), *GIS in law enforcement: Implementation issues and case studies* (pp. 211–227). New York: Taylor & Francis.

Chainey, S., & Tompson, L. (Eds.). (2008). *Crime mapping case studies: Practice and research*. Hoboken, NJ: Wiley.

Goldstein, H. (1979). Improving policing: A problem-oriented approach. *Crime and Delinquency, 25*, 236–258.

Paul, B. (2001). *Using crime mapping to measure the negative secondary effects of adult businesses in Fort Wayne, Indiana: A quasi-experimental methodology*. Paper presented to National Institute of Justice, Crime Mapping Research Center, 2001 International Crime Mapping Research Conference, Dallas, TX.

■ Questions for Review

1. Based on the literature review, what areas are promising for GIS to assist law enforcement in the investigating and prosecuting of crime? What weaknesses in utilizing GIS in law enforcement still need to be addressed?
2. Why would automated maps and reports be useful to police agencies? What types of reports or maps do you think would be most useful in an automated format? Why?

■ Chapter Exercises

Exercise 9: Spatial Deviation

Data Needed for This Exercise
- Everything needed for this exercise is located in EX9.mxd.

Lesson Objectives
- Identify a crime series based on modus operandi (MO) factors.
- Describe peculiarities of a robbery crime series and determine what may be important to the analysis.

- Introduce several key elements in determining suspect patterns of activity in a crime series based on crime theory.
- Discuss physical elements of the environment that can assist in analysis of some crime series.

Task Description
- In this exercise we will look at a robbery series and walk through the steps and key elements we should look for when doing an analysis of robbery or any other crime we may be dealing with.

1. Open EX9.mxd.
2. When you open this exercise you should be looking at a make-believe robbery series consisting of nine total robberies spread out over the Gateway patrol division in Glendale. Open the attribute table for the TypicalRobberySeries layer.
3. Notice that on the map there are only eight labels, but in the table there are nine crimes. The labels are number based on the FID number, which is a field that ArcGIS automatically creates. This field starts with 0 and goes to 8 instead of starting at 1 and going to 9.
4. Now look at the dates and times these crimes occurred. Are the dates in sequential order?_____. Notice that there are two offenses that occurred around 1500 hours, two that occurred around 0100, and five that occurred between 2000 and 2300 or so hours. Based on this observation, when would you advise undercover units to do surveillance in this region? _____. Congratulations; if you said 2000 hours to midnight or something similar you just did a temporal analysis of a robbery series. It really is as simple as that with 10 or fewer incidents where you can visually see the pattern (if there is one).
5. Go down to the field called MPDOW and see which days of the week these crimes occurred on. Does there seem to be any pattern? _____. Which day(s) of week would be best for the surveillance team?_____.

    If you indicated that Saturday or Wednesday would be good days, then you would be indicating that 67% or so of the robberies occurred on those days of week (six out of nine).
6. Close the table and look at the points on the map. Do you see any pattern of travel between hits? _____.
7. In a typical robbery (or other crimes) series you will see that the offender keeps crossing back and forth across a mean center of the crimes where there are no incidents. When you see this pattern

and the blank hole between several incidents, this often means that the offender lives near the center of the crime series or in the locations where there are no crimes in this series. This doesn't always have to be in the dead center, but keep this theory in mind when doing your analysis. If you see this type of general pattern, you are probably dealing with a marauder-like (poacher) offender, and he has some connection to the area where no crimes in the series are being committed.

8. Use the Line Drawing tool and connect the dots (see DVD Figure 9–1).

9. If you follow this line back and forth in the crime order sequence, you can see that the offender keeps moving back and forth across the blank middle of this series where there are no crimes, and perhaps he may be starting to move a little east from where he started. Another bad habit we sometimes get into is zooming into the robbery series and then ignoring everything else outside of the extents of it. It is *always* a good idea to zoom out once or twice and then see how the city landscape may affect our crime series. When you have zoomed out twice, turn on the Bike Path layer. Is there a chance that the offender could be using these bike paths to get to and from the crime scenes? Yes, especially if one of the tidbits of information we got from this crime series is that the suspect was seen leaving several robberies on a bicycle. Don't get tunnel vision when viewing your analysis, and be sure to take a look at some reasonable pieces of landscape information that could assist you with some insight into your offender's behavior. Another good example of this is if there is a huge lake right in the middle of this crime series. Instead of the suspect living in the lake (believe me, some analyst someplace has advised patrol of just such a thing), the suspect would probably be a commuter (hunter) type that was prowling the lake edge for victims for some reason.

10. For another piece of the puzzle for predicting where the bad guy will go next, let's measure the distance between each crime in the series. Using the Measure tool, measure the distance between each crime in the series to the next crime in the series and provide the average distance traveled here _____.

11. You should have computed about 1.6 miles as an average.

12. Using what you can rationally derive from the travel paths between crimes and the average distance, draw a pie-shaped wedge from the last crime (#8 label) in the direction you think the offender will go for about 1.6 miles (see DVD Figure 9–2). Your pie-shaped wedge

may be tilted more to the east or west, but you get the idea. You have just completed one of the standard ways that a prediction for a new crime in a crime series is performed by analysts.

13. There are several tools, like CrimeStat III and the old Animal Movement extension from USGS (United States Geological Survey), that attempt to predict the actual angle or bearing an offender may travel from the last crime based on the previous crimes; however, my research has not shown any of these to be highly accurate in a typical crime series. The CrimeStat III Correlated Walk Analysis method does a good job when the offender travels in a very recognizable linear, or clockwise/counterclockwise, fashion during the commission of crimes. Unfortunately, analysts do not see these patterns very often, so your best guess and judgment is as good of a predictor of directionality as other tools that are currently available.

14. When we start looking at crimes we need to look at the victimology as well. What kind of target was hit? If this series involved video stores, for example, and all of the video stores in the area were already hit, do you think that would change the suspect's thinking? _____. If the only stores that had not been robbed were east of the stores that had been robbed, do you think your directional analysis might be different? _____. If the offender had hit several stores multiple times and got a large sum of cash at each robbery at those locations and minor collections at the other stores, do you think that may influence his decision and your analysis? _____. We cannot think of every possibility, of course, but the ones that make sense and would directly relate to robberies you might commit if you were this robber could be useful and may need to be measured or considered. Don't overthink the analysis, but by reading the reports, knowing the details of each incident, and then applying that knowledge with statistical analysis, you can greatly assist in creating an operationally useful bulletin that can be provided to investigators.

15. In their book *Crime Analysis: From First Report to Final Arrest*, Gottlieb et al. describe a method of predicting the next location in a crime series called a spatial deviation rectangle (Gottlieb, Arenberg, & Singh, 1994). This method involves finding the average or mean of the X coordinate and the average or mean of the Y coordinate and the standard deviation of both.

16. Open the attribute table of the TypicalRobberySeries theme. Find the X and Y Coordinates fields.

17. Right click on the X field name and choose **Statistics** from the drop-down menu.
18. Write down the X field's mean and standard deviation here _____, _____.
19. Repeat this process for the Y field: _____, _____.
20. Now we just need to do some adding and subtracting to get the corners of our 68% and 95% rectangles. (A discussion of why we chose 68% and 95% is included in Chapter 8; for now see DVD Table 9–1 to check your results.)
21. Now we just need to add each of these points to the map using the Draw tool again, but this time use the Draw Point tool. Click on the Draw Point tool, and click in the approximate middle of the crime series. Then right click the point, choose **Properties**, and then size, position, and change the X and Y coordinates to the mean. Repeat this for the mean-1stdev, mean+1stdev, and so on until you have five points on the map (see DVD Figure 9–3). See what I mean by zooming out at least twice?
22. See DVD Figure 9–4 to compare your results. The middle green square is the mean center or what is sometimes referred to as the mathematical center of the crime series.
23. The other four squares represent the corners of the 68% and 95% rectangles, which predict the area where the offender is most likely to hit next.
24. If we connect the dots using the Draw tool again and the Draw Rectangle tool, we get our standard deviation rectangles.
25. For some analysis efforts, this may be sufficient and we should stop here. After all, our goal is to give the undercover units places to watch and wait for the suspect to hit again. We would like to give them the specific store the offender will hit, but although this is possible, it isn't always likely or easy to do. We could combine our directional analysis and our standard deviation rectangles into one area that is most likely for a new hit, which would significantly reduce the total area we are predicting. In this case, the pie-shaped directional polygon we created actually is within the larger rectangle, so we could just use it as our high-risk area for a new hit, or we could use the rectangles if the MO information is shaky or we think we are missing several robberies in the series that have not been identified yet. There are two really key things we need to do with a tactical analysis: predict where we think the offender will most likely hit again, and determine our confidence level in

that prediction. Many analysts do this by simply using the larger rectangle, which encompasses the area in which the offender has already hit. In this case that is about 10 square miles (right click on the graphic, choose **Properties** and **Area** from the drop-downs, and change the area measurement type to miles).

26. Imagine you are a detective assigned to the robbery detail. Tell me what your response would be if an analyst provided you with a map that says the suspect will hit somewhere where they already have hit before? Yep, you got it, insane laughter as they tear up the map in front of you! Therefore, in this series the directional pie-shaped wedge we created would probably be the best bet because the area is not as large as the entire 10 square miles, and it does use the travel patterns of the offender. So many analysts who are currently in the field are often unwilling to be wrong and therefore take the safe prediction or just show where the current robberies in the series have been committed when they present a map. Pardon me, but your job is to analyze, so why not give it your best shot? Yes, you will be wrong now and then, but you are also going to be right, and catching robbers is a very exhilarating thing to do even when you are not standing there with the handcuffs. Out of several hundred tactical analyses that I have performed, I have been statistically correct about 70% of the time (the offender did hit in a predicted area). Does that mean the officers arrested someone on 70% of my predictions? No, but about 20% of those predictions have resulted in an arrest of an offender or the identification of one, and that is a good thing.

27. Turn off the layers in the Robbery Layer 1–Typical data frame, right click on the Robbery Layer 2 data frame, and choose **Activate** from the drop-down menu.

28. Open the attribute table for the LinearRobberySeries layer.

29. Notice that we have 18 incidents. Look at the RPTNUM field and see that there are duplicate records for a few reports. We need to get rid of these extra records before we can continue.

30. To edit a layer and remove records, we need to put the layer in edit mode.

31. Right click on the **LinearRobberySeries** layer, choose **Data → Export**, and create a copy of this layer in your student folder.

32. Add the new layer into your data frame and edit it instead of the original file.

33. To edit it, go to **View → Toolbars** in the menu bar. Turn on the Editor toolbar (see DVD Figure 9–5).

34. Click on the word **Editor** at the left of the toolbar, and choose **Start Editing**.
35. In the Start Editing dialog box, make sure you choose the source directory where you put the copy of the robbery series data. In the bottom box, you will see the name of your file when you have chosen the correct source directory (see DVD Figure 9–6).
36. If you have not already done so, close the attribute table for the original robbery series file and open the table for your copy. Notice that all of the field names are now white instead of the typical grayish brown. This indicates this table and the layer are in edit mode.
37. At the left margin of the table, click on the box next to the three duplicate records you need to delete (Rpt numbers 3–5382, 3–13272, and 1–53397). Make sure you delete only one of the duplicates and not all six records.
38. When you have selected these records (highlighted in light blue), hit the **DEL** key on your keyboard.
39. Close the table, then click on **Editor** on the Editor toolbar again, and choose **Stop Editing**. Choose **Save Edits** in the pop-up menu and close the Editor toolbar.
40. Notice that we now have only 15 records and that the FID field has numbers 0–14 in it for the records. They also appear to be sorted in order of occurrence.
41. We will recreate the analysis we performed for the typical robbery series with this table of 15 robberies, although we will make it a bit easier for you with built-in tools for part of it through the Spatial Statistics toolbox.
42. Label each crime using the FID as the label field. Use the banner text type from the text symbol list, select size 6 and bold font, and click on the **Placement Properties** button in the symbol menu. At the bottom of the Conflict Detection tab, click on the check box that says **Place Overlapping Labels**, and apply the legend.
43. Notice that all of these robberies run along W Camelback Rd. They move back and forth from east to west during the crime series. This type of behavior is mostly seen with a crime spree, although in those incidents the offender typically travels in one direction and not back and forth.
44. If you follow the crime sequence with your mouse, you can tell the offender started on the far east, traveled west, then east again and hit several businesses near the middle of the street segments

that are involved, but he is still moving east and west during the remaining parts of the robbery series.

45. You may have to zoom in on the center crimes to see the labels sufficiently.
46. In this case, the directional pattern we would show would probably be a buffer around the entire section of Camelback Rd. We might even restrict it to the smaller section where the last several hits have been.
47. To create a standard deviation rectangle for this series, we can use the Spatial Statistics toolbox. Click on the **red toolbox** in the menu bar and find the Spatial Statistics tools section at the bottom.
48. Click on the **Measuring Geographic Distributions** toolbox to expand it.
49. Double click on the **Mean Center** tool to launch it.
50. Your input feature class is the copy of the linear robbery series you made and edited. The output feature class (to store the mean center point) will be saved to your student folder and called linearmeanctr.shp.
51. Click **OK** to do the analysis, and then add the shapefile to the map (you need to add data to do this because it does not automatically get added).
52. You may need to change the symbol or turn off the labels for the crime series to see where the mean center point was placed.
53. Now double click on the **Directional Distribution** (standard deviation ellipses) tool and launch it.
54. The input feature class will be the same as step 50. Save the new file in your student folder as linstdevellipse1. Set the ellipse size to 1 standard deviation and run this tool.
55. Do steps 53 and 54 again, but this time choose an ellipse size of 2 standard deviations and name it linstdevellipse2.
56. Load both new shapefiles into the project and close the Arc toolbox.
57. Notice that the ellipses are following the street rather closely. You have completed this prediction. In this case, the pattern where the offender begins to concentrate on the middle of the street segment at the end is important and should be given higher emphasis if the MO information from the case review indicates it as well.
58. Minimize the Robbery Layer 2 data frame and activate the Robbery Layer 3 data frame.

59. In this layer we have a set of crimes that follow one another counterclockwise. Do you think the directional analysis for this series would be easy to predict? As I indicated before, most tools do not do a very good job of predicting the next direction or bearing of a new crime unless you have a series such as this. We can use another tool within the Spatial Statistics toolbox to help us figure out what angle would be optimum for a new hit in this series because of the regular nature of the travel pattern.
60. Choose the Linear Directional mean tool from the Spatial Statistics → Measuring Geographic Distributions tools.
61. The input feature class will be the Path1 theme.
62. Save the output file to your student folder as path1dir.
63. Add the new file to the project, and open the attribute table.
64. At the end you can see that it tells us the average length is ~6029 feet. That is good information. 6029/5280 = _____ miles.
65. The fields called AveX and AveY also tell us that the mean center is located there. Click on **Help** in the menu bar, and type in *Linear Directional Mean* to read the help item for the other fields. As you can see, this tool seems to indicate that the offender will travel north and east from the last incident, which is probably not too accurate in this series.
66. Create standard deviation ellipses for this crime series and save them in your student folder.
67. Combine what you think the directional analysis should be in this series, and the StDev Ellipses you create into one prediction by drawing an area where the two derived locations overlap and show the instructor what you came up with.
68. When you are done, close the Robbery Layer 3 and activate the Robbery Layer 4.
69. What is a key element in this robbery series? _____.
70. Create standard deviation ellipses for this robbery series using the spatial statistics tools.
71. Notice that the standard deviation ellipses take up a lot of territory where there are no crimes. This is one of those issues when you use a spatial statistics method that deals with the mean center and the XY coordinates of the occurrences. These statistical methods do not take into consideration any cultural landscape involved with the robbery series. In this case, it pretty much totally ignores the coincidence that every robbery is right next to the Loop 101 freeway. We can see that only crime number 8 (FID #7) is over a mile

from the freeway, and all the others are within a mile or less. We need to take advantage of this fact when we create our prediction in this case. The ellipses start us there, but we now need to create a buffer of 1 mile around the freeway and use it and the ellipses to predict the next hit location.

72. First, use a select by Attributes query on the street centerline file of "TYPE_12" = 'Freeways' to select only those street segments that are freeways.
73. In the Arc toolbox, go to the Analysis toolbox (top) and choose **Proximity** tool group and expand it. Choose the **Buffer** tool and double click it.
74. The input feature class is the street centerline file.
75. Change the output location to your student folder and call this file streetbuffer.
76. In the Distance Value section change the options available to say "1" and "miles."
77. In the dissolve type box, choose **All**.
78. Click **OK** to run this tool.
79. Add the new layer to the project.
80. Now we need just the area where the 2 StDev Ellipse and the buffer overlap.
81. In the Arc toolbox find the tool called **Analysis tools → Overlay → Intersect**.
82. Double click this tool to launch it.
83. The input feature classes need to be the buffer of the Freeways theme and the 2 Stdev ellipse polygons. Add them to the list on top.
84. Name the output feature class Fwy_Intersect and place it in your student folder.
85. The output type should be Input.
86. Click **OK** to run this tool. When it's done, add the layer to the map.
87. Notice that only the freeway buffer that intersected the 2 StDev ellipse remains. Can we follow the directional pattern between the hits manually? I had a hard time as well. To accomplish this we would need to use CrimeStat III or another extension, such as Hawth's tools, to create this line between crimes in the series. We will do this in a later lesson.
88. Activate Robbery Layer 5 now.

89. In this case we have an even more troubling type of series that standard deviation analysis is going to do very poorly at helping us identify the next hit location. In this series the suspect appears to be hopping back and forth between several cluster areas. This could be due to the type of places he is hitting (convenience stores on corners or strip malls) and probably indicates that he is commuting in these areas for the target-rich environment rather than living in the area and crimes of opportunity. Of course, we would need to try to verify our assumptions by reading the cases, but this spatial pattern indicates this could be true. In these circumstances, a prediction may be impossible.

90. Create standard deviation ellipses and a mean center and see if you can manually follow the path the offender is taking in this series. Don't spend hours on this, but do complete the first two layers using the toolbox tools we've discussed. Discuss with your fellow students how this series could be analyzed instead of using mean center and standard deviation spatial statistics.

91. Close and do not save the project.

Exercise 10: Time to Practice on Your Own

Data Needed for This Exercise
- Create a new project and save it in your student folder.
- Load the files you feel would best help you in your analysis.
- Load the indecent exposure series shapefile from C:\CIA\Exercises\Exercise_10\IndExp.shp.

Lesson Objectives
- Create a tactical analysis project and analyze an indecent exposure series using CrimeStat III or other available tools.

Task Description
- You have been assigned an indecent exposure crime series on which to make a tactical prediction. All of the cases have been determined to be related to the same offender based on careful review of the case files.
- You are to create a new ArcGIS project, add the indecent exposure layer you received from another analyst, and use CrimeStat III to create a convex hull polygon, standard deviation ellipses, center of minimum distance, and an interpolation density layer for this analysis.
- By combining these analytical efforts, create a prediction of the most likely areas the offender will be prowling for his next indecent exposure incident and notify detectives.

- If possible, consider directional analysis as you have time, and review the series to see if you can provide a best guess estimate of the probable location where the offender lives in relation to these offenses.
- The final analysis should look something like the map in DVD Figure 10–1 if you combined the various layers into one prediction.

Exercise 11: A Discussion of Victimology

   Data Needed for This Exercise
   - All data is contained in EX11.mxd.

   Lesson Objectives
   - In this lesson we will learn to make use of information about our victims of crime. Victims and where they are and/or what behaviors they are often involved in can lead them to be victims. We can learn from this information and find the similarities among victims in our series to help us predict the next location for a new hit with a greater degree of accuracy. Even if the victims are of different types, ages, and genders, we can sometimes use demographic data to assist us with finding similar areas of our city that could give us similar victim characteristics and use that to our advantage.

   Task Description
   - We will revisit our robbery series from exercise 9 and see how victimology can play a part in greatly reducing the possible locations our surveillance teams need to cover and how to use journey to crime information to help place roving vehicles in place to watch for the offender.

1. Open EX11.mxd and save a copy of it in your student folder.
2. When you open this exercise you will see a graphic prediction area, which is a combination of the standard deviation ellipses, the last hit buffer, and a directional analysis. This area is the most likely area for a new hit in this robbery series (see DVD Figure 11–1).
3. Open the attribute table for the robbery series.
4. Notice that several of our victims are Circle K, 7-Eleven, Safeway, and video stores in this series. Although the suspect has chosen different types of locations, he has chosen some pretty specific ones. We can also see that the suspect made the most money from his video store and Safeway store robberies (see the DolPDay field).
5. Close the attribute table.

6. Using this information we can search the business records we can obtain from our city tax license department for all businesses that have these names or similar names. This gives us the point shapefile called BIZ in the TOC.
7. Turn this theme on, then open its attribute table.
8. There are 122 such businesses in Glendale. It is not realistic and not operationally effective to tell the surveillance officers that they need to watch 122 different stores because most of those stores are not in our target area.
9. Turning on the All Businesses theme is not a great idea either because there are just too many small dots to make the map much use.
10. Because the prediction area is a simple graphic, click on it to get the small handles around it, and then click on the **BIZ** theme in the TOC.
11. Now go to Selection in the menu bar and choose **Select by Graphics** from the pop-up menu.
12. Only 17 of our possible target locations are within the prediction area. This is still a lot of targets, but at least it gives the surveillance units a few specific targets they could watch, or you could also do some proactive target hardening at a few of those locations by warning the business managers that they could be targeted and ask them to increase security measures.
13. You can then export this data to an Excel spreadsheet and use this information in your bulletin.
14. In some cases we don't have the luxury of having a few specific types of businesses that are being hit. An example would be crimes that may target elderly white females or only happen in areas that are zoned for commercial purposes.
15. We can use data layers from the US Census Bureau where we have the age, sex, or race of a victim that is being targeted every time in a series. The general process is to first use the Select by Location query to find out what census blocks, block groups, or tracts the current crimes are being committed in and look for patterns for those areas based on the census tract data. Some common variables found in robberies might be vacant properties, a higher number of rental properties, or female heads of household.
16. Another layer we can use sometimes is the Zoning layer that your city might have developed.

17. This layer often has data on the type of land use your city is divided up into. For example, turn on the Zoning layer, and then use the **Selection → Select by Location** query to select the features of the Zoning layer that intersect the robbery series data.
18. You should have found 11 records in the Zoning layer. The majority of these selected records in the zoning attribute table have a base_ zone of R1–6 or C-2, meaning residential 1–6 or commercial 2.
19. Now that we know these zoning areas are most likely to be hit, we can use a Select by Graphics query to select all the zoning areas within or intersecting our prediction area. Then we can limit these to just the C-2 and R1–6 areas by doing a regular Select by Attributes query (select from current selection setting) to select all of the zoning polygons that are either C-2 or R1–6 (see DVD Figure 11–2).
20. The light blue selection lines are the zoning polygons that intersected the prediction area. They are either C–2 or R1–6.
21. You can use this same basic select process to eliminate lakes, canals, freeways, and other unlikely spots from your prediction area that could not be a target just out of common sense.
22. It is always better to have the actual name of a business or some very specific target to identify for a new hit, but if the entire process of predicting an offender's behavior were easy, they wouldn't need a crime analyst.
23. You may find it necessary to drive out to the crime scenes to get a better feel for the area and an idea of why the suspect chose this business or this neighborhood to target. Sometimes they are easy to see on the map:
    a. All businesses are along major roadways or near freeway on/off ramps.
    b. All businesses are within strip malls.
    c. Residences are all along a nature trail, forested area, or bordering a waterway of some sort.
24. Sometimes they are not as evident without some hunting in census or zoning layers:
    a. The age of the victim is 60 years or older.
    b. The gender of the victim is always female and at least 55 years of age.
    c. The victims are always black, white, or Asian families.
    d. The locations of the businesses that were hit are always near vacant businesses within a strip mall, etc.

25. Victimology is always an important thing to look at when doing a tactical prediction, but it is often overlooked unless it is very obvious. Who your victim is and why the suspect chose that person or business can often be more important than the specific MO actions the offender took during the commission of the offense and should always be considered in your analysis.
26. Save and close your project.

References

Adams, T. M. (2001). *Historical homicide hot spots: The case of three cities.* Ann Arbor, MI: UMI.

Arizona Criminal Justice Commission. (2002). Crime mapping in Arizona report. Retrieved March 2009, from http://azcjc.gov/pubs/home/092502_CrimeMapping.pdf

Bitner, L., Gardner, J., & Caldwell, R. (2000). Analyzing gun violence. In N. Lavigne & J. Wartell (Eds.), *Crime mapping case studies: Successes in the field* (Vol. 2, pp. 13–18). Washington, DC: Police Executive Research Forum.

Block, R., & Block, C. R. (2000). The Bronx and Chicago: Street robbery in the environs of rapid transit stations. In V. Goldsmith, P. G. McGuire, J. H. Mollenkopf, & T. A. Ross (Eds.), *Analyzing crime patterns: Frontiers of practice* (pp. 137–152). Thousand Oaks, CA: Sage.

Canter, P. (1998). Baltimore County's autodialer system. In N. Lavigne & J. Wartell (Eds.), *Crime mapping case studies: Successes in the field* (Vol. 1, pp. 81–91). Washington, DC: Police Executive Research Forum.

Casady, T. (2003). Lincoln Police Department—specific examples of GIS successes. In M. R. Leipnik & D. P. Albert (Eds.), *GIS in law enforcement: Implementation issues and case studies* (pp. 146–158). New York: Taylor & Francis.

Casady, T. (2008a). Automating briefings for police officers. In S. Chainey & L. Tompson (Eds.), *Crime mapping case studies: Practice and research* (pp. 27–32). Hoboken, NJ: Wiley.

Casady, T. (2008b). Rolling the dice: The arrest of Roosevelt Erving in Lincoln, Nebraska. In S. Chainey and L. Tompson (Eds.), *Crime mapping case studies: Practice and research* (pp. 63–68). Hoboken, NJ: Wiley.

Cook, P. (1998). Mapping a murderer's path. In N. Lavigne & J. Wartell (Eds.), *Crime mapping case studies: Successes in the field* (Vol. 1, pp. 123–128). Washington, DC: Police Executive Research Forum.

Cordner, G., & Biebel, E.P. (2005). Problem-oriented policing in practice. *Criminology & Public Policy*, 4(2), 155–180.

Eck, J. E., & Spelman, W. (1987). *Problem solving: Problem oriented policing in Newport News*. Washington, DC: Police Executive Research Forum.

Eisenberg, D., & Schmerler, K. (2004, May 19). The Chula Vista Motel project. Presentation at the Massachusetts Association of Crime Analysts, Hyannis, MA.

Fagan, J., & Davies, G. (2000). Crime in public housing: Two-way diffusion effects in surrounding neighborhoods. In V. Goldsmith, P. G. McGuire, J. H. Mollenkopf, & T. A. Ross (Eds.), *Analyzing crime patterns: Frontiers of practice* (pp. 121–135). Thousand Oaks, CA: Sage.

Freisthier, B., Gruenewald, P. J., Remer, L. G., Lery, B., & Needell, B. (2007). Exploring the spatial dynamics of alcohol outlets and child protective services referrals, substantiations, and foster care entries. *Child Maltreatment*, 12(2), 114–124.

Frost, G. A. (2005). Integrating GIS, GPS, and MIS on the web: EMPACT in Florida. In F. Wang (Ed.), *Geographic information systems and crime analysis* (pp. 183–195). Hershey, PA: IDEA Group.

Gersh, J. S., & Beardsley, K. C. (2000). Evaluating the impact of a drug crackdown. In N. Lavigne & J. Wartell (Eds.), *Crime mapping case studies: Successes in the field* (Vol. 2, pp. 19–27). Washington, DC: Police Executive Research Forum.

Gottleib, S., Arenberg, S., & Singh, R. (1994). *Crime analysis: From first report to final arrest*. Montclair, CA: Alpha Group.

Goldstein, H. (1990). *Problem oriented policing*. New York: McGraw-Hill.

Hendersen, K., & Lowell, Lt. R. (2000). Reducing campus crime through high definition mapping. N. Lavigne & J. Wartell (Eds.), *Crime mapping case studies: Successes in the field* (Vol. 2, pp. 3–12). Washington, DC: Police Executive Research Forum.

Hill, B. (2003). Operationalizing GIS to investigate serial robberies in Phoenix, Arizona. In M. R. Leipnik & D. P. Albert (Eds.), *GIS in law enforcement: Implementation issues and case studies* (pp. 146–158). New York: Taylor & Francis.

Holzman, H. R., Hyatt, R. A., & Dempster, J. M. (2001). Patterns of aggravated assault in public housing: Mapping the nexus of offense, place, gender, and race. *Violence Against Women*, 7(6), 662–684.

Holzman, H. R., Hyatt, R. A., & Kudrick, T. R. (2005). Measuring crime in and around public housing using GIS. In F. Wang (Ed.),

Geographic information systems and crime analysis (pp. 311–329). Hershey, PA: IDEA Group.

Hubbs, R. (1998). The Greenway Rapist case: Matching repeat offenders with crime locations. In N. Lavigne & J. Wartell (Eds.), *Crime mapping case studies: Successes in the field* (Vol. 1, pp. 93–98). Washington, DC: Police Executive Research Forum.

Hubbs, R. (2003). Mapping crime and community problems in Knoxville, Tennessee. In M. R. Leipnik & D. P. Albert (Eds.), *GIS in law enforcement: Implementation issues and case studies* (pp. 127–145). New York: Taylor & Francis.

Kennedy, D. M., Braga, A. A., & Piehl, A. M. (1998). The unknown universe: Mapping gangs and gang violence in Boston. In D. Weisburd & T. McEwen (Eds.), *Crime mapping and crime prevention: Crime prevention studies* (Vol. 8, pp. 219–262). Monsey, NY: Criminal Justice Press.

Leipnik, M. R., & Albert, D. P. (2003). How law enforcement agencies can make geographic information technologies work for them. In M. R. Leipnik & D. P. Albert (Eds.), *GIS in law enforcement: Implementation issues and case studies* (pp. 3–8). New York: Taylor & Francis.

Leipnik, M., Botelli, J., Von Essen, I., Schmidt, A., Anderson, L., & Cooper, T. (2003). Apprehending murderers in Spokane, Washington using GIS and GPS. In M. R. Leipnik & D. P. Albert (Eds.), *GIS in law enforcement: Implementation issues and case studies* (pp. 167–183). New York: Taylor & Francis.

Lockwood, D. (2007). Mapping crime in Savannah: Social disadvantage, land use, and violent crimes reported to the police. *Social Science Computer Review, 25*(2), 194–209.

Martin, D., Barnes, E., & Britt, D. (1998). The multiple impacts of mapping it out: Police, geographic information systems (GIS) and community mobilization during devil's night in Detroit, Michigan. In N. Lavigne & J. Wartell (Eds.), *Crime mapping case studies: Successes in the field* (Vol. 1, pp. 3–13). Washington, DC: Police Executive Research Forum.

McGuire, P. G. (2000). The New York Police Department COMPSTAT process: Mapping for analysis, evaluation, and accountability. In V. Goldsmith, P. G. McGuire, J. H. Mollenkopf, & T. A. Ross (Eds.), *Analyzing crime patterns: Frontiers of practice* (pp. 11–22). Thousand Oaks, CA: Sage.

Moland, R. S. (1998). Graphical display of murder trial evidence. In N. Lavigne & J. Wartell (Eds.), *Crime mapping case studies: Successes in the field* (Vol. 1, pp. 69–79). Washington, DC: Police Executive Research Forum.

Newman, O. (1972). *Defensible space: people and design in the violent city.* New York: Macmillan.

Otto, A. C., Maly, K. W., & Schismenos, D. (2000). Cracking down on gangs with GIS. In N. Lavigne & J. Wartell (Eds.), *Crime mapping case studies: Successes in the field* (Vol. 2, pp. 101–107). Washington, DC: Police Executive Research Forum.

Overall, C., & Day, G. (2008). The Hammer Gang: An exercise in the spatial analysis of an armed robbery series using the probability grid method. In S. Chainey & L. Tompson (Eds.), *Crime mapping case studies: Practice and research* (pp. 55–62). Hoboken, NJ: Wiley.

Paulsen, D. J. (2003). Comparing actual hot spots v. media hot spots: Houston, Texas, homicides 1986–94. In M. R. Leipnik & D. P Albert (Eds.), *GIS in law enforcement: Implementation issues and case studies* (pp. 98–102). New York: Taylor & Francis.

Polk, J., Hammond, M. M., Yoder, C. S., & Burke, M. L. (2000). Enforcing civil gang injunctions. In N. Lavigne & J. Wartell (Eds.), *Crime mapping case studies: Successes in the field* (Vol. 2, pp. 123–131). Washington, DC: Police Executive Research Forum.

Ratcliffe, J. H. (2004). The Hotspot Matrix: A framework for the spatiotemporal targeting of crime reduction. *Police Practice and Research, 5*(1), 5–23.

Ratcliffe, J. H., & McCullough, M. J. (1998). The perception of crime hot spots: A spatial study in Nottingham, U. K. In N. Lavigne & J. Wartell (Eds.), *Crime mapping case studies: Successes in the field* (Vol. 1, pp. 45–51). Washington, DC: Police Executive Research Forum.

Reno, S. (1998). Using crime mapping to address residential burglary. In N. Lavigne & J. Wartell (Eds.), *Crime mapping case studies: Successes in the field* (Vol. 1, pp. 15–21). Washington, DC: Police Executive Research Forum.

Rich, T. F. (1998). Crime mapping by community organizations: Initial successes in Hartford's Blue Hills neighborhood. In N. Lavigne & J. Wartell (Eds.), *Crime mapping case studies: Successes in the field* (Vol. 1, pp. 35–41). Washington, DC: Police Executive Research Forum.

Rieckenberg, E. J., & Grube, T. (1998). Reducing traffic accidents using geographic analysis. In N. Lavigne & J. Wartell (Eds.), *Crime mapping case studies: Successes in the field* (Vol. 1, pp. 23–26). Washington, DC: Police Executive Research Forum.

Robbin, C. A. (2000). Apprehending violent robbers through a crime series analysis. In N. Lavigne & J. Wartell (Eds.), *Crime mapping case studies: Successes in the field* (Vol. 2, pp. 73–79). Washington, DC: Police Executive Research Forum.

Romig, K., & Feidler, A. (2008). A therapeutic landscape? Contextualizing methamphetamine in North Dakota. In Y. F. Thomas, D. Richardson, & I. Cheung (Eds.), *Geography and drug addiction*. Berlin: Springer Science and Business Media.

Roncek, D. W. (2000). Schools and crime. In V. Goldsmith, P. G. McGuire, J. H. Mollenkopf, & T. A. Ross (Eds.), *Analyzing crime patterns: Frontiers of practice* (pp. 153–168). Thousand Oaks, CA: Sage.

Santiago, J. J. (1998). The problem of auto theft in Newark. In N. Lavigne & J. Wartell (Eds.), *Crime mapping case studies: Successes in the field* (Vol. 1, pp. 53–59). Washington, DC: Police Executive Research Forum.

Schmitz, P., Cooper, A., Davidson, A., & Roussow, K. (2000). Breaking alibis through cell phone mapping. In N. Lavigne & J. Wartell (Eds.), *Crime mapping case studies: Successes in the field* (Vol. 2, pp. 65–71). Washington, DC: Police Executive Research Forum.

Suresh, G., & Vito, G. F. (2007). The tragedy of public housing: Spatial analysis of hot spots of aggravated assaults in Louisville, KY (1989–1998). *American Journal of Criminal Justice, 32*(1–2), 99–115, 130. Retrieved March 2, 2009, from Criminal Justice Periodicals database. (Document ID: 1411351851).

Taylor, R. B. (1998). Crime and small-scale places: What we know, what we can prevent, and what else we need to know. In National Institute of Justice (Ed.), *Crime and place: Plenary Papers of the 1997 Conference on Criminal Justice Research and Evaluation* (pp. 1–22). Washington, DC: US Department of Justice.

Warden, J., & Shaw, J. (2000). Predicting a residential break-in pattern. In N. Lavigne & J. Wartell (Eds.), *Crime mapping case studies: Successes in the field* (Vol. 2, pp. 81–87). Washington, DC: Police Executive Research Forum.

Wernicke, S. (1998). Close the door on crime: A mapping project. In N. Lavigne & J. Wartell (Eds.), *Crime mapping case studies: Successes in the field* (Vol. 1, pp. 27–33). Washington, DC: Police Executive Research Forum.

Wernicke, S. (2000). Reducing construction site crime. Overland Park, Kansas. In N. Lavigne & J. Wartell (Eds.), *Crime mapping case studies: Successes in the field* (Vol. 2, pp. 29–36). Washington, DC: Police Executive Research Forum.

Winship, C. (2002). *End of a miracle? Crime, faith, and partnership in Boston in the 1990s*. Retrieved April 2008, from http://www.wjh.harvard.edu/soc/faculty/winship/End_of_a_Miracle.pdf

Woodby, K., & Johnson, A. (2000). Identifying a serial indecent exposure suspect. In N. Lavigne & J. Wartell (Eds.), *Crime mapping case studies: Successes in the field* (Vol. 2, pp. 55–63). Washington, DC: Police Executive Research Forum.

Crime Mapping and Analysis Data

CHAPTER 6

▶ ▶ LEARNING OBJECTIVES

This chapter is designed to provide the beginning crime analysis and mapping student with an awareness of the different types of data commonly used in crime analysis and mapping. Data for crime analysis is produced in several different forms and can be obtained from many different places. While it is important to understand *how* and *when* to use different data types (Chapter 8 will discuss this notion in greater detail), it is first important to understand the key sources of data available and their characteristics. In this chapter, key data sources and different types of data are identified and discussed. In addition, readers will learn the importance of clean data, and the strengths and weaknesses of different data sources are examined. After studying this chapter, you should be able to:

- Know what is meant by the term *clean* data.
- Identify and explain different *units of analysis* for incident characteristics, including time and space.
- Know the appropriate uses for both qualitative and quantitative data.
- Identify and explain the major sources of crime data in the United States and be able to recognize their strengths and weaknesses.
- Understand the *crime funnel* of the criminal justice system and how it affects the crime data that mappers and analysts use.
- Explain different types of *contextual* data used in crime analysis and mapping, and list various resources for finding this data.
- Have a basic understanding of various issues with crime mapping data, including data sharing and ethical and legal issues.
- Have a basic understanding of the various tools that crime mappers and analysts use to create and modify crime maps and the tools used to present their findings to others.

▶ ▶ KEY TERMS

Attribute Table
Contextual Data
Data Cleaning
Geocoding
Institutional Review Boards
National Crime Victimization Survey
National Incident Based Reporting System
Probability Sample
TIGER
Uniform Crime Reports

■ Introduction

Typical crime analysts will work with different types of data and data sources throughout their careers. The audiences these analytical efforts attract could be the public, fellow analysts, police officers, command staff, other law enforcement agencies, and even other governmental entities. In a large number of these analytical efforts, the question at hand will not involve forecasting or predicting but simple production of maps that describe a crime problem. An example might be creating a beat map that shows patrol officers where new beat boundaries have been drawn, or showing the hot spots of armed robbery incidents in the city. The second type of analytical effort involves forecasting or predicting where something is likely to happen again, how many crimes of a specific type we might expect in a given area, and even where an offender could potentially live in relation to a series of crimes committed.

Currently there are a large number of excellent data sources available for crime analysis and mapping, many of them free of charge. Police departments, social welfare agencies, and even the US Census Bureau are just a few of the data gold mines available to crime analysts. However, crime analysts must use a level of caution in the analysis process with all data, regardless of the source. The old adage "garbage in, garbage out" will serve you well. We must always address the data needs and analysis to be performed based on the audience to whom the analytic product will be directed. For example, we would likely need much less information about several specific incidents within a certain geographic boundary if we were presenting that information to a neighborhood watch group instead of a group of detectives within our own agency. The audience and the data that is available for the analysis requested will generally guide your mapping product to fruition. This works for not only the descriptive types of maps, but also those that are also developed for problem solving and prediction. When you wish to make meaningful assumptions or assertions based on data you've collected, then we need to take this concept a bit further and add the scientific process to the collection, collation, dissemination, and evaluation of our data we wish to use. Any analyst can make a map; the key is knowing your data well enough to make a map that turns data into information and information into actionable knowledge for your audience. There has been and will likely continue to be much argument about whether data collected by a police department is a sample or the entire population. From the police officer's perspective, the crime reports and information they write reports about is the true crime problem and represents the picture of crime within his or her jurisdiction. To the researcher, we know that not all crime is reported, and even policies within police departments may cause anomalies in the formal reporting of certain crimes (gas drive outs, very minor thefts, etc.). These policies are generally put into place to save time for officers to deal with more serious crimes and save money in the police department's budget. On the other hand, we also know that not all crime is actually reported to the police, so from a researcher's perspective, the data collected by a police department should be viewed as the entire population of *reported* crime and thus represented as such, keeping in mind that reported crime does not always represent true population characteristics of all committed crime.

When doing meaningful research, it is important to determine if the data was properly collected following the guidelines of the scientific process. This includes the requirement that the data be a scientifically selected **probability sample** (which can be difficult or even impossible in many types of crime analysis). A probability sample is a scientifically selected sample group chosen from a larger population where all cases within that population have an equal chance of being selected to be in the sample group. There are many different sampling techniques that can be applied toward the selection of both probability and non-probability samples, and any good research methodology book will have detailed descriptions of each technique. For our purposes, you need to be able to justify that the cases you include in your analysis include all or are representative of the larger population that you are trying to explain. In crime mapping, this typically means including all cases or units that fit the required characteristics in the analysis. For example, if your analysis or crime map seeks to understand robbery-motivated homicides within city limits for a given time period, *all* known robbery-motivated homicides must be included in the analysis. If any are excluded, the analysis will be limited in its power to explain or predict these crimes. The problem is that the complete information about a particular crime is not always available, and analysts are left to make assumptions that may or may not be correct. For example, there may be a series of homicides that were motivated by robbery, but the evidence left behind fails to show a complete enough picture to rule them as such. In addition, there may be more than one motivation for the offender to commit homicide and robbery, but the data entry person may only list the most apparent one. Thus, the more assumptions an analyst must make in his or her analysis, the more likely there will be errors.

A second issue of concern for crime analysis is **data cleaning**. This process involves eliminating information that is unnecessary to your analysis and making sure there is no bad information. It can also mean taking the coded data and normalizing it so that anyone would understand what the codes mean. For example, a records management system stores gender values as 0, 1, or 2, where 0 is unknown, 1 is male, and 2 is female. We would want to clean this field so that a layperson would understand what 0, 1, or 2 means when we get the data ready for a mapping production process. With quantitative analysis, most available data comes in a database management format (such as SPSS, Stata, SAS, or Access) or spreadsheet format (such as Excel). Cleaning the data simply involves running frequency reports or manually scanning the data for information that does not appear to be correct. This process, to many analysts, is a daily event. Any field within the data set's rows could have errors that are created from commission (incorrectly entering the wrong information) or omission (leaving the field blank or null). For example, if you come across someone who is 157 years of age when computing frequencies or scanning the column that holds information of victims' ages, you obviously need to correct or clean up this person's file. It is possible that the person who entered the data (you, in some cases) accidentally hit the "1" key before typing in the person's true age of 57 years. In this case, simply deleting the extra numeral is the solution. The data could also just be incorrect, and if there are too many errors within one record, it may have to be excluded. Another example might be a victim whose cause of death is listed as gunshot to the head, but under the weapon field heading, a knife may be listed. Obviously, this does not make a whole lot of sense. The analyst will need to go back to the original report (if available) and clean up the data entry mistakes. Depending on where the data comes from, this may not be possible, and in this case, the entry may have to be removed from your analysis. A very important point to make here is that you must never make assumptions about what the correct information is. In the first example, it may be possible that the 5 or 7 were incorrect, leaving the victim's true age to be 17 or 15 years—a far distance from 57. In the second example, it is possible that the perpetrator also used a knife, and thus simply replacing the knife with a gun would not present the entire picture. Another possible example would come from changes in policies, or even software and hardware, over time within a police agency. For example, when a department's records management system (RMS) was first created, its creators probably used the race variables of B (black), W (white), S (Spanish), A (Asian), and O (other) as the only allowable entries. Over time and due to racial profiling complaints, the RMS may have undergone changes that shifted these variables to A (African American), C (Caucasian), H (Hispanic), O (Oriental), and U (unknown or other). Chances are the data that was entered for older cases was not modified when the new format was made policy, and if you go back in time to do analysis, you may find the need to modify entries to match current policies and procedures to get good quality data to work with. Analysts need to be intimately familiar with their department's RMS to know what garbage needs to be sifted through in the cleaning process. There is a philosophical debate as to *who* in the department should be keepers of the data. In some departments, the belief is that correcting errors in the RMS is the responsibility of crime analysts. However, most analysts believe (and we agree) that this responsibility should be held by the records department or a similar custodial department, leaving the analyst the time required to perform crime analysis.

In addition to clean data, another topic often discussed with regards to crime mapping is the concept of scale. For example, if your analysis is general in nature and only covers a citywide scale, then chances are your detail and ultimate cleanliness of the data may not be as demanding as if you were analyzing a specific robbery series occurring within a geographic area of your city.

Continuing the topic of scale, a third issue of importance is choosing the appropriate unit of analysis. Analysts must carefully select the appropriate level of analysis. Choosing a unit of analysis or scale that is too small or too large may obscure important relationships. An illustration of this in crime analysis would be a comparison of two well-known studies, the first being the infamous Kansas City Preventive Patrol Experiment. Published in 1974 by the Police Foundation (Kelling et al., 1974), this study evaluated an experiment that was conducted by the Kansas City Police Department (KCPD). In this experiment, KCPD varied their patrol levels in 15 police beats. For approximately 1 year, five beats received no routine patrol (although police did respond to calls for service), five beats received at least double the amount of patrol they would normally receive, and five beats received the normal amount of patrol they had always received. Using multiple sources of data, including victimization surveys, reported crime rates, arrest data, a survey of local businesses, attitudinal surveys, and trained observers who monitored police–citizen interaction, the major findings were that crimes thought to be preventable by routine patrol, such as burglaries, auto thefts, robberies, and vandalism, were not changed. Based on these results, the utility of routine patrol, a key component of policing, was questioned. In addition to its inability to prevent crime, routine patrol was also determined in this study to have no impact on fear of crime or satisfaction with the police. This study's unit of analysis (the police beat) was based on political boundaries that were meaningless to crime control efforts. Recall from earlier discussions that caution must be used when selecting which unit of analysis to employ in analysis. This is very important in crime mapping and analysis. If the boundaries of an area that are being examined do not have spatial meaning to crime or the social context of crime, important relationships can be obscured. This point is supported by the results of another study of police patrol that was conducted almost 2 decades after the Kansas City experiment. In this study, the unit of analysis was a hot spot, a much smaller area than a police beat, whose borders are a function of crime itself and therefore are very meaningful to crime control efforts.

Sherman and Weisburd (1995) conducted their study of preventative patrol in the city of Minneapolis. This 1-year study also examined crime and disorder in 110 hot spots, 55 of which received increased levels of police patrol. Their findings indicated that hot spots that received a higher dosage of routine patrol showed modest reductions in crime and significant reductions in disorder. In this study, the unit of analysis, identified hot spots, intensifies the deterrent effects of preventative patrol.

As you can see, choosing the appropriate unit of analysis is extremely important in analyzing crime. We cannot assume that the Kansas City Preventive Patrol Experiment would have yielded different results had the researchers chosen a smaller unit of analysis, but it is plausible that this could be the case. The following sections of this chapter narrow our discussion to examine specific types of data that are commonly used in crime mapping and analysis and their respective strengths and weaknesses. As you read this chapter, keep in mind the importance of having representative, clean data collected at the appropriate unit of analysis.

■ Measuring Crime

Crime is measured in a variety of ways, utilizing a variety of methodologies, each with its own strengths and weaknesses. Whether using "official" data or self report measures, crime analysts must be intimately familiar with their data sources and include the limitations of these sources and how they impact the interpretation of their results. The following sections discuss some of the most common ways in which we currently measure crime, beginning with official crime data.

Provided in the discussions of commonly used data is also an examination of the strengths and weaknesses that impact interpretations of crime analysis performed using these data sources.

Official Crime Data

Official crime data can be classified into a few general categories: incidents, people, places, and things. Incident data typically includes data on reported offenses, calls for service, arrests, traffic citations, and accidents. People data involves persons listed in reports, corrections data, probation information, parole records, and sex offender registry information. Places data includes such things as locations (addresses) or boundaries for businesses, schools, beats, and council districts, and information about locations within the city that the analyst might use. Data surrounding things includes property taken or recovered in a criminal incident, evidence found at the scene, stolen vehicles, weapons, and other items of interest to the analyst.

Let us begin with the most well-known source of crime data. The **Uniform Crime Reports** (UCR) is published annually by the Federal Bureau of Investigation (FBI). Participating law enforcement agencies keep track of known offenses and report monthly totals to the FBI, who then analyzes this information to produce the Crime Index. Thus, the UCR is *summary* data. The UCR collects information on Part I index crimes, which consist of murder and nonnegligent manslaughter, forcible rape, robbery, aggravated assault, burglary, larceny–theft, motor vehicle theft, and arson. In addition, arrest data for Part II crimes (less serious crimes, including simple assaults and driving under the influence) is also collected. A major strength of the UCR is that it has been in operation since 1930 and thus is available for trend analysis. The UCR is a great resource for analysts who are performing large-scale analyses. However, because the information that is collected represents crime totals for grouped categories and is measured at the agency level, comparisons based on place can only be made at the precinct or larger level (this is important for the analyst to keep in mind if he or she needs a smaller scale).

Other weaknesses of the UCR data for crime analysis include:

- The program is voluntary, and therefore not all agencies participate.
- Included offenses are only those known to police. Crimes not reported or discovered by police are not included in the UCR.
- The definitions for crimes differ from state to state, so what one agency considers to be a burglary based on its local criminal

statutes may not be a burglary under UCR definitions. This often causes some confusion and potential weaknesses in the UCR data collection.
- UCR data is based on the primary or most serious offense that occurred during the commission of a crime. For example, if a suspect kidnapped, raped, and then murdered a victim, under UCR hierarchy rules, only the murder would be reported.
- UCR counts some crimes by the number of victims, not the number of incidents.

Another source of data, the **National Incident Based Reporting System** (NIBRS), a component of the UCR, provides greater detail in its crime reporting ability. In contrast to the UCR, it provides *incident*-level data. It was developed because the law enforcement community recognized a need for more detailed information about crime. While the UCR collects most of its crime data in the form of category totals, NIBRS breaks down data into specific subcategories. For example, the fraud offenses category contains the crimes of false pretenses/swindle/confidence game, credit card/automatic teller machine fraud, impersonation, welfare fraud, and wire fraud. NIBRS collects information at the incident level for 22 different offense categories (Group A offenses) and records incident and arrest data about the incident, including victim and offender characteristics and types and value of property stolen and recovered (see **TABLE 6-1**). In addition, NIBRS collects arrest data (only) on 11 other offense categories (Group B offenses, see **TABLE 6-2**). The NIBRS program does not have the hierarchy rule and thus will provide information for *all* crimes a criminal commits within an incident. Furthermore, NIBRS includes updated definitions for crimes such as rape (the UCR does not account for male victims of rape) and differentiates between attempted and completed crimes (the UCR Summary system does not). Last, NIBRS accounts for victimless crimes (e.g., drug offenses), whereas the UCR Summary reporting system only counts crimes against persons and crimes against property.

Although the NIBRS is a wonderful addition to the UCR and greatly enhances our abilities in crime analysis, it suffers from the same limitations as the UCR. In addition, NIBRS has not been around as long as the UCR, and not every state reports NIBRS data. The FBI reports that as of 2004, approximately 5271 law enforcement agencies reported NIBRS data to the UCR program and there are 26 state programs certified for NIBRS participation. The Federal Bureau of

TABLE 6-1 NIBRS Group A Offenses

1. Arson
2. Assault offenses: Aggravated assault, simple assault, intimidation
3. Bribery
4. Burglary/breaking and entering
5. Counterfeiting/forgery
6. Destruction/damage/vandalism of property
7. Drug/narcotic offenses: Drug/narcotic violations, drug equipment violations
8. Embezzlement
9. Extortion/blackmail
10. Fraud offenses
11. Gambling offenses
12. Homicide offenses
13. Kidnapping/abduction
14. Larceny/theft offenses
15. Motor vehicle theft
16. Pornography/obscene material
17. Prostitution offenses
18. Robbery
19. Sex offenses, forcible: Forcible rape, forcible sodomy, sexual assault with an object, forcible fondling
20. Sex offenses, nonforcible: Incest, statutory rape
21. Stolen property offenses (receiving, etc.)
22. Weapon law violations

Investigation (n.d.) says that "the data from those agencies represent 20 percent of the U.S. population and 16 percent of the crime statistics collected by the UCR program."

Both the UCR and NIBRS data can be accessed online and free of charge through the National Archive of Criminal Justice Data (NACJD) held by the Inter-University Consortium for Political and Social Research (ICPSR) housed at the University of Michigan. These data files can be easily downloaded to a personal computer, or the user can perform a limited, but useful, number of analyses online using the provided Survey Documentation and Analysis (SDA)

TABLE 6-2 NIBRS Group B Offenses

1. Bad checks
2. Curfew/loitering/vagrancy violations
3. Disorderly conduct
4. Driving under the influence
5. Drunkenness
6. Family offenses, nonviolent
7. Liquor law violations
8. Peeping Tom
9. Runaway
10. Trespass of real property
11. All other offenses

program. In addition to the UCR and NIBRS, the NACJD houses a Web site dedicated to GIS data sets that are organized into the following categories:

- Geographic Data in Mapping Software Files
- FIPS County Codes
- Collection-Specific County Codes
- Census Tract Codes
- Zip Codes
- Block Groups
- Police Precincts
- Police Beats
- Addresses
- XY Coordinates
- Reference Collections to Link Geographic Identifiers

Crime analysts may also access data from local law enforcement agencies. Depending on the agency, this may be a very simple or very complicated process. Typically the analyst is employed by an agency, and thus data access is a nonissue. Students, academic researchers, and outside analysts must usually go through a public records request process from the agency. Depending on how the data is collected and stored within the police department, the data may be furnished in a variety of electronic formats and may even be supplied in hard copy form. In this case, the analyst must enter the data into a database prior to analysis. A crime incidents database typically houses a wealth of information, including a record number, date of reported incident and occurrence (if known), type, method, and location of crime, and disposition. In some cases, characteristics about the victim and offender are also included. When an arrest is made in the case, an arrest report is completed. This information is entered into an arrests database and contains basic information on the charge, date and location of the arrest, arrestee's address, and characteristics of the arrestee.

Calls for service data is generated from citizen- and officer-initiated calls for service. Typically, 911 dispatchers record information from a call, including the date and time of the call (and the time of police arrival), the type of call (what the complaint is), the location of the call (sometimes this is an exact address, sometimes it is a street segment or intersection), and the disposition(s) of a call. Calls for service records are useful to analysts, especially in performing evaluation of department performance measures, such as response times (which are often more important to citizen satisfaction than solving crimes). However, there are limitations that must be understood when using this data. First, when a citizen calls for police assistance, their depiction of the situation may be misleading or inaccurate, leading the dispatcher to enter an incorrect call type. However, many CAD systems distinguish between an initial call type and a verified or final call type, or the initial call type may be verified by a detective after the call is investigated further. Second, many people may call police regarding the same incident. For example, several people may hear gunshots or a woman screaming. Third, the address that the citizen calls from may not always be the address where the incident or crime occurred. A good example is a theft from a vehicle where the victim discovers the theft at his or her place of business, but the victim drives home and calls police to make a report from there. In crime mapping, if each caller to an event receives a point on the map, analysis can count each caller as an incident, potentially producing artificially inflated results. Last, errors in inputting addresses can lead to difficulty in **geocoding** (the process of taking street addresses from a tabular file and matching them to a reference file to produce the points on a map). In addition to data and information values that could need a good cleaning, we also need to make sure that the address is in such a format that will ensure that it geocodes properly when brought into a GIS system for mapping. For example, the person who entered the data may have typed the address 123 Main St when in fact it is 123 Main *Rd*. This is a threat to validity because unmatched addresses must either be matched manually (which can be time consuming) or removed from the analysis.

Police departments also keep track of traffic and accident information. Accident databases typically house information on the time and date of accidents, road conditions at the time of the accident, and violations of traffic laws, including whether or not alcohol played a role in a crash. Some states also collect information about whether or not one or more parties were using a cell phone at the time of the accident. Traffic databases hold information on traffic citations and vehicle stops. This data has a variety of uses. For example, in more recent years, this data has been used to perform racial profiling research.

A persons database contains information about people involved in a criminal incident. Usually linked to the criminal incident database, files contain information on witnesses, victims, and suspects. Probation and parole departments also maintain records about their parolees and probationers. These records are vital to investigations in crime mapping and can be used to prioritize a list of suspects who live or work near a criminal incident.

In addition to parolees and probationers, police departments maintain information on sex offenders who live within their jurisdiction. Known addresses of registered sex offenders can be examined in light of regulations that outline where this type of offender can live and work (most cities, for example, have laws forbidding sex offenders to live or work within 2000 or 2500 feet of a school, park, or day care facility). Mapping capabilities make this analysis much more efficient. There may be challenges in court to distances that are calculated using mapping software, so analysts should verify the distance accuracy of their software and layers, which is spatial cleaning of the data.

Police departments may also maintain records on property and vehicles that have been stolen or recovered or have been used in the commission of a crime. These databases may also be linked to the criminal incident database through a record number. The primary information held by these databases is physical descriptions of the vehicle or property, VIN and serial numbers, or other identifying information. Somewhat related to this category are files and information that are collected in reference to the location of evidence collected during an investigation. This could be such things as locations where a murder victim was taken, murdered, and his or her body dumped, where evidence was found, or where cell phone towers from which phone calls were received or made from a suspect's phone are located.

Self-Report

The most widely known self-report data used in crime analysis is the **National Crime Victimization Survey** (NCVS). Launched in

1873 and housed by the NACJD, approximately 77,290 households (see Bureau of Justice Statistics Web site at http://www.ojp.usdoj.gov/bjs/pub/ascii/ntcm.txt for complete methodology) are selected in which household members (above the age of 12 years) are asked to report their victimization experiences, if any. The NCVS is especially interesting because it captures those crimes that are unreported to police and elicits detailed information about crime victims and their experiences. However, the NCVS is not without its faults, which include:

- There is no way to verify if a respondent is providing truthful and accurate information.
- Information on homicides is not collected. ("Dead men can tell no tales.")
- Crime victims under the age of 12 years are unaccounted for.
- Victimless crime is unaccounted for.

As shown in this section's discussion of official and self-report crime data, analysts must take caution to ensure their data is the correct data, is clean, and is measured at the appropriate scale prior to performing any analyses. Remember, every data set has its strengths and weaknesses and must be used appropriately if analysts wish to get accurate and usable results from their analyses.

The Crime Funnel

The crime funnel is an important model of crime in the criminal justice system and provides an important lesson for crime analysts. The basic premise is that as criminals progress through the criminal justice process (beginning with a crime being committed and ending with sentencing of the offender), the number of criminals who are processed within the formal criminal justice system becomes progressively smaller (see **Figure 6–1**).

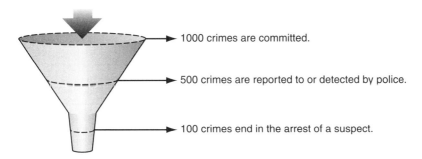

Figure 6–1 Criminal Justice Funnel Model

Based on yearly comparisons of the UCR and NCVS data, it is argued that approximately half of crime is not reported to or detected by police. Of the 50% of crime that is known to police, only about 20% ends in the arrest of a suspect. This means that a suspect is arrested in only about 10% of all crimes committed. The likelihood of arrest is not evenly distributed across all crime types, however. Homicides, for example, generally obtain the attention of law enforcement, and murderers usually leave behind important physical evidence that helps aid in the successful investigation of the crime. On the other hand, residential burglary, if detected long after the burglars leave and if there are no witnesses to the crime, rarely ends in an arrest. Homicide, a crime that is most often reported to or detected by police, has a high clearance rate. Burglary, also often reported to police (because one needs a police report to submit to insurance agencies) has a low clearance rate.

What does this mean to crime analysis? It means that the data crime analysts use to explain crime and make predictions about crime is often incomplete. This is a common limitation in crime mapping and analysis. For example, when examining a robbery series, an analyst may not have information on *all* of the robberies committed by an offender because some of them may not have been reported or another nearby agency had one or more currently unknown related robberies that met the series' modus operandi. Information about these unreported or unknown robberies may be important, even critical, in predicting where the next robbery is likely to take place or where the offender is likely to live or work.

The next section continues this chapter's discussion of data sources available to police. Here we will examine noncrime data that is important in crime analysis. Contextual data—information about the people or places involved in crimes—is just as important in understanding the larger picture of crime as the crime data itself.

■ Contextual Data

Contextual data contains information about people or places that is not crime related. For example, in crime mapping, analysts often include information about population characteristics, such as median income or the percentage of single-female-headed households. In addition, analysts often include places, such as bars or pawn shops, in their maps to provide context to the crimes they are analyzing. Contextual data is data that is typically supplied by other public and private organizations outside of the criminal justice system. Census data is most commonly used in crime analysis. In crime mapping, locations of places such as bars and pawn shops, and areas such as school and voting districts, are often used. This data can come from a city sales tax licensing department or commercial software designed for businesses that contains this information. In some cases you can even collect this data from Google Maps, Yahoo! Maps, or other types of Internet applications and then geocode the results.

US Census Bureau

The US Census Bureau is a gold mine for crime analysts and crime mappers who are looking for contextual data. According to the Census Bureau Web site, the first US census (after America's independence) was taken in 1790. At this time, people were simply counted by US Marshalls on horseback. In addition to the population census taken every 10 years, the US Census Bureau has grown and expanded to include gathering information on fisheries, manufacturing, poverty, and crime, and today it collects information year round from households and businesses on a variety of important issues. Many of these issues have important implications in crime analysis. The following is a list of some categories of data that the Census Bureau collects, which are often used in crime research and analysis.

- Age
- Race/ethnicity
- Gender
- Marital status
- Education
- Income
- Poverty status
- Employment
- House value
- Own/rent

The Census Bureau Web site makes it very easy to download files in a variety of formats. In addition, they have an online mapping component that allows users to quickly make basic maps of crime and other social variables that are important in understanding crime. The Topologically Integrated Geographic Encoding and Referencing system (**TIGER**) "is a Census Bureau computer database that contains all census-required map features and attributes for the United States and its possessions, plus the specifications, procedures, computer programs, and related input materials required to build and use it" (U.S. Census Bureau, 2004). Several states have direct links to these files from

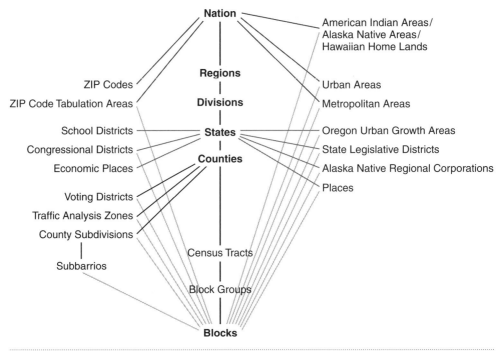

Figure 6–2 Census Data
Source: From census.gov

their own GIS Web sites, making it even easier for crime analysts to access locally relevant information.

Census information is organized by different units of analysis, beginning at the block level. **Figure 6–2** is a diagram developed by the Census Bureau to explain all levels of analysis for which data is available.

However, not *all* data is available at *all* levels of analysis. There are several reasons for this, the primary reason being the protection of privacy. Most information comes in the form of a mean or median (these measures of central tendency are discussed in Chapter 8) or a percentage. For example, crime mappers could access the median household income or the percentage of persons who do not hold a high school diploma or equivalent at various levels (block, block group, tract, etc.).

Local Social Agencies

If the analyst does not find the contextual data he or she is looking for in the Census Bureau data, he or she may want to contact local government and/or social agencies. One of the best places to acquire current, local GIS files is at city and/or county planning departments. Some states even have a dedicated GIS agency. Massachusetts, for example, has the Office of Geographic and Environmental Information, which maintains a very accessible Web site with links to state-specific data. It is extremely rare that a crime analyst would have the skills and training necessary to create the base map files necessary for geocoding and analyzing crime. Today, however, these skills are not required for the analyst because these files are almost always easily accessible from other agencies.

In addition to GIS files, other public agencies collect a variety of data that is useful to the crime analyst. Information about establishments with liquor licenses, public assistance and housing, abandoned buildings, and even graffiti cleanup are collected by various agencies and are great resources for crime analysts. However, it is up to the analysts to perform the legwork necessary to access these data files. Sometimes this may mean a simple phone call or e-mail, but at other times, the process may be more lengthy or complicated. Regardless, this data can be very valuable to crime analysts.

Universities also store data that analysts can use. Earlier in this chapter we discussed ICPSR, which is housed within the University of Michigan. Many universities hold GIS-related data as part of their responsibility as data repositories. They also employ researchers who, in addition to data sharing, often collaborate with practitioners on a variety of research projects related to crime prevention, and these collaborations have often produced interesting and innovative research. In addition to universities, there are multiple Web sites (www.gislounge.com and www.geographynetwork.com, to name a few) that hold GIS data free for the taking. Of course, caution must be employed when using these free sources of data because (depending on its original source) the data may contain errors that will impede accurate analysis.

■ Tools for Crime Analysts and Mappers

Aside from GIS software, crime mappers must also be proficient in a variety of other computer software programs. This book will only discuss a few of these; however, as with crime mapping software, there are a variety of options to choose from depending on the applications the analyst wishes to complete. Remember, the programs vary widely in their applications and cost. In this text, we will briefly discuss Microsoft programs because they are the most commonly used applications for crime analysts. These programs are relatively inexpensive but have powerful applications for crime analysts.

Microsoft

Microsoft Access is a desktop database application commonly used to store, query, and analyze data. Access is also capable of creating reports by using fairly simple functions. It is a relational database system, which allows an analyst to link related files and organize and analyze the data. It allows for a multidimensional view of tabular data and has tremendous capabilities in law enforcement and in crime analysis. Hicks (2007), for example, used Access and Microsoft Outlook (an e-mail system) to develop a system whereby automated probationer contacts were created and compiled into a relational database. Every 24 hours this system checks police reports for the names and dates of birth of approximately 120,000 probationers. If an exact match is found, an e-mail alert is generated and sent to Hicks and the assigned probation officer, giving a brief summary of what transpired (Suspicious person? Arrest?). This information can then be imported into Microsoft Excel for future quantitative analysis, or it can be geocoded onto a map for spatial analysis.

Excel is a spreadsheet application that allows the user to create multiple charts and graphs and to perform quantitative analyses as simple as counting or as complex as multivariate regression. It is the computational engine behind Access. One can also create an **attribute table** in Excel, which can be geocoded as a map layer. An attribute table contains the information that makes up a map's layer. For example, an attribute table of homicides would contain key information in columns (address, crime type, weapon used, victim and offender characteristics, etc.) for individual cases (listed in rows). A brief tutorial on Excel is contained in the appendix of this text.

Microsoft Word and Microsoft PowerPoint are software programs used primarily for presentation purposes. In Word, one can type a professional-looking report by importing charts, graphs, tables, and even maps. It also has merging applications so that the user can quickly and easily print labels and envelopes for mailing. PowerPoint is often used for oral presentations, allowing the presenter to write text or import pictures, charts, tables, and even audio and video clips to slide presentations. It is also used to create flowcharts and other diagrams used in professional presentations.

Statistical Package for the Social Sciences

Statistical Package for the Social Sciences (SPSS) is commonly used in academia, government, and the private sector to enter, store, organize, and analyze data. Although it is much more expensive than Excel, SPSS is easier to use than Excel due to its point-and-click application. In a few short keystrokes, SPSS can perform complicated computations. While not part of its base software package, users also have the option to purchase a variety of extensions, including a mapping component.

Other Software Tools

There are many different analysis tools to choose from, and it matters little which package is ultimately selected. The decision should be based on how user friendly the software is, what applications the analyst needs it to perform, and how much money the analyst is prepared to spend. Microsoft, SPSS, and many other computer tools allow an analyst to perform many different types of analyses at varying degrees of ease and cost. Regardless of which tools the analyst selects for his or her toolbox, it is important to remember the lessons learned about the limitations of different types of statistical analyses and the weaknesses of different types of data that are discussed throughout this chapter.

■ Data Sharing Issues

The last section of this chapter finishes our discussion of data by examining issues of data sharing, including ethical issues and legal issues. There are a number of entities and persons who might request data from crime analysts and mappers. First, other agencies may wish to access data for their own analyses. Cities, for example, are not islands. They have borders that touch other towns or cities. Criminals often pay no attention to these borders and commit crimes in multiple jurisdictions. An analysis that is conducted in only one town and ignores the white space of bordering jurisdictions will provide an incomplete picture of crime. Another issue regarding interagency data sharing has to do with the compatibility of computer programs. Many records management systems (RMS) were custom built for individual departments and may not interface well with other systems. Many law enforcement regional data sharing systems are in the process of being developed nationwide, but the cost is substantial and often requires changes within the participating agencies that are difficult to implement.

Second, academic researchers are often interested in acquiring data for their own research projects. Traditionally, this relationship between police agencies and academics has been collegial and productive. However, it is important that both parties be clear about what the data will be used for and how issues of confidentiality will be maintained. In addition, **institutional review boards** (IRBs) should review the research proposal to identify problems in methodology and ethics of the research design.

Third, citizens and community groups may also make periodic requests of crime analysts. In fact, many agencies have crime mapping availability for citizens on their department Web page. (A list of many of these departments can be found at http://www.ojp.usdoj.gov/nij/maps/links.htm.) Some departments publish monthly or quarterly maps, while others allow interested citizens to interact with an online mapping component by entering an address or landmark to retrieve crime information. Upon doing so, aggregated crime data is presented in map form for the area surrounding the entered address or landmark. In many states, the addresses and names of persons involved in a criminal event are considered to be public information (except in the case of juveniles or sexual assault victims, where it can be argued that a victim's right to privacy outweighs the public's right to know), and thus maps that depict crime incidents at the address level would not normally pose privacy issues. However, with the widespread use of the Internet and increased mapping abilities, one can argue that the public's access to crime data is greater now than it has ever been. This may pose potential problems depending on who is accessing the crime maps and for what reasons (Wartell & Thomas, 2001).

A fourth issue has to do with the commercial use of crime maps that are made available by law enforcement. Private businesses, such as mortgage brokers, insurance agents, or businesses dealing in private security, may use crime maps in devious and discriminatory ways. Mortgage brokers and insurance agents, for example, could use crime information to refuse loans or insurance (or require overpriced premiums) to persons living in high-risk neighborhoods (a process called *redlining*; Zenou & Boccard, 1999). In addition, the people who live in these neighborhoods may be exploited by businesses that use crime maps to scare people into buying security measures.

Last, as will be emphasized in Chapter 8, the misuse of crime data or the analyst's failure to follow statistical rules in analysis (intentionally or unintentionally) may lead to maps that are difficult to interpret. For example, a map that displays crime information at the census tract level could be misinterpreted to mean that the *entire* census tract suffers from high rates of crime when it may only be one small section of the tract that has a large crime problem. People who view this information may decide not to conduct business in this tract, including purchasing homes or starting businesses.

■ Conclusion

As you can see, knowing which data sources to use and how to share the completed analysis (whether in report or map form) is diverse and complex. Understanding and clearly articulating the limitations of analyses is critical to limiting the possibility of misinterpretations by your audience. Furthermore, understanding the legal and ethical issues involved with distributing crime maps, data, and analysis is imperative for crime analysts.

The Massachusetts Association of Crime Analysts' Web site contains an article entitled "The Ten Commandments of Crime Analysis" (Bruce, 1999), which provides a useful framework for summarizing the main points presented in this chapter and reinforces the points to be discussed in Chapter 8. Several of these commandments are useful to our current discussion. First, commandment three states "thou shalt take responsibility for thine own data." Crime analysts must be accountable in using appropriate, clean, timely, and accurate data in their analyses. Second, (commandment five) "thou shalt never present statistics (or maps) by themselves." Analysts should always provide qualitative descriptions of the maps and analyses they present, making sure to include any limitations of the analyses. Commandments six and seven, "thou shalt know thy jurisdiction from one end unto the other" and "thou shalt not stop crime analysis at thy jurisdiction's borders" reinforce the notion that many jurisdictional or municipal boundaries commonly used in mapping files have no real meaning to crime and/or criminals. It is important for analysts to have an understanding of the hot spots in their jurisdiction as well as an understanding of how crime in bordering jurisdictions affects one another. The eighth commandment of crime analysis is thou "shalt focus equal attention on the six Ws." Thus, the analyst should attempt to answer the who, what, where, when, how, and why of a crime series, patterns, and hot spots. Finally, commandment nine asks that analysts "remember thy community, and keep it holy." This includes prioritizing crime analysis activities and providing as much information as you can without violating the privacy of crime victims and others or causing harm to ongoing investigations.

■ Recommended Reading

Stallo, M. A. (2003). *Using Microsoft Office to improve law enforcement operations: Crime analysis, community policing, and investigations.* Analysis Consulting and Training for Law Enforcement, Inc. http://www.actnowinc.org/

Stallo, M. A., & Bruce, C. (2008). *Better policing with Microsoft Office.* Charleston, SC: BookSurge. ISBN-10: 1419609483

■ Questions for Review

1. What are the limitations of using department crime data in crime analysis?
2. Why is the selection of scale or unit of analysis so important in creating crime maps or performing crime analysis?

3. What are the strengths and weaknesses of the UCR?
4. What are the strengths and weakness of the NIBRS?
5. What are the limitations in using calls for service data?
6. What is contextual data, and how is it useful to crime mapping and analysis? Provide an example of contextual data in your discussion.
7. What is geocoding?
8. What are the ethical and legal considerations in sharing data with other agencies or the public?
9. What are the "ten commandments" of crime analysis?

■ Chapter Glossary

Attribute Table An attribute table contains the information that makes up a map's layer. It is contained in a spreadsheet or database application, such as Excel or Access. For example, an attribute table of homicides would contain key information in columns (address, crime type, weapon used, victim and offender characteristics, etc.) for individual cases (listed in rows).

Contextual Data Contextual data is data that includes information on variables related to crime, but not crime itself. Contextual data often includes information about population or spatial characteristics.

Data Cleaning The process of eliminating information unnecessary to your analysis and making sure there is no bad information. It may also mean creating or computing new variables needed for an analysis or recoding existing variables into a format that is more useful or easier to use in the analysis.

Geocoding Geocoding is the process of taking street addresses from a tabular file and matching them to a reference file to produce the points on a map.

Institutional Review Board An internal review system typically found in colleges and universities that is concerned with the potential ethical issues in research proposed by faculty, staff, and students.

National Crime Victimization Survey A self-report survey of crime in the United States, launched in 1873 and housed by the National Archive of Criminal Justice Data (NACJD). Approximately 77,290 households are selected in which household members (above the age of 12 years) are asked to report their victimization experiences, if any. The NCVS captures those crimes that are unreported to police and elicits detailed information about crime victims and their experiences.

National Incident Based Reporting System NIBRS collects information at the incident level for 22 different offense categories (Group A offenses) and records incident and arrest data about the incident, including victim and offender characteristics, and types and value of property stolen and recovered. In addition, NIBRS collects arrest data (only) on 11 other offense categories (Group B offenses). NIBRS collects incident-level data and breaks down the data into specific subcategories. Not every state participates in the NIBRS data collection system.

Probability Sample A probability sample is a scientifically selected sample group chosen from a larger population where all cases within that population have an equal chance of being selected to be in the sample group.

TIGER The Topologically Integrated Geographic Encoding and Referencing system is a computer database operated by the US Census Bureau that contains all of the census-required map features and attributes for the United States, including the specifications, procedures, computer programs, and the related input materials that are necessary to develop and utilize it.

Uniform Crime Reports Participating law enforcement agencies keep track of known offenses (primarily Part I index crimes) and report monthly totals to the Federal Bureau of Investigation (FBI), who then analyzes this information to produce the Crime Index. The UCR data is published annually by the FBI.

Chapter Exercises

Exercise 12: Joins and Relates

Data Needed for This Exercise
- All needed data is included in EX12.mxd.

Lesson Objectives
- Tabular joins are helpful when combined with summary tables.
- Use and create spatial joins to select data one more way.
- Use a relation to select records when tabular joins are just not good enough.

Task Description
- Take steps to create a choropleth map by creating a summary table by grids for total UCR Part I offenses reported, and then

add the summary table data to the grid polygon theme to create a rudimentary shaded grid (choropleth) map.
- A spatial join will be used to add attributes from the grid table to the calls for service table so that each call for service knows what grid it is in.
- Create a relationship between data in a table and another summary table to help in selecting records of a specific type.

1. You have to develop a hot spot map for the monthly Compstat meeting that shows the total number of Part I crimes. You've decided to use a simple choropleth map to show the hot spots and are going to use the count, or the total number of UCR Part I crimes within each quarter square mile grid in this project.
2. Open EX12a.mxd in the C:\CIA\Exercises\Exercise_12\ folder.
3. Part I crimes have already been selected where «UCRGROUP» LIKE 'PART 1 %' and a new layer was created using this selection.
4. You should have around 36,769 records in the attribute table, and one of the fields at the right should be Grid_name.
5. We need a summary table or count by each unique grid name. Right click the **Grid_name** field and choose **Summarize** from the pop-up menu.
6. Save the summary table in your student folder as P1GridSum.dbf and add it into the project when it is finished.
7. Close the attribute table and open the summary table.
8. This summary table contains grid names and the count of the total Part I UCR crimes that occurred within it.
9. We can now join this data to the grid layer so that each grid will know how many Part I crimes occurred within it as well. This is called aggregation of the data.
10. Right click on the **Grid** theme and choose **Joins and Relates**. In the next pop-up choose **Join**.
11. The top box will have "Join attributes from a table" (see DVD Figure 12–1).
12. The second box will say "grid_name."
13. The third box will say "P1GridSum."
14. The last box will say "Grid_name" as well.
15. We are saying that the Grid theme and the P1GridSum.dbf summary table have a field that has similar and matchable data in the form of the field called "grid_name."

16. To perform an attribute join, you must have such a connection between tables. The field names do not have to be identical; however, the data in the field does need to be identical or the join will not work.
17. Click **OK** to apply the join. You should see that you are back at the map display, and it seems as if nothing happened, right? Wrong!
18. Open the attribute table for the Grid theme and then scroll all the way to the right. Observe that a new field named "Count_GRID_NAME" has been added. As your browse through the records in the table you will see number values in this field for several of the grids. Why do you also see null values as well? _____.
19. You see null values because there was no record in the P1GridSum table for that grid, so there was no value to bring into this polygon's attributes. This means that the grids with null in them did not have any Part I crimes. Close the table.
20. Now we want to change the legend and the name of our Grid theme to reflect what we are going to show on the map (hot spots of Part I crime). Right click the **Grid** theme and go to the **General** tab. In the Name field type *Hotspots for Part 1 Crimes*.
21. Now go to the **Symbology** tab, and at the left choose **Quantities** from the Show window area. Choose **Graduated Colors** and then go to the **Value** box section.
22. Use the drop-down box to find **Count_grid_Name**, and choose it for this box.
23. Click **Apply** and view your hot spot map!
24. Notice that the grids with a <null> value are not showing at all in the map display.
25. The easy fix for this is to just bring in another copy of the grid layer, make it hollow with a thin gray line, and make sure it is above this theme in the TOC. This will appear as if all the grids are there, but actually two themes are creating the look you need. The harder, but more permanent, solution is to calculate new fields in the Grid theme's attribute table for the fields called "gridname" and "count", and when you calculate the counts, make any grid with a null value = 0 instead of null.
26. Close EX12a.mxd and open EX12b.mxd.
27. Open the attribute table for the BURGCFS2004 theme and notice that there is no grid field or grid name field anywhere in the table.

28. Close the attribute table.
29. Right click the **BURGCFS2004** theme. Choose **Joins and Relates**, and then choose **Join**.
30. In the dialog menu, make the following settings:
    a. Select **Join Data from Another Layer Based on a Spatial Join** in the top box.
    b. Select **Grid** in the second box.
    c. Check the **If Falls Inside** box.
    d. Name the file BURGJOIN.shp and put it in your student folder.
31. When the join process is done running, it will add the new theme to the TOC.
32. Open the attribute table and go to the right of the table. Remember what the last field name used to be? Now there are several others that came from the grid layer there. Find the **Grid_name** field and right click it. Click on **Summarize** and create a summary table called BurgSum.dbf and save it to your student folder.
33. Load the BurgSum.dbf table into ArcMap by answering **Yes** to the pop-up question.
34. Open the table. What do you see? _____.
35. See if you remember how to do a tabular join from this table to the Grid theme's table and create a shaded grid map of the results. Your map should look like the map shown in DVD Figure 12–2.
36. Save your project!
37. Close EX12b.mxd and open EX12c.mxd.
38. In this next section we will learn a little bit about relates and how they differ from joins. Unlike joins, attribute data is not added to a theme or a new theme (spatial join). Instead, data in one table can be used to search or query data in another table.
39. Our task is to find the recovery locations of the Hondas that have been stolen out of beat 46 in the past several years. We are looking for possible patterns.
40. You can see where the stolen vehicles we are dealing with are located. Turn on the Recovery theme to see that the data stretches out across all of Maricopa County and includes all stolen cars, so we need to limit our data to just those recovered vehicles that are in the Stolen Vehicles theme.
41. To create a new relate, we have to decide which theme is going to be the theme that is related, or the master theme for lack of a better

description. In this case we want the Stolen Vehicles theme to be the master theme. Click on this theme to select it, and then right click it and choose **Joins and Relates**, and then choose **Relate**.

42. Complete the Relate dialog menu as follows (see DVD Figure 12–3):
    1. *UNIQUE_*
    2. *Recovered_Stolen_Vehicles*
    3. *UNIQUE_*
    4. *Stolen_Beat46_to_Recovery*
43. Click the **OK** button to create the relate.
44. Did anything happen? Apparently not, but in fact a relationship between the UNIQUE_ field names in each theme was created for the next step.
45. Our objective here is to find all the records out of all recoveries that were related to the Honda thefts in beat 46.
46. Right click on the **Stolen Vehicle** theme and open the attribute table.
47. Click on the **Options** button and choose **Select All**.
48. Click on the **Options** button again, but this time choose **Related Tables** and select the Stolen_Beat46_to_Recovery relationship we created.
49. ArcMap will pause for a few moments, and then it opens the recovered vehicle attribute table for you and has 163 records selected. Click the **Selected** button and see that they are all Hondas that are directly related to the stolen vehicles.
50. Close the tables and make sure you do *not* clear the selection.
51. Right click on the **Recovered Vehicles** theme and choose **Data → Export**. Save the selected records as RecVeh.shp in your student folder, and then add it into the project.
52. Turn off the original Recovered Vehicle theme and the Stolen Vehicles theme, and then zoom to the extents of the Recovered Vehicles theme you just added.
53. Use the Identify tool to check out those points way up northwest and southeast of Glendale. Make sure to set the selectable layers so you don't get data from other layers as well.
54. Wow, there are Honda recoveries all the way to Las Vegas and down into Douglas, Arizona!
55. Do you see any patterns or clusters of recoveries in any place?

56. I did a summary table on the Recovery City field and found that 102 vehicles were recovered someplace in Phoenix, with 39 recovered in Glendale (mostly around where they were taken from, it seems) and 11 in Peoria. Might this indicate that Glendale should be doing auto theft prevention with Phoenix and Peoria?

57. Save and close your project.

■ References

Bruce, C. (1999). *The ten commandments of crime analysis.* http://www.macrimeanalysts.com/articles/tencommandments.pdf

Federal Bureau of Investigation. (n.d.). *National incident-based reporting system: Frequently asked questions.* Retrieved March 2009, from Federal Bureau of Investigations Web site: http://www.fbi.gov/ucr/faqs.htm

Hicks, D. (2007). *Spatial analysis of probationer contacts and automated email alerts: NIJ workshop.* Presented at the 9th Crime Mapping Research Conference, March 28–31, 2007, Pittsburgh, PA.

Kelling, G. L., Pate, T., Dieckman, D., & Brown, C. E. (1974). *The Kansas City preventive patrol experiment: A summary report.* Washington, DC: Police Foundation.

Sherman, L. W., & Weisburd, D. (1995). General deterrent effects of police patrol in crime "hot spots": A randomized, controlled trial. *Justice Quarterly, 12*(4), 625.

U.S. Census Bureau. (2004). *Maps in American FactFinder.* Retrieved March 2009, from U. S. Census Bureau Web site: http://factfinder.census.gov/jsp/saff/SAFFInfo.jsp?_pageID=gn7_maps

Wartell, J., & Thomas, J. T. (2001). *Privacy in the information age: A guide for sharing crime maps and spatial data.* Washington, DC: National Institute of Justice. http://www.ncjrs.gov/txtfiles1/nij/188739.txt

Zenou, Y., & Boccard, N. (1999). Racial discrimination and redlining in cities. *Journal of Urban Economics, 48,* 260–285.

People and Places: Current Crime Trends

CHAPTER 7

▶ ▶ **LEARNING OBJECTIVES**

Students from every major discipline can acknowledge that crime is not randomly or equally distributed across population, time, or space. Crime rates can be affected by such things as geographic considerations and/or changes within an infrastructure, weather conditions and major weather events, economic conditions and political changes, isolated events such as a mass killing or terrorist attacks, the use of the death penalty, and many others. While many of these conditions change with some level of predictability (like the change of season), others are spontaneous events that create a long-lasting wave of change that can be both unanticipated and very long lasting (such as hurricane Katrina, 9/11, and the Virginia Tech shootings). This chapter provides a snapshot view of what crime looks like today and how it has changed over the last few decades. We will also discuss crime trends and their changes over time. Current trends in criminal offending and victimization, offender and victim characteristics, and how crime trends exist in time and space will be emphasized. After studying this chapter, you should be able to:

- Identify who is at the greatest risk of becoming a victim of various crimes based on age, gender, and race.
- Identify who is most likely to commit various crimes based on age, gender, and race.
- Identify different motivations for offending.
- Understand the nature of victim and offender relationships.
- Explain patterns in group crimes.
- Understand the role victims play in their own victimization.
- Understand the role of drugs and alcohol in crime.
- Identify the relationship between geographic size, location, time, and crime.

■ Introduction

In the next chapter, we will discuss several different types of statistical analyses that are useful to crime analysts. The previous chapter provided an examination of the types of crime data (including the limitations of these data) that analysts use in crime analysis and crime mapping, including calls for service, arrest data, victimization surveys, and information collected by the US Census Bureau. This chapter examines that data to identify the trends and patterns found in criminal offending and victimization. This chapter also includes a discussion of any changes in trends or patterns that have occurred within the last few decades. Students should view the information contained in this chapter with a critical eye using what they learned in previous chapters (specifically Chapter 6) about the strengths and weaknesses of crime measurement and statistical analysis (to be discussed more in depth in the next chapter). The majority of the information discussed in this chapter comes from the data sources examined in Chapter 6, primarily the UCR and the NCVS.

The overwhelming majority of calls to police are not crime related. Durose, Schmitt, and Langan (2005) reported for the Bureau of Justice Statistics (BJS) that of the 45.3 million people that had personal contact with police in 2002, only 2.9% of those people were arrested by police. In addition, of that 45.3 million people, only 25% indicated that their contact was initiated to report a crime. These numbers suggest that a large majority (approximately 33 million) of police contacts are initiated around activities that are not directly related to crime. This is an important notion for the analyst to understand because he or she will most likely be called upon to complete analyses for the department that focus on quality of life issues, such as barking dogs, loud parties, etc. In the sections that follow, a description of a current snapshot of crime in the United States is provided, organized by trends across people and places.

■ People

The primary people involved in criminal incidents are, of course, victims and offenders. When these two types of people interact, you have a criminal event. Witnesses, responsible parties, investigative leads, and other persons help us to collect information about how the event happened, where it happened, and who did what to whom. The following sections discuss characteristics of both victims and offenders based on crime data from the UCR and the NCVS. When reading this chapter, keep in mind the discussion from Chapter 6 regarding the strengths and weaknesses of each of these data sources. Also, try to imagine why and how characteristics of people interact with time and space for a more comprehensive picture of crime.

Victims

Recall from Chapter 6 that the UCR measures Part I index offenses and is a voluntary reporting system for law enforcement agencies.

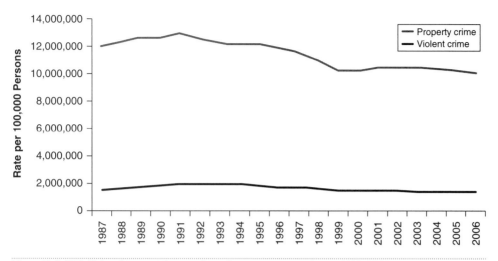

Figure 7–1 UCR: Violent Crime and Property Crime Rates in the United States, 1987–2006

(Remember, the UCR crime rates are reported as rates per 100,000 people. This is done to make the statistics more meaningful to the reader and is used as a way to make the data comparable across jurisdictions of various sizes.) According to the FBI, an estimated 1,417,745 crimes of violence (murder, forcible rape, robbery, and aggravated assault) were reported to the UCR in the United States during 2006. This is up approximately 1.9% from 2005 but down 13.3% from 1997. (Crime trends often show minor fluctuations over short time periods while reflecting more consistent patterns over longer time periods.) In 2006, reported property crimes (burglary, larceny–theft, motor vehicle theft, and arson) totaled 9,983,568. This is a decrease of 1.9% from 2005 and a decrease of 13.6% from 1997 (see **Figure 7–1**).

In 2005, murder was the least frequent form of violent victimization (approximately six murders per 100,000 persons). While males of all races accounted for approximately 77% of murdered victims between 1976 and 2005, blacks were more than six times more likely than whites to be murdered during this time period. Persons aged 18–24 years were at highest risk of being murdered during this same time period (17 per 100,000).

The NCVS indicates similar crime trends over time. However, it consistently estimates that there is approximately twice the number of crimes committed as are reported to the UCR, supporting the general assertion that this report may be a better indicator of victimizations in the United States (see **Figure 7–2**).

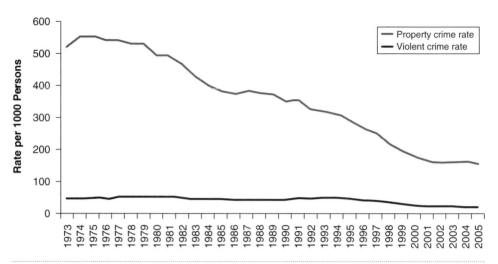

Figure 7–2 NCVS: Violent Crime and Property Crime Rates in the United States, 1973–2005

According to the NCVS (recall that the NCVS reports crime rates per 1000 people, only measures crimes perpetrated against victims aged 12 years and older, and excludes victims of murder), 25 million crimes were committed against US residents 12 years of age or older in 2006. Of these crimes, an estimated 19 million were property crimes, 6 million were violent crimes, and 174,000 were personal thefts. Of the approximate 6 million violent crimes during this period, slightly over 5 million included both aggravated and simple assaults (Klaus, 2007). In 2005, the national violent crime rate was down 58% from what it was in 1993, and for the most part, it has been steadily decreasing since about 1993. This does not mean, however, that every city in America has been enjoying lower crime rates for the last decade and a half. As will be discussed later in this chapter, while crime has been steadily declining for the nation as a whole, it has been increasing for different demographic groups and in different areas of the country.

According to the NCVS, males were at a slightly higher risk for violent victimization in 2006, at 26.5 per 1000 males aged 12 years and older (this higher risk for males is true for all violent crime categories with the exception of sexual assaults). Females were victims of violent crimes at a rate of 22.9 per 1000 females aged 12 years and older. When this crime is further separated by age, it appears that females aged 34 years and older have higher rates of this type of victimization than males in the same age group. Males aged 24 years and younger, however, have slightly higher rates of this type of victimization than females in the same age group.

In 2006, the NCVS reported that persons between the ages of 16 and 19 years held the dubious honor of being the group with the highest rate of violent victimization: 52.3 per 1000 persons within this age group. Persons in the age groups 50 to 64 years and 65 years or older held the lowest risks of violent victimization. This trend remains fairly consistent over time, with the highest risk of being a victim of a violent crime belonging to persons under the age of 24 years. Thus, the risk of violent victimization progressively decreases with age.

This crime trend presents an excellent opportunity for the budding analyst to ponder a few questions about this victimization trend and what influences it. When considering the victimization rate previously shown for those over age 50 years, how do you think the following considerations affect that rate, and why does it appear to remain so stable?

- Activities of the victim
- Location of the victim
- Environment of the victim
- Age of the victim
- The number of potential victims in that community or particular location
- The extent of the violence perpetrated upon those victims

Now consider what is known about the victim population. In general, we can say with some certainty that this population (people aged 50 years and older) in general is more likely to perform their daily activities during the day and tends to curtail their movements in the evening hours. We can also say that this population tends to live in one location for longer periods and probably has a sense of its community. We can also say that this population, by and large, tends to be physically weaker than the typical offender population and may live alone, as with the case of widows and widowers. How do you think these factors combine to create or thwart victimization? As the baby boomer population bubble enters the retirement years (beginning in 2008), do you think this trend will remain consistent? If one considers the cities where large segments of the population are in these age groups (such as Phoenix, Arizona), how might that affect the mapping process, and what data characteristics might the analyst screen when performing their analyses?

For property crimes, persons younger than age 20 years had the highest risk of victimization. Persons between the ages of 50 and 64 years and those aged 65 years or older were the least likely to experience property victimization. However, while the elderly were least likely to be victimized of a property crime; more than 90% of crimes against the elderly between 1993 and 2002 were property crimes. (Is this a matter of better reporting, being easier targets, or more property ownership? Again, these are things to consider when analyzing crime trends.) Teens and young adults experienced the highest rates of victimization for both property and violent crimes. Approximately 40% of crimes against persons aged 12 to 24 years between 1993 and 2002 were property crimes.

When examining the impact of race on victimization, the NCVS data show that "black victims experienced higher rates of violence than whites or persons of other races" (Rand & Catalano, 2007, p. 4). The overall rate of violent victimization for black individuals was 33 per 1000 black persons. For white individuals, the overall rate of violent victimization was 23 per 1000 whites. Property crime rates for Hispanics (210 per 1000 households) were higher than property crime rates for non-Hispanics (148 per 1000 households; BJS, 2005).

According to NCVS data, in 2005 black women were at a higher risk to be victims of a personal crime (with a rate of 25.5 victimizations per 1000 black women) than white women (16.5 victims of a personal crime per 1000 white women). Black women were also at almost three times greater risk for being the victim of a rape or sexual assault. The risk of being a victim of a rape or sexual assault for Hispanic women (1.1 per 1000) was slightly higher than for non-Hispanic women (.7 per 1000; BJS, 2005).

When examining variations in victimization rates based on socioeconomic status, persons in households with an annual income under $7500 (in 2005) had higher rates for both burglary and robbery victimization. This is true for both black and white victims. Persons in this income category also endured higher rates of assaults than persons in higher income brackets. However, when separating for race, the income bracket for black victims that held the highest risk of assault was between $15,000 and $24,000. For property crime overall, persons living in households with less than $7500 total income have consistently held the highest rate of victimization since 1999 (BJS, 2005).

When examining the impact of marital status on victimization, a few interesting trends emerge. According to NCVS data, persons who never married have the highest rate of victimization for personal crimes (39 per 1000 persons never married). This group is followed in risk by persons who are divorced or separated (32.8 per 1000 divorced or separated persons), then persons who are married (10.8 per 1000 married persons), followed by widowed persons, who are at the lowest risk of being victimized of a personal crime (6.9 per 1000 widowed persons; BJS, 2005). These trends follow common sense in that a large number of persons who have not yet married are young (hence the highest risk), and a large number of persons who are widowed are older (hence the lowest risk). Much of the victimization risk of divorced or separated persons can also be explained by other factors, such as age and socioeconomic status because divorces are expensive and often leave one partner in worse financial condition than he or she was during the marriage. Divorced or separated females, for example, have slightly higher rates of victimization for crimes of violence overall and are at higher risk to be victimized of completed acts of violence, rape and sexual assault, and simple assault (BJS, 2005).

When looking at the number of persons living within a household, it appears that the more people living under one roof, the higher rate of victimization the household suffers. For example, the rate of household victimization for a property crime in 2005 was 100 per 1000 households for households with only one occupant. This rate increases progressively, and households with six or more persons suffer a rate of 324.5 property crimes per 1000 households with six or more occupants (BJS, 2005).

Last, in examining the notion of time, in 2005 approximately 53% of violent victimizations (excluding murder) occurred during the daytime (6:00 a.m. to 6:00 p.m.). Thirty-five percent of violent crimes occurred between the hours of 6:00 p.m. and midnight. For property crimes, this pattern changes modestly. In 2005, approximately 36% of property crimes were committed between the hours of 6:00 a.m. and 6:00 p.m., and 44% of the crimes were committed during the nighttime (14% between 6:00 p.m. and midnight, and 16.5% between midnight and 6:00 a.m.; BJS, 2005).

Offenders

Many of the same trends and patterns of victimization also appear in patterns of offending. This is because *most* crime is largely intra-age, intrarace, and intraclass, meaning that offender and victim populations tend to be the same people. That said, a young, white male is most likely to be victimized by a young, white male. For example, in 2005, according to NCVS data, male victims perceived 78.5% of their offenders to be male for overall crimes of violence. Consistently, white victims of violent crimes perceived their offenders to be white 64% of the time and perceived their offenders to be black only 17% of the time. Black victims of violent crimes perceived the race of their offenders to be black 73% of the time and white only 10% of the time.

Overall, approximately 56% of offenders were perceived by their victims to be white compared with 25.3% who were perceived to be black. However, when examining individual crime categories, victims perceived their offenders to be white a higher percentage of the time for the categories of attempting to take property with injury (attempted robbery) and for both categories of assault (simple and aggravated). For the categories of rape and sexual assault and robbery (completed and attempted without injury), victims perceived the race of their offenders to be black a higher percentage of the time.

The FBI estimated that 14,380,370 arrests for all offenses (excluding traffic violations) were made in 2006. Approximately 1,540,297 of these arrests were for property crimes and 611,523 were for violent crimes. Drug abuse violations accounted for 1,889,810 of total arrests. Approximately 84.5% of the total arrests in 2006 were of persons aged 18 years and over. Arrests of juveniles in 2006 for violent crimes, including murder and robbery, increased from 2005 (by 3.4% for murder and 18.9% for robbery).

Males accounted for a larger percentage of the distribution of arrested persons (76.3% for total crimes) for every crime category measured by the UCR except embezzlement, prostitution, and runaways. (Females accounted for 52.7%, 64.2%, and 56.6%, respectively.) Females also accounted for 10.9% of persons arrested for murder and nonnegligent manslaughter.

For total offenses, whites accounted for 69.7% of arrested persons, and when compared to blacks on various crime types, whites accounted for more of the arrests for all other categories except for murder and nonnegligent homicide, robbery, and gambling (BJS, 2005).

Motivations for Offending

Depending on the type of crime, offender motivations may not be completely known. Oftentimes they are assumed by clues left at a crime scene or by victim and witness accounts of the crime. Sometimes the offenders themselves offer their motivations for committing crimes. For most property crimes, such as larceny and burglary, and violent crimes, such as robbery, the motivation is usually assumed to revolve around economics. However, motivations become more difficult to ascertain for crimes such as vandalizing property, auto theft, and arson. In some instances of auto theft, for example, the offender may have simply needed to get somewhere; other times, especially with juveniles, the thrill of stealing a car is the motivation. In addition, the motivations for burglary are thought to be much more complex than simple economic gain, and self-report studies have identified several motivations beyond monetary gain, including the excitement of the job, the challenge, and using one's intelligence to beat the police (Wood, Gove, & Cochran, 1994). Research also suggests that criminals typically do not specialize in any one crime type and that a majority of crime is opportunistic, negating the need for high levels of forethought and planning (Farrington, 1992; Gottfredson & Hirschi, 1990). Wood et al. (1994), in their study of incarcerated adults in Oklahoma, found that a variety of motivations existed for the crimes of shoplifting, burglary, robbery, rape, and assault, including:

- Ease of opportunity
- Thrill or excitement
- Money and property rewards
- Difficulty and challenge
- Anger, frustration, and rage
- Power and control
- Sexual relief and/or satisfaction
- Revenge/hatred/payback
- Sudden impulse or whim
- Unintentional/accident
- Offender was on drugs
- Respect and admiration of others
- To buy or steal drugs or alcohol
- Need money for food, rent, or bills
- Peer pressure or group behavior

For shoplifting, offenders in their study ranked money and excitement as the most important motivations for their crime. For burglary, monetary motivations were ranked as very important by most offenders, while the ease of opportunity and the thrill or excitement were also very important. This is also true for robbers who indicated that power and control were important motivations behind their crimes. For the crime of rape, offenders ranked the ease of opportunity, the thrill or "high" of performing the act, anger/frustration/rage, power and control, sexual satisfaction, revenge/hatred/payback, and a sudden impulse as very important motivations for their crimes. Last, for assault, offenders noted the motivations of the thrill of excitement and the "high," anger/frustration/rage, power and control, revenge/hatred/payback, a sudden impulse, being on drugs, admiration and respect of others, and peer or group pressure as being important.

Determining the motivations for crimes is often a difficult task because we are left to make these assumptions based on the information left at crime scenes and the self-reports of victims, witnesses, and offenders involved in the crimes (which can be suspect). Information about bias, however, is typically more readily available in hate crimes. For example, the FBI reports that in 2006 there were 9652 victims of hate crimes. Of these, 52.1% were targeted because of their race, 18.1% were targeted because of their religion, and 15.3% were targeted because of their sexual orientation. In addition, 13.5% of hate crime victims were targeted because of their ethnicity or national origin, and 1% were targeted because of their disability. For hate crimes motivated by racial bias, 66.4% were targeted because of antiblack sentiment, 21% because of antiwhite bias. Approximately 65.4% of religious-based hate crimes were anti-Jewish attacks, and 11.9% were attacks based on anti-Islamic bias. For hate crimes motivated by a bias against sexual orientation, the majority of attacks were credited to anti-male-homosexual bias. When examining ethnicity, 62.8% of attacks were anti-Hispanic. Finally, of the 95 persons targeted because of their disability, 74 were attacked because of a mental disability.

Victim–Offender Relationships

For personal crimes in 2005, approximately 52% of victimizations were committed by persons who were strangers to their victims. Most homicide victims knew their murderers in some capacity (BJS, 2005). The number of homicides where the relationship between the victim and offender is unknown has been decreasing since 1999 (BJS, 2005).

The nature of victim and offender relationships varies by crime type. For example, when examining rape and sexual assault, only about 35% of offenders were strangers to their victims. For robbery, only 20% of offenders were known to their victims (BJS, 2005). When examining relationships between victims and offenders, some interesting patterns emerge. For example, the NCVS indicates that for 2006, approximately 5% of the offenders who victimize males were described as intimates, and 47% were strangers (down from 54% in 2005). For female victims, however, 21% of their offenders were described as intimates, and 29% (down from 34% in 2005) were strangers. In 2006, for both male and female victims, 40% of their offenders were considered to be a friend or acquaintance (Rand & Catalano, 2007). "Though the rate and level of homicide change from year to year the relationship between victim characteristics in homicide tends to remain the same" (Catalano, 2006a, p. 2). Each year most murder victims are male. In

2004 approximately 78% of murder victims were male. Of the cases where the relationship between the victim and offender is known, the majority of victims knew their murderer. In 2004 approximately 77% of the victims knew their offenders. These offenders are overwhelmingly males aged 18 years or older.

Only one in 320 households in 2005 experienced intimate partner violence (Klaus, 2007). Nonfatal intimate partner violence has been declining since 1993 (Catalano, 2006b). Females aged 20–24 years were at the highest risk of nonfatal intimate partner violence. Between 2001 and 2005, approximately 96% of female victims of intimate partner violence were victimized by males. Another 3% reported that their abuser was a female. For male victims of intimate partner violence, 82% reported that their abuser was female, and 16% reported that their abuser was another male. Married females reported the lowest rates of nonfatal intimate partner violence, while separated females reported the highest rates. Between 2004 and 2005, black females aged 12 years and older had a higher risk of being the victim of intimate partner violence (4.6 per 1000 persons of the same group) than white females aged 12 years and older (3.1 per 1000 persons of the same group; Catalano, 2006b). Rates of intimate partner violence for Hispanic females and non-Hispanic females have been similar (4.3 per 1000 for Hispanic females and 4.2 per 1000 for non-Hispanic females) for the years 2001 to 2005 (Catalano, 2006b). Alcohol or drugs were present in approximately 42% of all incidents of nonfatal intimate partner violence.

Group and Other Miscellaneous Patterns

The proceeding sections discuss group and other patterns in offending including weapon use and gang-related crime. General patterns of behavior that occur in criminal events within your own geographic region can be specific to that region, or more global in nature. Understanding the who, what, when, where, and how of criminals in your region can aide you in developing better analysis products for your agency. It is important for the analysts to have an understanding of these patterns when performing analyses and creating crime maps as they can help inform the analyses. As with individual characteristics, these patterns are dynamic and are not absolute, but may be useful in understanding criminal incidents in a broader context.

Group Offending Patterns

In 2005, for all crimes of violence, approximately 79% of incidents involved only one offender. When comparing crimes committed by strangers with crimes committed by nonstrangers, 68%

of stranger crimes were committed by only one person, and 90% of nonstranger crimes were committed by only one person. However, when multiple offenders were present, offenders aged 12–20 years accounted for 41% of these crimes (based on victims' perceptions of the age of their offenders; BJS, 2005). When examining the perceived race of offenders for crimes committed by more than one person, victims perceived their offenders to be black 40% of the time and white 29% of the time.

Weapon Use

For 2006 approximately 67% of all violent incidents (excluding murder) were committed without the use of a weapon. In 9% of these incidents, the presence of a firearm was indicated (Rand & Catalano, 2007). The presence of a weapon was indicated in 48.3% of robbery victimizations and 21.7% of assaults, and firearms were used in 26.3% of robberies and only 6.7% of all assaults. Crimes of violence were more likely to be perpetrated without a weapon when the victim and offender knew each other (17.7%) than when the victim and offender were strangers (30.4%). For intimate partner violence, a weapon of some type was present during the offense 15.6% of the time for incidents with female victims and 28.1% for male victims. Firearms were present in 3.6% of the incidents with female victims and 0.5% of incidents with male victims.

Gang-Related Crime

It is difficult to separate crimes that are gang related from crimes that, while they are committed by known gang members, are not gang related. Thus, while victims' perceptions about the gang status of their offenders are important, this does not necessarily mean that the crimes themselves were gang related.

According to NCVS data, between 1998 and 2003 approximately 6% of violent victimizations were perpetrated by persons believed to be gang members (this is down from 9% during the years 1993 to 1996). This estimate is based on the perceptions of the crime victims themselves, so the true figure could be higher or lower. According to the FBI's supplementary homicide reports, the percentage of homicides that were gang related between 1993 and 2003 ranged from just under 5% to a peak of 6.9% (Harrell, 2005).

Reporting Crime

As discussed in Chapter 6, the crimes that are reported to or detected by law enforcement only represent about 50% of total crime. Therefore, a discussion of *why* people decide to report their victimizations (or why they do not) is important.

According to the Bureau of Justice Statistics (BJS), the reporting of crime to the police across crime types or crime victims is neither random nor equal. The BJS estimates that 49% of violent crimes and 38% of property crime were reported to the police during 2006. For violent crime, robbery and aggravated assaults were more likely to be reported to police than rape and sexual assaults. For property crime, motor vehicle thefts were most likely to be reported to police. For personal crimes, victims of purse snatching or pocket picking and victims of rape or sexual assault were least likely to report their victimizations to police (35.2% and 38.3%, respectively). Victims of robbery were most likely to report the crimes to police (especially victims of completed robberies).

Approximately 39.6% of victims of property crimes reported their victimizations to police in 2005. Victims of motor vehicle thefts were most likely to report their victimizations to authorities (92.4% for completed motor vehicle thefts). Female victims of violence were more likely to report their victimizations than were males (54.6% versus 42.4%). Females were more likely to report victimizations for every violent crime category concerning both stranger and non-stranger offenders. There were no substantive or significant differences in reporting across gender for victims of property crimes. Black victims of violent crimes were only slightly more likely to report their victimizations (49.4%) than were white victims (47.2%). Black victims were slightly more inclined to report being the victim of a property crime than were white victims (39.6% of white males versus 44% of black males; 38.8% of white females versus 44.7% of black females). Hispanic females were more likely to report being victims of violent crimes (60.3%) than were non-Hispanic females (53.5%).

Victims of violent crimes aged 65 years and older were most likely to report their victimizations (66.1%). Victims aged 12–19 years were least likely to report their victimizations of violent crimes (34.5%). Victims of property crimes who reported less than $7500 in annual family income were least likely to report their victimization to police (36.8%). Victims reporting their annual family income to be between $7500 and $14,999 were most likely to report their property crime victimization to police (40.8%).

The most often cited reason for not reporting personal crime victimization was that the offender was unsuccessful or that the object was recovered. The second most often cited reason was that the incident was a private or personal matter. For victims of property crimes, 26.7% of those who did not report their victimizations to police cited that the object(s) were recovered or that the offender was unsuccessful.

Victim Precipitation

According to NCVS data, 9.4% of violent crime victims reported being the first to use or threaten physical force during the incident. A victim being the first to use or threaten physical violence was more likely in assault crimes as well. Approximately 70% of respondents who were victims of a violent crime in 2005 reported using self-protective measures of some kind. Self-protective measures by victims were highest for attempted robberies (90.8%) and lowest for completed robberies (46.8%). An estimated 81.6% of rape and sexual assault victims reported taking self-protective measures prior to the offense occurring. Interestingly, there does not appear to be significant differences across race or gender, with the exception of robbery, where white victims were far more likely to report taking self-protective measures before being victimized (67.9% white versus 51.9% black). Persons aged 20–34 years and aged 65 years and older were more likely to report using self-protective measures than any other age group for all crimes of violence. Persons aged 12–19 years and 50–64 years were least likely to take self-protective measures.

Resisting or capturing the offender was the most frequently cited self-protective measure for victims of violence in 2005. This held true for every violent crime category (rape and sexual assault, robbery, and assault). Running away and getting help or giving an alarm were cited as the second and third most employed measure of self-protection used. Male victims of violence were more likely to cite resisting or capturing the offender (33.4%) than were female victims of violence (15.8%). Female victims of violence were most likely to get help or give alarm (18.9%) or run away and hide (17.9%). There were subtle, but no significant, differences across race. For example, white victims (15%) were more likely to run away or hide than black victims (11.9%), and black victims were more likely to scare or warn the offender (13.9%) than were white victims (10.1%).

In every crime category, victims who took self-protective measures were far more likely to report their behaviors as helpful to the situation. Approximately 42% of violent crime victims reported that their self-protective measures helped them to avoid injury or avoid greater injury. When victims reported that their self-protective measures hurt the situation, the majority (65.5%) indicated that their behavior made the offender angrier and/or more aggressive. Analysts might consider this when considering crime rates across their jurisdictions. For example, they might consider if the presence of a college community and the attendant housing decreases the rate of crime or causes criminal events to be dispersed to different parts of the community. Could the use of protective measures entice criminals to do harm to other, less defensive populations, such as the elderly we discussed earlier, or are there equal attempted victimizations across the jurisdiction but more successful attempts upon the elderly population? Could this trend be affected by the administration of the college through the addition of facilities or majors of study and the enhancement of the college's public safety department? How might you consider this information when producing a crime map, and how would you present this information to your chief or sheriff? As you can imagine, a crime analyst could elicit several conclusions from the same data, all of which could be based on firm analysis.

Alcohol and Drug Use

In 2005, for crimes of violence, 27.5% of victims perceived their offenders to be under the influence of drugs or alcohol during the commission of the crime. For rape and sexual assault, this figure increased to 35.5%. The category with the lowest percentage of victims who perceived their offenders to be under the influence of drugs or alcohol was robbery (22.5% of victims).

In 2002 approximately "68% of [jail] inmates reported symptoms in the year before their admission to jail that met substance dependence or abuse criteria" (Karberg & James, 2005, p. 1). Approximately 16% of inmates reported that the reason they committed their offense was to get money for drugs, and approximately 50% of all convicted inmates reported being under the influence of drugs or alcohol at the time of the offense. Of these offenders, 71% were dependent upon or abused drugs or alcohol.

A larger percentage of female inmates reported substance dependence (51.8%) than did their male counterparts (44.3%). White inmates reported more substance dependence (55.4%) than did black (40.4%) or Hispanic inmates (35.7%). Female and white inmates were more likely to report using drugs at the time of their offense. Convicted females were more likely to report being under the influence of drugs (34%) than alcohol (22%) at the time of their offense. Males were more likely to report using alcohol at the time of their offense. An astounding 41.6% of inmates convicted of homicide reported using alcohol at the time of the offense. Marijuana and cocaine or crack were most often cited as the inmate's substance of choice. This example again presents another opportunity to discuss mapping considerations. Knowing that a large proportion of violent offenders uses the drugs mentioned, an analyst might consider the trends in drug arrests and the geographic location of those arrests when predicting future violent crimes. For example, can you see how a chief might appreciate knowing that a division in his jurisdiction is about to undergo an increase in the homicide rate due to drug activity? When analysts can appreciate these connections and nuances, they can then assist the administration with resource allocation and the creation of crime prevention measures.

Self-report studies of school age students also provide information about the drug and alcohol abuse of minors. "In 2005, one quarter of all students in grades 9–12 reported that someone had offered, sold, or given them an illegal drug on school property in the past twelve months" (Dinkes, Cataldi, Kena, Baum, & Snyder, 2006, p. vi). In 2005 approximately 43% of students in grades 9 through 12 reported consuming at least one drink of alcohol during the past 30 days, and 20% of students reported using marijuana during the past 30 days.

■ Places

As this book's main focus is the importance of spatial relationships to crime, this chapter would not be complete without a discussion of "place." Just as crime is unevenly distributed across people, it is also unevenly distributed across place. The following section provides a cursory review of crime distribution based on geography and basic spatial characteristics (primarily looking at the differences between urban and rural areas).

Location

According to the NCVS, urban and suburban victimization rates remained stable from 2005 to 2006. Persons living in urban areas are far more likely to be a victim of a violent crime, while persons living in suburban areas are far less likely than urban residents to be a victim of a violent crime but slightly more likely than rural residents. Between 2004 and 2005 the rate of violent crimes per 1000 persons aged 12 years or older was 29.4 in urban areas, 18.3 in suburban areas, and 18.1 in rural areas (population below 50,000 persons).

Rates vary significantly across urban areas. For example, in 2006, the UCR reported a rate of violent crime for the Detroit-Livonia-Dearborn metropolitan division to be 1277 per 100,000 persons. In the Washington, DC-Arlington-Alexandria metropolitan statistical area, the violent crime rate is a much lower 538 per 100,000. The FBI cautions, however, not to make comparisons based solely on population and lists the following "known" variables that can cause crime to vary from place to place (some of which will be discussed in the following two chapters). They are as follows:

- Population density and degree of urbanization
- Variations in composition of the population, particularly youth concentration
- Stability of the population with respect to residents' mobility, commuting patterns, and transient factors
- Modes of transportation and highway system
- Economic conditions, including median income, poverty level, and job availability
- Cultural factors and educational, recreational, and religious characteristics
- Family conditions with respect to divorce and family cohesiveness
- Climate
- Effective strength of law enforcement agencies
- Administrative and investigative emphases of law enforcement
- Policies of other components of the criminal justice system (i.e., prosecution, judicial, corrections, and probation)
- Citizens' attitudes toward crime
- Crime reporting practices of the citizenry

In 2005 the western region of the United States held the highest rate of violent victimization at 25.9 violent crimes per 1000 persons. The south had the lowest rate of violent crime with 18.5 violent crimes per 1000 persons. However, remember that NCVS data does not count homicides (because dead people cannot report their own victimization). According to UCR data, in 2006 the violent crime rate (comprised of murder, forcible rape, robbery and aggravated assault) was highest in the south (548 per 100,000) and lowest in the northeast (392 per 100,000). When examining property crime rates, the west again holds the highest rate of victimization with 206 property crimes per 1000 households. The lowest property crime rate is attributed to the northeast with approximately 104 property crimes per 1000 households.

Type of Location

For crimes of violence (excluding murder) in 2005, approximately 15% of victimizations occurred at or in the victim's home. This location is second only to the percentage of violent crimes committed on a street not near the victim's home (18.6% of violent incidents). When crimes committed by strangers are compared with crimes committed by nonstrangers, it is not surprising that crimes of violence more often occurred at home when perpetrated by a nonstranger. For the crimes of rape and sexual assault, 36% of incidents occurred at or in the victim's residence, and 24% of incidents occurred at, in, or near a relative or neighbor's home. Robbery and assault by armed offenders most often transpired on a street not near the victim's home.

When examining what victims were doing when victimized, for crimes of violence in 2005, the largest categories included victims involved in leisure activities away from home (22.3%) and victims involved in other activities in the home (21.5%). For property crimes, the largest percentage of victims reported sleeping while being victimized (27%).

The number of units in a housing structure is also an important variable in understanding victimization rates. It appears that for all property crime, households occupying housing structures with two units hold the highest rate of victimization (195 per 1000 households of the same). NCVS data indicates that housing structures with three units held the lowest risk for property crime (BJS, 2005).

Schools

In 2005 12.3% of violent victimizations occurred inside a school building or on school property. During the period of July 1, 2004, through June 30, 2005, 28 youth from the ages of 5 through 18 years were victims of school-associated deaths (seven of these were suicides; Dinkes, et al., 2006). "In 2004, students ages 12–18 were more likely to be victims of theft at school than away from school" (p. iv). Victimization rates at our nation's schools were lower in 2004 (55 victimizations per 1000 students) than in 2003 (73 victimizations per 1000 students). Victimization rates for students away from school also declined between 2003 (32 victimizations per 1000 persons aged 12–18 years) and 2004 (21 victimizations per 1000 persons aged 12–18 years). However, "the percentage of public schools experiencing one or more violent incidents increased between the 1999–2000 and 2003–04 school years"(p. vi). Male students were more likely than female students to report being threatened or injured with a weapon on school property in the past year. In 2005 10% of Hispanic students reported being threatened with a weapon on school property compared with 7% of white students. School enrollment size was an important variable in understanding the frequency of discipline problems.

In 2005 approximately 24% of urban school students aged 12 to 18 years indicated there were gangs in their schools. This is up from

21% in 2003. Students enrolled in suburban and rural schools were less likely to report the presence of gangs at their schools.

In 2005 approximately 19% of students in the 9th through 12th grades reported carrying a weapon anywhere, while approximately 6% reported that during the previous 30 days they had carried a weapon on school property. In 2005 black and Hispanic students reported being more fearful of their safety than white students.

Own or Rent?

People who rent their homes are far more likely to be victims of property crime than people who own their homes. In 2005 the rates of property crime victimization for renters was 192 per 1000 households. The property crime rate for people who own their home was approximately 136 per 1000 households (Catalano, 2006a).

Residential Mobility

Residential mobility is also important in understanding victimization risks. In 2005, for crimes of violence, people aged 12 years and older who reported living in their homes for a period of less than 6 months held the highest rate of victimization at 51.3 per 1000 persons of the same group. This risk declines as the time spent living in a home increases, while people living in their home for 5 or more years held the lowest risk at 14.6 crimes of violence per 1000 people. This trend also holds true for property crimes. People living in their homes for less than 6 months suffer a property victimization rate of 297 per 1000, while people living in their homes for 5 or more years had a significantly reduced rate of property victimization of 137 per 1000 (BJS, 2005).

Urban Versus Rural: Is There Really a Difference?

The short answer to this question is yes. In general, people living in urban areas are at greater risk of victimization than people living in suburban and rural areas, with rural inhabitants typically having the lowest risk. However, this pattern is not constant across regions. For example, in the Northeast, persons living in rural areas have only a slightly lower risk of being assaulted (18.9 per 1000) than those living in urban areas (20.8 per 1000), while suburban residents have the lowest assault rate of 13 per 1000 persons. In the Midwest rural residents have a higher risk of being victimized of a violent crime (21.5 per 1000 compared with 17.6 per 1000), and assault victimizations in the Midwest are also higher in rural areas than they are in suburban areas. In the South, rural residents enjoy the lowest risk of violent crime in every category. In the West, rural residents have a higher risk of being a victim of a violent crime than either urban or suburban residents, most notably for aggravated and simple assault (BJS, 2005).

In examining the risk of property victimization, rural residents have slightly higher risks for household burglary than suburban residents. This in true in all regions except for the West, where rural residents have lower rates of household burglary than their urban and suburban counterparts (BJS, 2005).

■ Online Resources

- Reports and data on crime and victimization in the United States: http://www.ojp.usdoj.gov/bjs/
- Sourcebook of Criminal Justice Statistics: http://www.albany.edu/sourcebook/
- Uniform Crime Reports (FBI): http://www.fbi.gov/ucr/ucr.htm

■ Conclusion

So now that your head is spinning with crime statistics, you may be wondering: What does any of this matter to a crime mapper? The answer to your question is this: Knowing that some people and some places are more affected by crime than others is the first step when identifying those factors important in understanding when, where, how, and why crime is distributed the way it appears to be. Understanding the factors associated with criminal events allows us to make better predictions about where and when crime is most likely to occur. In addition, the ability to predict crime provides us with an opportunity to prevent crime—the ultimate goal of law enforcement.

Recall that the first section of this text discussed key theories that offer various explanations to both individual and environmental causes of crime. While reading this chapter and recalling these discussions of crime theory, you should begin to make the connections between *how* crime is distributed and *why* it is distributed in some manner. Now that you have a solid understanding of the foundations of crime mapping and analysis, including theoretical and empirical research in crime mapping, Section 3 will provide an introductory examination of the basic statistical and spatial analyses that crime analysts frequently perform.

■ Questions for Review

1. Based on the crime statistics discussed throughout this chapter, *who* is most likely to be a victim of a violent crime?
2. Who is *least* likely to be victim of a violent crime?
3. How do the crime theories discussed in the first section of this textbook explain some of the crime trends that appear in the UCR and NCVS statistics?

Chapter Exercises

Exercise 13: Practical Administrative Exercise 1—The Media

Data Needed for This Exercise
- _____ (Crime points)
- _____ (School districts)
- _____ (Schools)
- _____ (Streets)
- _____ (City polygon)

Lesson Objectives
- Use the GIS skills you have developed to complete an assignment for the media.

Task Description

You are a crime analyst working for the City of Glendale, and you have received a public records request from a local news agency that says the following:

Steven Conrad
Chief of Police
City of Glendale
6835 N. 57th Dr.
Glendale, AZ 85301

January 31, 2006

RE: Public Records Request

Dear Chief Conrad,

The Channel 12 news anchor office would like to ask the Glendale Police Department's help in addressing an issue of community concern. The PUSD Governing Board has recently requested that staff investigate the issue of closing *high school* campuses within the Peoria District. A committee, made up of citizens, staff, and district administrators, will be examining various issues relating to daily campus activities and student traffic around each of the district's seven high schools. The news staff has been asked to analyze crime data and information within the vicinity of each Peoria District high school for a news report we are doing on the issue of closed campuses. We would like to request data on all reported crimes during the calendar year 2004. Information that would be of great help would include the time of day of incident, type of incident, and address of violation. If this data cannot be released, then an analysis of the total crimes surrounding Peoria schools in your jurisdiction in the form of a GIS product would be acceptable.

I can be contacted at (623) 412-5274 should you have any questions. Thank you very much for your consideration of this application.

Sincerely,

Caroline L. Ruiz
Research and GIS Specialist
Channel 12 News
1645 N. 16th St.
Phoenix, AZ 85003

The police chief has decided that the individual crime data will not be released and tells you that you will need to complete the analysis that was requested in the form of maps, tables, or charts and respond to Ms. Ruiz in the next day or two with the results of your analysis.

Think about the process involved in answering this request before proceeding. You should likely have at least five data sets you have listed in the Data Needed for This Exercise section to complete this request. You are going to need a point layer that has crime data included within it and *not* calls for service. You are going to need some way of identifying where the Peoria District is located and which schools are in Glendale that you can analyze. You will have to decide how far away from the school is a reasonable distance to analyze because Ms. Ruiz indicates she wants crime data "around" Peoria District schools. How far out from each school should this be? Should it be a 3 mile, 0.25 mile, or 0.13 mile radius (the average distance a juvenile travels in Glendale to commit a crime)? This may require some experimentation when you know which schools we are talking about. How many maps do you need to create: one or more? What kind of supporting documentation should you create: a table and a chart from Excel or just a chart on the map layout from ArcGIS 9.2?

When you have thought about this process and the decisions you need to make, you then need to create a new ArcMap project and load the appropriate data from shapefiles or a personal geodatabase. Depending on what data sets and layers you have added to your project, you should have an ArcMap project that looks something like DVD Figure 13 1.

Remember that more than one layer of GIS data could be used to answer the question and that there may be more than one source in the geodatabase or the GIS layers folder for the data you need. You may wish to review what is contained in each layer you are thinking about using through ArcCatalog first. If the data contains all the information that you think is important and covers the areas you feel need to be covered, then you can drag and drop them from ArcCatalog directly into your map. Hopefully, you can see how important it is to *really know your data* before you begin an analysis or a map. It really pays to learn as much about what the source is for the data, when it was last updated (metadata), and how accurate it may be. If needed, ask questions about which layer is the most accurate from the other analysts who have used those layers before. What is the City of Peoria doing in their analysis? Test your analysis with both data sets, and see how the results compare.

If you decided to use the layers that I used (see DVD Figure 13–2), then you will find that the Area Schools layer and the School Points layers actually contain different data and cover different geographic areas.

The School Points layer was created by the Glendale Crime Analysis Unit on February 3, 2006 per the metadata information in ArcCatalog. It is also apparent that it only covers public schools within the City of Glendale city boundaries. The Area Schools layer covers Glendale and surrounding cities and includes charter schools and other locations like that as well as the public schools. Which one do you think you will use? _____. Why do you think you will use this layer over the other one? _____
_____
_____
_____.

I have decided to use the Area Schools layer for my analysis, and now I need to find out which schools are within the Peoria School District. Because the request only discusses "high schools," I need to show only the high schools on my map. I can accomplish this by just adding a marker symbol for high schools in the legend (under categories, unique values).I can also create a theme definition so that I only see high schools on the map. I am going to choose the slightly more complicated method of creating a theme definition. Another way to accomplish this would be to select all of the high schools and then create a new shapefile of just the high schools. This is recommended if you think you will revisit this analysis effort again later.

Here we set up a theme definition to show just the high schools:

1. Open the Layer Properties menu by double clicking the **Area Schools** theme.
2. Click on the **Definition Query** tab and enter the following formula in the window using the Query Builder button. Make sure to check the data and fields first to see if this is a logical query for

your data. In addition, be mindful of the single quotation marks and other characters you use so that they match the values in the layer you are using.

[SCH_TYPE] = 'HIGH SCHOOL'

3. Now you should see only the high schools on the map display when you have applied this definition query.
4. Now you need to turn on the High School Districts theme and identify which polygon boundary is the Peoria Unified School District. This can be accomplished in the three following ways:
   a. _____(Legend)
   b. _____(Query)
   c. _____(Identify)
5. The names of the two high schools that are inside Glendale and within the Peoria District are:
   a. _____
   b. _____
6. Based on where the high schools are located and your search radius, do you think your final analysis should have a disclaimer? This may depend on the area around the schools that we choose for our analysis parameters. Think about this. Why? _____.
7. Now we can select the crimes "around" each school, create one or two maps as needed for clarity, and use this data to create a chart and/or a table of the data that we will send to Ms. Ruiz.
8. There are at least three ways to select the total number of crimes around these two schools, but first we need to select just these two schools so that anything we do will only use these schools in the final result.
9. Select the two schools by using the Select tool (white arrow in standard toolbar). Hold the CTRL key down after selecting the first school, and then select the other. Now, right click on the **Area Schools** layer name and choose **Data** from the drop-down menu. Choose **Export Data** from the second drop-down menu, and then save your data as a shapefile in your student folder. Call the theme PeoriaHS.shp, and then load this layer back into your project. Turn off the area schools layer and show only the PeoriaHS theme.

10. I previously said that there are at least three different ways to select all crimes within the distance you have chosen. Two of the obvious ways are:
   a. _____(Select)
   b. _____(Draw)

11. Again, the way we are going to do it is the most difficult method. We are going to use the Buffer wizard to create buffers for us around both points. To learn where we can find the Buffer wizard and what it does, let's open the Help dialog by clicking on **Help** in the file menu and then typing in *Buffer Wizard* under the Search section. You should get a help reference that says a lot about different tools for doing buffering as shown on DVD Figure 13–3. We want to look up how to add the Buffer wizard to the Tools menu bar. Read this section of the help text, and if you have questions about how to accomplish this task, ask the instructor *after* reading the help text (see DVD Figure 13–3). When you have accomplished adding the Buffer wizard to the toolbar, you should see a new tool icon on the toolbar (see DVD Figure 13–4). You can use the Buffer wizard to add buffer points, lines, or polygons. (Remember this for future exercises.)

12. I used the Measure tool and found that these two schools are about 8500 feet apart from each other. Converted to miles, this would mean that they are _____ miles apart (8500 ÷ 5280 = ?).

13. I am going to choose a distance of 1320 feet or _____ miles as my buffer distance, but you can use whichever you think is the most appropriate. This is *your* analysis, not mine. You have to be ethically responsible for the decisions you make during your analysis process if your agency does not have specific policies that are this detailed. Most people are not going to question your analysis (maps are always true, remember?), but if they do, you want to make sure, through your own code of ethics, that the data is accurate, reliable, and properly portrays the information that it should.

14. I am going to pick the PeoriaHS theme as the layer I want to create buffers around, and I want to pick all features so that I get two buffers at the end. I will choose *feet* in the Buffer wizard pop-up

and set the distance to *1320 feet*. In the next dialog, I will set Dissolve Barriers Between to *No* and then create a new shapefile called Buffer of PeoriaHS and save it in the StudentSave folder. I will then add it to my project and turn it on. I want to change the outline to red with a thickness of 2 pt. and no fill color.

15. I will now use the Selection menu item to Select by Location all crimes within each of the buffer polygons that were created. Do I want to do this query for both polygons at the same time or do them separately? If I do them separately (by first selecting the buffer polygon for Cactus HS and use the Select by Graphics routine to select all the crimes within that buffer, then make a new shapefile of just those crimes and name it CactusHS.shp), I may be better off. I can then do the same for the other buffer polygon and have two sets of crime points that I can further analyze separately. It is smarter to do them separately rather than at the same time and make shapefiles of each set of crimes you can use later. You can also do them at the same time and later use a few selection steps and edit the attribute table to add the two school names to the crime point layer.

16. Our goal will determine which way we pull our data together. If you are going to bring this data into Excel, for example, it doesn't matter if the crime points are all in the same file or not, or if they're in two separate files, as long as the name of the school is associated with the right data (because I added a field in ArcMap and added the names). We can then make tables and charts as needed in Excel and finish this assignment after we have made our map pretty.

17. If we decide to use ArcMap and add a table to the layout and a graph, for example, we will need to summarize all of the crimes by one of the fields that describe the crime type (like UCR_Group in this data). Each school will need to be selected first, one at a time. Then you would use the summary tables to add them to a graph or a table in the layout, depending on what would make the most sense for your project.

18. When you have completed your summary table, you can right click the top of the table and add it to your layout. Notice you have very little control over the appearance of the table, however. Another option is to create the summary table in Excel and then copy and paste (special) a bitmap into your map. It is now a graphic and not an actual table, but it gives you more control over the appearance of the table on the layout.

19. When you have completed your map and analysis, please write a short letter to Ms. Ruiz explaining what findings you think are important. You can use the letter template for Glendale located in the C:\CIA\documents\word_templates folder. Be brief and to the point with your analysis results, print the letter and associated tables or charts, and share them with your classmates to get their opinions on what you have created. Critiques by your peers can be a very valuable tool for you to learn more about mapping and analysis that speaks to the issues.

References

Bureau of Justice Statistics. (2005). *Criminal victimization in the United States, 2005 statistical tables*. US Department of Justice, Office of Justice Programs, Bureau of Justice Statistics.

Catalano, S. (2006a). *Criminal victimization, 2005*. US Department of Justice, Office of Justice Programs, Bureau of Justice Statistics.

Catalano, S. (2006b). *Intimate partner violence in the United States*. US Department of Justice, Office of Justice Programs, Bureau of Justice Statistics.

Dinkes, R., Cataldi, E. F., Kena, G., Baum, K., & Snyder, T. (2006). *Indicators of school crime and safety: 2006* (NCES 2007-003/NCJ 214262). US Departments of Education and Justice Statistics. Washington, DC: US Government Printing Office.

Durose, M. R., Schmitt, E. L., & Langan, P. A. (2005). *Contacts between police and the public: Findings from the 2002 national survey*. US Department of Justice, Office of Justice Programs, Bureau of Justice Statistics.

Farrington, D. P. (1992). Explaining the beginning, progress, and ending of antisocial behavior from birth to adulthood. In J. McCord (Ed.), *Facts, frameworks, and forecasts: Advances in criminological theory* (Vol. 3, pp. 253–286). New Brunswick, NJ: Transaction.

Gottfredson, M. R., & Hirschi, T. (1990). *A general theory of crime*. Stanford, CA: Stanford University Press.

Harrell, E. (2005). *Bureau of Justice Statistics crime data brief: Violence by gang members, 1993–2003*. US Department of Justice, Office of Justice Programs.

Karberg, J. C., & James, D. J. (2005). *Bureau of Justice special report: Substance dependence, abuse, and treatment of jail inmates, 2002*. Retrieved February 2009, from http://www.ojp.usdoj.gov/bjs/abstract/sdatji02.htm

Klaus, P. (2007). *Crime and the nation's households, 2005.* US Department of Justice, Office of Justice Programs, Bureau of Justice Statistics.

Rand, M., & Catalano, S. (2007). *Criminal victimization, 2006.* US Department of Justice, Office of Justice Programs, Bureau of Justice Statistics.

Wood, P. B., Gove, W. R., & Cochran, J. K. (1994). Motivations for violent crime among incarcerated adults: A consideration of reinforcement processes. *Oklahoma Criminal Justice Research Consortium Journal.* Retrieved March 2008, from http://www.doc.state.ok.us/offenders/ocjrc/94/940650G.HTM

Statistics and Analyses

SECTION III

CHAPTER 8   A Brief Review of Statistics
CHAPTER 9   Distance Analysis
CHAPTER 10 Hot Spot Analysis

A Brief Review of Statistics

CHAPTER 8

▶ ▶ LEARNING OBJECTIVES

This chapter is designed to provide students with a basic understanding of data and how data is manipulated by crime mappers and analysts. Basic concepts of statistical analysis and their applications in crime mapping will also be discussed. After studying this chapter, you should be able to:

- Understand the difference between *qualitative* and *quantitative* data.
- Define and explain levels of measurement including *nominal*, *ordinal*, *interval*, and *ratio*.
- Understand the difference between *discrete* and *continuous* variables.
- Understand *descriptive* statistics, including typical measures of *central tendency* and *dispersion*.
- Understand *inferential* statistics, including typical tests of *significance* and measures of *association*.
- Understand what a *regression* model is and how it works.
- Understand the limitations of statistics and how their improper application can yield misleading results.
- Define and explain *classification* in crime mapping and be able to identify strengths and weaknesses of each method.

▶ ▶ KEY TERMS

Antecedent Variable
Bimodal
Causal Relationship
Choropleth Map
Coefficient
Contingent Variable
Continuous
Dependent Variable
Dichotomous
Discrete
Exhaustive
Frequency Distribution
Histogram
Independent Variable
Interquartile Range
Interval
Intervening Variable
Inverse Relationship
Linear Relationship
Mean
Mean Center
Measures of Association
Median
Mode
Multicollinearity
Mutually Exclusive
Nominal
Normal Curve
Operationalize
Ordinal
(continued)

■ Introduction

This chapter explains basic statistical terms and concepts that are necessary for the beginning crime analyst to understand when performing statistical analyses and making crime maps. Our intent here is not to overload the student with a comprehensive and detailed discussion of formulas and equations, but to provide enough information for the beginning analyst to understand the strengths and weaknesses of statistical procedures commonly used in crime analysis and crime mapping. This chapter also discusses various characteristics of data. These characteristics are important in understanding how maps and analyses can be interpreted.

283

KEY TERMS

Positive Relationship
Qualitative
Quantitative
Range
Ratio
Reliability
Skewed
Spatial Autocorrelation
Spurious Relationship
Standard Deviation
Standard Deviation Ellipses
Tests of Significance
Unimodal
Unit of Analysis
Validity
Variables
Variance

A Crash Course in Statistics

As with any scientific approach to finding solutions, we need to analyze data. Crime analysis is no exception. With these analyses comes the responsibility to effectively (and ethically) use statistics to derive adequate answers to questions posed by command officers and others that we can depend on to make decisions. Having a basic understanding of statistics and statistical methods enables us to analyze our data and avoid mistakes or misinterpretations of the data.

Types of Data

Before we can begin examining the how and why of data analysis, we must first examine the different types of data and how each can be manipulated. Essentially, there are two types of research: **qualitative** and **quantitative**. In the simplest terms, qualitative research typically yields narrative-oriented information (such as the categories "hospital" or "park"), while quantitative research generally produces number-oriented information (usually coined "data"). Of course, as with virtually everything in the social sciences, this concept is not as simple as it appears. For example, suppose an analyst or researcher wanted to know why a certain intersection within a city experienced a higher level of drive-by shootings than anywhere else in the city over the last year and why these types of shootings continue to increase. He or she may choose any number of research designs to study the problem, including a qualitative or quantitative approach or a hybrid of the two. One qualitative design might be to conduct interviews of residents and business owners who live and work in the immediate vicinity of the intersection. Narratives gathered from these interviews might suggest that the neighborhood has recently undergone a transition characterized by high levels of instability and diversity. (For example, a participant might state, "Since they changed the bus schedule, this neighborhood has really fallen apart.") Upon further investigation, the researcher might discover that a nearby factory where many of the residents were employed was recently closed, resulting in high rates of unemployment, which in turn may have caused many families to move out of the neighborhood. The subsequent cheap real estate attracted recent immigrants to move into the neighborhood. This resulted in a culture clash amongst old and new residents in an already strained neighborhood. This situation also increased the tension between rivaling gangs (for example, the battle over drug territories), which increased the violent incidents between them, including drive-by shootings.

Using the same hypothetical problem, let's suppose the same researcher employed a quantitative design to study the violent intersection. First, he would collect official crime statistics for the area immediately surrounding the intersection of interest. In addition, he would gather information from secondary sources about the neighborhood and its residents, including real estate values, employment rates, residential mobility rates, and income and education levels. This data may tell the researcher that there are high unemployment levels, low real estate values, and high transition rates within the neighborhood. Statistical analysis may then determine that these and other factors contributed significantly to the number of drive-by shootings. However, while statistical analysis provides some explanation of the key factors or **variables** (any trait that can change values from case to case; individual variables are what data sets are made of) that impact the amount of violence in a neighborhood, it cannot explain how and why the neighborhood was transformed to begin with and why it is continually declining. This information comes from the qualitative methods discussed earlier, including resident interviews.

As you can see from this hypothetical example, qualitative approaches are excellent in their ability to collect very rich information. This information is usually in the form of narratives or stories that can provide a great deal of insight to the researcher. In crime analysis and mapping, qualitative information is invaluable to understanding maps and statistical outputs. However, qualitative information is impossible to analyze unless it is converted to a numerical output, which can sometimes be difficult. For example, in mapping gang activity and territory, Kennedy, Braga, and Piehl (1998) collected information from police officers, probation officers, and street workers about gang membership, activity, and territory to create maps as part of a multi-agency collaborative effort geared toward reducing gun violence among Boston youth. The qualitative information that was gathered was then combined with Boston police department data on gun assaults, weapons offenses, drug offenses, armed robbery, youth homicide, and calls for service for "shots fired" (Kennedy, et al., 1998, p. 245) and mapped to illustrate gang rivalries, gang boundaries, and locations of violent incidents. The resulting analysis produced a better understanding of the links between gangs and crime in the city of Boston.

Qualitative and quantitative research designs have their own strengths and weaknesses. Research designs that incorporate elements of both yield the most comprehensive and accurate information and will have a better chance of providing the answers that researchers seek.

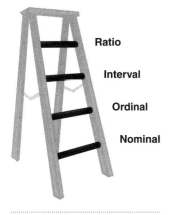

Figure 8–1 Levels of Measurement

Now that the basic differences between qualitative and quantitative research data have been discussed, the next step is to gain an understanding of the levels of measurement. Variables can be measured using four different levels of measurement. Visualizing the levels of measurement as a four-step ladder is helpful (see Figure 8–1).

The lowest step represents **nominal**-level data, which is often referred to as qualitative data. Nominal-level data is that which is measured in categories, or categorical data (in this text, categorical data is defined as data measured in word-based categories, such as "yes" or "no"), and cannot be ranked. Race, for example, is a nominal-level variable. Typical categories for race include African American, Caucasian, Hispanic, and Asian. Nominal variables are often collapsed into the smallest number of categories possible so that they may be used in different types of analyses. For example, the variable race on a survey or data collection instrument may be originally separated into the following categories:

- African American
- Native American
- Caucasian
- Hispanic
- Asian
- Other

These six categories can also be collapsed into two categories: Caucasian and non-Caucasian. A variable in this form (**dichotomous**) can be more versatile. Collapsing smaller categories into larger ones makes statistical analysis easier. Categories that hold too few cases are difficult to use in comparisons and thus need to be combined with other categories. An important thing to remember, however, is that collapsing categories into larger ones can hide important relationships between them. For example, the difference between African Americans and Native Americans on a given variable may be significant, but if they are both collapsed into the category of non-Caucasian, any differences between the two groups will be obscured. Similarly, if smaller areas (such as intersections or street segments) are collapsed into larger areas (such as census blocks or block groups), important crime patterns may be hidden. However, a dichotomous or *dummy* variable (one with only two categories as in the earlier race example) is a very powerful form, statisticswise, for a nominal variable to be in. Last, categories, both numeric and alphanumeric, must be **mutually exclusive** (a case should not be able to fit into more than one category) and completely **exhaustive** (every possible answer must be included in the answer set). This makes variables such as race very tricky to measure. It is easy to envision instances where persons could fit into more than one racial category. When categories are not mutually exclusive, this introduces an error to any analysis that is performed using these variables.

Examples of nominal-level variables in crime maps might be city parks or cemeteries. They cannot be ranked as variables themselves, and depending on what data is available about the park or cemetery (the facilities that are located there, different physical attributes), they may be able to be placed into separate categories. However, on the map, they are just shown with different colors that represent the various nominal variables available. Obviously they can be ranked on other variables, for example location, size, or criminal activity, but these are separate measures for other variables that can be ranked.

The second step in the ladder represents **ordinal**-level data, which contain categorical or numerical data that can be ranked, but the precise value is not known. The most often used example for categorical ordinal data are variables constructed using the Likert scale. The following survey question, designed as a partial measure of fear of crime, is a useful example.

I feel safe walking in my neighborhood alone at night.
1. Strongly agree   4. Disagree
2. Agree   5. Strongly disagree
3. Neutral   6. Don't know

Note that the answer sets can be ranked in terms of the level of agreement but that there is no meaningful or quantifiable difference between "strongly agree" and "agree." Another point that must be noted regarding any variable is that traits, concepts, and ideas in criminal justice can be difficult to **operationalize**, or measure. Using the previous example of fear, the statement, "I feel safe walking in my neighborhood alone at night" might be interpreted by respondents in several ways. Someone reading this statement may answer "strongly disagree," not because he is fearful of crime, but he is fearful of other things. The road that he lives on may have a lot of traffic, so he might be fearful of getting hit by a car, or his neighborhood may have a chronic problem with loose dogs, and he is fearful of getting attacked by a dog. Neither of these interpretations involves fear of crime, yet if he answers that he strongly agrees with the statement, he will be counted as a respondent who is fearful of crime.

When measuring a complex idea such as fear of crime, it is important to construct questions as carefully as possible and to ask multiple questions about the concept to achieve **validity** and **reliability** in the variable's operationalization. (Validity means that a variable accurately reflects the trait or concept it is measuring; reliability means that the measure is representative consistently across people, places, and time.) This is a very important point because available data for crime analysis and crime mapping does not always contain variables that are measured perfectly. When this is the case, maps can describe and illustrate, but analysis is limited in its interpretation or its ability to provide inferences.

A second example of an ordinal measure uses numerical data in ranges:

What is your annual household income?
1. Less than $20,000
2. Between $20,000 and $40,000
3. Between $40,001 and $60,000
4. Between $60,001 and $80,000
5. More than $80,000

Note that while the answer sets have clear numeric ranges and can be ranked in order of income level, the analyst does not know the exact amount of household income attributable to the respondent. Also note that the class intervals are equal (each range within the answer set is $20,000). Numeric ranges for ordinal data must either contain answer sets with equal interval widths or contain ranges that are logically or theoretically based. For example, suppose the analyst needed to create a map at the beat level showing reported burglaries for the past week for a large city. The map will be a **choropleth map** (uses color intensity to show the value of a given variable), and because this is a very large city with numerous police beats, the analyst cannot use the precise numbers for each beat or the map, while very colorful, will be difficult to read. Therefore, the analyst needs to group the number of burglaries into ordinal ranges. These ranges should be based on equal intervals (1–5, 6–10 … 20+) or on ranges based on some logic (perhaps in ranges based on natural breaks within the data).

The third step in the level of measurement ladder represents **interval**-level data. With interval-level data, the precise value of a measure is known and thus can also be ranked. For example, suppose the previous survey question regarding income was changed to:

What is your annual household income? _____.

The analyst would know exactly how much money in household income is attributed to the respondent and could rank it from lowest to highest. A mapping example might be to use the total number of burglaries at the beat level for a large city such as Baltimore. An analyst would not want to use the actual total number of burglaries if he or she were creating a map at the beat level of Baltimore because at that unit of analysis, the map would be very difficult to read.

The last step in the ladder represents **ratio**-level data. For most statistical analysis purposes, interval- and ratio-level data are treated the same (all ratio-level data is also interval-level data). The example of income used in the discussion of interval-level variables is also a ratio-level variable. Ratio level data has a true zero point and continues infinitely to the positive and to the negative. Therefore, some analyses that rely on the assumption of ratio-level data will provide results that do not exist in the real world—for example, predicting that 111.23 crimes will occur within a given area. In real life, this will never happen. Similarly, a prediction of a negative number of crimes cannot exist in the real world (we can reduce crimes by five, but we cannot have a negative number of crimes as a real value for an area).

Using the ladder analogy, the top level (ratio) assumes all the characteristics of lower levels in addition to its own unique characteristics. However, lower levels can only assume characteristics of the levels beneath them. For example, one can take an interval-level variable, such as income, and collapse it into a **frequency distribution** that displays the number of respondents with a range of household incomes (see **Figure 8–2**).

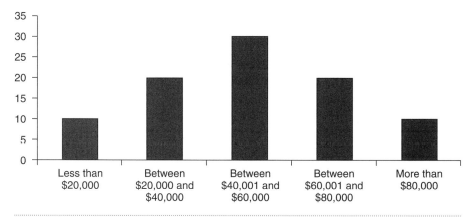

Figure 8–2 Annual Household Income Frequency Distribution, Example 1

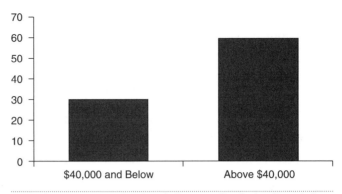

Figure 8–3 Annual Household Income Frequency Distribution, Example 2

This frequency distribution can also be collapsed further into a distribution that illustrates those above and below a given household income (see **Figure 8–3**).

The reverse, however, is not true. One cannot take information gleaned from an ordinal-level measure and turn it into an interval-level measure. For example, if on a survey, a respondent indicates that he or she commits between 10 and 20 crimes per month, we cannot assume that this person commits 12 crimes per month, only that the number is somewhere between 10 and 20 crimes per month. This is why it is best, when possible, to collect interval- or ratio-level data. This notion will become more and more apparent as you progress in your studies of crime analysis and crime mapping.

There are two more concepts that must be addressed before we move into a discussion of statistics: *discrete* and *continuous*. These terms are used in other chapters when we discuss the utility of various spatial analyses (specifically when performing hot spot analyses), so it is important to have a basic understanding of what they mean. **Discrete** variables are variables that cannot be subdivided. The number of persons living in a household is a discrete variable. For example, there cannot be 2.3 persons living in a household. There can be 2, or there can be 3, but not 2.3. Similarly, as discussed earlier, an area cannot have a total of 111.23 crimes. It is either 111 or 112 because crime is a discrete variable. Another example is a variable expressed in rate form. Often, when discrete variables are converted into rates, they will look like continuous variables. For example, 412 burglaries in a city of 300,000, converted to rate form, computes to be 13.73 burglaries per 10,000 persons. **Continuous** variables, on the other hand, *can* be subdivided—theoretically they can be subdivided an infinite number of times. Time is a good example of a continuous variable. Response times could be measured in minutes, seconds, milliseconds, nanoseconds, and so on. Usually, these variables are rounded to the unit that makes the most sense or is the most convenient; for example, there is little sense in comparing response times to the nanosecond (billionth of a second). Response times are usually rounded to the unit of minute or second. Knowing whether a variable is discrete or continuous is important in understanding which statistical analyses can be performed and how they can be interpreted.

Variables can be presented in a variety of ways. For example, as will be discussed in the next section, variables can be presented using measures of central tendency, such as the most frequent value, the middle value, or the average value in a distribution. Raw values are also often used but are usually standardized in some way to allow for meaningful comparisons. For example, when comparing the number of crimes across cities in eastern Massachusetts, it is necessary to compute a *rate* based on population so that the number of crime incidents for heavily populated cities can be compared to less populated cities in a meaningful way. As was discussed in the previous chapter, rates can be calculated for any number of persons but is typically done in crime analysis per 1000 persons, 10,000 persons, or 100,000 persons (usually some power of 10 to eliminate decimal points). The computation is simple: Divide the number of crimes in an area by its population, and multiply this value by whatever number of persons you would like the rate to be based upon. Recall from the previous chapter that crime rates as presented in the UCR were per 100,000 persons, whereas victimization rates presented by the NCVS were per 1000 persons.

Rate = $[(f_1) \div (f_2)] \times$ (some power of 10 to eliminate decimal points)

Ratios are often used to compare the frequency of values in one category to another. To compute a ratio, simply use the value of category one as the numerator and the value of category two as the denominator, and divide to compute the ratio. For example, if the frequency of property crimes is 300 and the frequency of violent crimes is 10, the resultant ratio is 300 ÷ 10, which equals 30. This is interpreted by stating that for every one violent crime, there are 30 property crimes.

$$\text{Ratio} = (f_1) \div (f_2)$$

Percentage changes are also useful for comparisons, typically from a given time period to another. The computation for computing a percentage change is not difficult; simply put, the analyst first takes the value at the later time (f_2) and subtracts the value at the earlier time (f_1) and divides this number by the value at the earlier time (f_1). This computed value is then multiplied by 100 to express the change in the form of a percentage.

$$\text{Percentage change} = [(f_2 - f_1) \div (f_1)] \times 100$$

Descriptive Statistics

Descriptive statistics are used primarily to summarize large amounts of information. The frequency example used earlier (in Figures 8–2 and 8–3) summarized the household incomes of 90 persons by grouping them into ranges. The most common descriptive statistics are measures of central tendency and measures of variation (or dispersion). The most often used measures of central tendency include the *mode*, *median*, and *mean*.

The **mode** (Mo) is simply the most frequent score that appears in a distribution. Using the earlier example (Figure 8–2) the modal category for household income is "Between $40,001 and $60,000." That is, the highest number of people (30) fall into this category. Another example using age might be as follows: There are 10 members in the fictitious Reservoir Street gang. Bob, Bill, and Jason are 17 years of age; Ted and Tim are 16 years of age; Kevin and Tyler are 14 years of age; and Steve, George, and Ed are 15 years of age. This data ($n = 10$) has a **bimodal** distribution. That is, there are two modes: 17 and 15 (three persons are 17 years of age, and three persons are 15 years of age). Another example of a bimodal distribution is a sample of a robbery series where an offender hits most frequently between 1700 and 1800 hours and also between 0700 and 0800 hours (each the same number of times) in a series of hits.

The mode can be used for variables measures at all levels of measurement, but it is the only measure of central tendency that can be used for nominal and categorical data. That is, the mode requires only frequencies (how many cases are in each category) and thus can be used with variables that are measured categorically or in word-based categories (such as race or gender). For example, in mapping places that sell alcohol in a large city, there may be 100 restaurants that sell alcohol, 80 bars, 15 grocery stores that sell beer and wine, and 10 liquor stores, producing a total of 205 establishments that sell alcohol, with restaurant being the modal category. Other measures of central tendency require at least midpoints (median) or numerical scores (mean) in their computations.

The **median** (Md) is simply the middle score within a distribution. That said, the variable must be represented numerically and thus be able to be ranked from lowest to highest. (We will not go into the mathematics of it in this book, but there are formulas available that use midpoints to compute the median for ordinal data that is numeric. When the data available for crime analysis is measured at the ordinal level, it is necessary to compute an approximate median that can be used for further analysis.) The median cannot be used on nominal or categorical data. In odd-numbered distributions, the median is always the middle score. In even-numbered distributions, there are two middle cases, so the median is defined as the score exactly halfway between the two middle cases. Using the age example again, and ranking the ages from lowest to highest, the median for this distribution is 15.5.

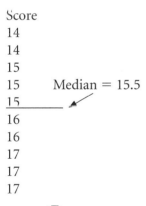

The **mean** ($\bar{X}$), or average, represents the arithmetic average of a distribution. It is easy to compute. One must simply add up all the scores in the distribution and then divide this sum (Σ) by the number of scores in a distribution. (Using midpoints, there are formulas that will compute an estimated mean for numeric, ordinal-level data, but it is beyond the scope of this textbook. Most good textbooks on statistics will contain these formulas. Again, if the only data available is measured at the ordinal level, it may be necessary to estimate the mean so that further statistical analysis can be conducted.) Using the age example again, the average is computed to be 15.6. The mean is the most powerful measure of central tendency and thus the most often used. However, because it is computed using every score within a distribution, it is sensitive to extreme values and should not be used for distributions that are highly **skewed**.

```
Scores
14
14
15
15
15      n = 10
16
16      X̄ = Σxi ÷ n
17
17      X̄ = 156 ÷ 10 = 15.6
17
Σ = 156
```

An example of a skewed distribution might be best illustrated using the concept of response times for a police department. The computed average response time for a department may be higher than the median response time because extreme scores skew the distribution. The median is a more useful measure of central tendency to use in skewed distributions. This is an important point because most of the data that crime analysts use in their analyses and maps is skewed. This concept will be explained later in this chapter, but essentially a distribution is skewed if the majority of cases are in either the low end or high end of the distribution. (However, despite this, many police commanders are accustomed to hearing the average of crime variables and may not readily accept median values without some explanation.) For example, when analyzing calls for service data over 1 year, the majority of addresses will only be represented once within the data set. That is, the overwhelming number of people who call police within a 1-year period will only call once. However, a select few of the addresses will show multiple calls to the police. This distribution will be skewed, and thus the median and mode number of calls for service should be used instead of the mean to describe the data.

A **mean center** analysis is one of the most common descriptive analyses used in crime mapping. The procedure is similar to computing a mean for a series of scores. The mean center is the approximate center of a series of events (typically crime incidents). It is primarily used to compare the center points of different types of crimes or similar crimes that occurred at different times. As with the mean of a distribution, the mean center is sensitive to outliers. In addition, if a distribution is oddly shaped or has more than one mode (for example, has two clusters within a distribution), the mean center (while in the center of the distribution) is deceiving.

A useful hint is to compare the mode, median, and mean of a distribution to get a better understanding of a variable. If the mode, median, and mean are fairly close, the distribution is not skewed, and reporting the mean on its own is appropriate. If the measures are vastly different (indicating a skewed distribution), it might be more appropriate to report just the median. TABLE 8-1 provides some helpful reminders of when it is appropriate to use the mode, median, and mean.

The next subject to tackle in this crash course on statistics is measures of dispersion (also called measures of variability). Measures of central tendency report the most common, the middle, and the typical (or average) scores. Measures of dispersion take us further in our understanding by reporting the shape of the distribution. There are several simple and familiar statistics that report dispersion. The most common include the **range** (the distance between the lowest and highest score), **interquartile range** (the distance between the 25th and 75th percentile), **variance** (the average squared distance of each score in a

TABLE 8-1 When to Use the Mode, Median, and Mean

Measure	When to Use
Mode	• With nominal-level variables.
	• You want to report the most frequent, or common, score.
Median	• Variables are numeric and measured at the ordinal level or higher.
	• You have interval- or ratio-level variables with highly skewed distributions.
	• You want to report the central score.
Mean	• You have variables measured at the interval or ratio level (unless highly skewed).
	• You want to report the average score.

distribution from the mean of the distribution), and **standard deviation** (the average distance of each score from the mean). Reporting the range provides the audience with an understanding of the lowest and highest point, providing perspective to where an individual score may lie in a distribution. For example, if a neighborhood experienced 15 burglaries in a week, knowing the lowest number of burglaries any neighborhood experienced that week and the highest number of burglaries any neighborhood experienced that week provides more meaning to the number 15 burglaries. Using the interquartile range provides the audience with an even better understanding of where an individual score lies in relation to other scores in a distribution. For example, knowing the interquartile range allows for understanding whether or not a neighborhood is in the top 25% or 50%, or the lower 25% or 50%, for burglaries. There are several other measures of dispersion that are used less often, but for our purposes we will focus our efforts on the most important measure of dispersion to crime analysis and mapping: the standard deviation.

The standard deviation requires that a mean be computed and thus is only appropriate for interval-level variables. (Again, there are formulas that allow you to compute a standard deviation for ordinal-level numeric data, but these computations, found in most good statistics books, are beyond the scope of this textbook.) The computation for standard deviation is simple enough, and the information it provides is invaluable to crime analysts. As defined, the standard deviation is the square root of the squares of the deviation. It gives us the average distance of each score from the mean, which tells us the shape of the distribution and whether or not a distribution is tight or disperse. (That is, how far away from the mean any given score is likely to be.) The problem is that if the mean is subtracted from each score in the distribution, the sum of these values will be zero. This is because the mean represents the balancing point of the distribution. Thus, dividing the sum of these values by the number of scores in the distribution (to find the average deviation from the mean) will also be zero. Therefore, one must square each of the distances (the individual scores minus the mean) and *then* add them up and divide them by our n. This will provide the variance (S^2) of the distribution. To compute the standard deviation (S), one simply needs to take the square root of the variance. (When using a random sample instead of the whole population, the formula uses $n - 1$ instead of n to account for error when using smaller samples.) The following is how to compute the variance and standard deviation for the age example we have been using:

Scores (x_i)	$x_i - \bar{X}$	$(x_i - \bar{X})^2$	
14	−1.6	2.56	$S^2 = \Sigma(x_i - \bar{X})2 \div n - 1$
14	−1.6	2.56	$S^2 = 12.4 \div 9$
15	−0.6	0.36	$S^2 = 1.38$
15	−0.6	0.36	
15	−0.6	0.36	$S = \sqrt{1.38}$
16	0.4	0.16	$S = 1.17$
16	0.4	0.16	
17	1.4	1.96	
17	1.4	1.96	
17	1.4	1.96	
$\Sigma = 156$	$\Sigma = 0$	$\Sigma = 12.4$	

$n = 10$
$\bar{X} = 15.6$

The average age of this sample is 15.6 years. The standard deviation of this sample indicates that for the variable age, the standard deviation from the mean is 1.17 years. Later in this chapter, distributions that are based on the bell-shaped, or normal, curve will be discussed. Understanding the meaning of a standard deviation is vital to understanding *normal* distributions.

In crime mapping, the standard deviation is used in several ways to illustrate levels of crime dispersion. For example, an analyst may compute and compare the standard deviation distances for crimes within a large city. The larger the standard deviation distances, the more dispersed the crimes are. **Standard deviation ellipses** (oval and circular shaped) are used to illustrate the size and shape of the distribution. The smaller the ellipse, the closer together crime events are located. A tall and skinny ellipse indicates a distribution that is more tightly clustered from west to east and more dispersed from north to south.

Inferential Statistics

The goal of inferential statistics is to study a sample (again, randomly selected) to learn about a larger population. Often in crime analysis, the populations or areas we are interested in studying are much too large to properly survey. For example, if a researcher wanted to know the thought process of burglars in choosing their intended targets and needed to interview all burglars, he would have a difficult time to say the least. Aside from the time and resources involved in such a study (the US Department of Justice estimates that approximately 298,835 arrests for burglary were made nationwide in 2005), not all burglars get caught and thus are not known (the US Department of Justice estimates that 2,154,126 burglaries were reported in 2005), and it is plausible to assume that not all burglars would agree to be involved in any research study. That said, it is important to be able to *infer* that what is known about a sample adequately represents the larger population from which it was drawn. To do this, we must rely on the rules of probability.

Probability theory is the foundation of inferential statistics. Probability is defined as the number of times any given outcome will occur if the event is repeated many times. For example, if a random coin was flipped 100 times, and there was nothing wrong with the coin (for example, it was not two headed or weighted on one side), one would expect to see the coin land on heads about 0.5 or 50% of the time. Of course, rarely will the result be exactly 50 heads and 50 tails, but it is usually close, 48 to 52 for example. What is interesting about random samples is that they usually reflect what is called the normal curve or bell-shaped distribution. This means that if you perform the 100-flip coin toss, most of the results should be very close to what you would find if you performed the coin toss 1000 times, 10,000 times, or more. That is, the results should be close to a 50–50 distribution of heads and tails.

The **normal curve** is a concept that is of great importance to statistics. Not only is it key to understanding inferential statistics, but it is a powerful tool that allows us to form precise statements about empirical distributions and to infer sample statistics to a larger population. The normal curve is a theoretical model that is **unimodal** (it has only one mode) and is flawlessly smooth and symmetrical. It is bell shaped, has tails that continue infinitely in both directions, and the value of its peak is the mode, median, and mean for the distribution. The most important point about the normal curve is that the distances along the horizontal axis, when measured in standard deviations, always contain the same proportion of the total area under the curve. That said, if the mean and standard deviation of a distribution are known, they can be used to describe the distribution more precisely, and analysts gain the ability to determine where any given score lies within a distribution. **Figure 8–4** illustrates the normal curve.

Notice that the mean, median, and mode are all equal. And, because the median is the exact middle of a distribution, the median splits the distribution exactly in half. Also notice that the numbers along the axis represent standard deviations from the mean. Note that the mean is represented by a zero. This is because the mean itself

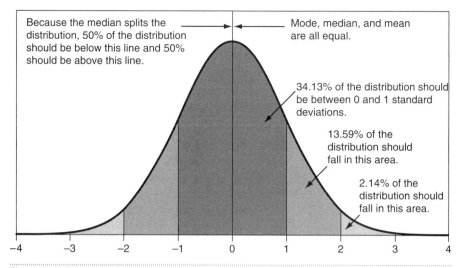

Figure 8–4 The Normal Curve, Example 1

cannot deviate from the mean. Thus it is 0 standard deviations away from the mean.

The standard deviations along the axis also correspond to a proportionate area under the total curve. Note that the area between 0 and 1 standard deviations encompasses 34.13% of the area under the normal curve. This is also true for the area between 0 and –1 because the normal curve is symmetrical. Essentially, what this means is that if an empirical distribution is normally distributed, we would expect to find 68.26% of the scores to be within one standard deviation above and below the mean. Along the same lines, we would expect 99.72% of the distribution to fall within three standard deviations above and below the mean. What this means is that given the assumption of normality, it is highly unlikely that we would find very many scores that are farther away than three standard deviations from the mean. **Figure 8–5** illustrates the normal curve for a distribution that has a mean of 100 and a standard deviation of 10.

Note that the range of scores within three standard deviations below and above the mean correspond to a range of values between 70 and 130. Thus, 99.72% of the distribution should fall between 70 and 130, with 95.44% of the distribution falling between 80 and 120, and with 68.26% of the scores falling between 90 and 110. We would only expect a score above 130 or below 70 less than 0.28% of the time.

In the world of criminals and their crimes, what this means is that analysts do not have to sample all criminals and all crimes to

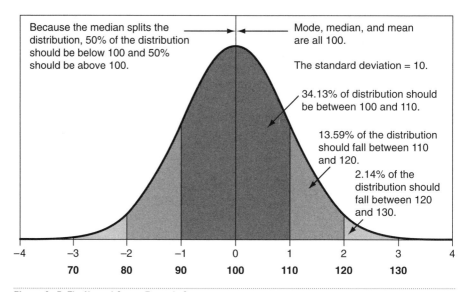

Figure 8–5 The Normal Curve, Example 2

get a better understanding of criminal behavior. They can randomly sample a proportion of them and be able to *infer* that what they learn about the sample is also true for the entire population. Essentially, if analysts correctly selected a random sample, 99.72% of the time the sample average on any given variable should be within three standard deviations of the true population (the entire population the sample was selected to represent).

Suppose analysts wanted to know the average number of arrests for inmates in a given state. (This would typically be a known parameter of this population, so we could use it to perform a check on our sample to see if it is representative of the population of all inmates in the state.) If the average number of arrests for this population is five with a standard deviation of one, and analysts randomly sampled 1000 inmates, they should find that approximately 683 (68.34%) of the inmates have between four and six arrests (one standard deviation from the mean). Approximately 954 (95.44%) should have been arrested between three and seven times (two standard deviations from the mean). Finally, approximately 997 (99.72%) should have between two and eight arrests (three standard deviations from the mean). Three or fewer persons should fall into the category of less than two arrests or more than eight arrests. If the parameters of this variable are applied to the entire population of inmates, the distribution should be very similar. That is, less than 1% (0.28% to be more precise) of the inmates should have less than two arrests or more than eight arrests.

The shape of the normal curve can be tall, short, wide or narrow, as long as it retains all of the characteristics of the normal curve (unimodal, symmetrical, and smooth). A disperse distribution will be short and wide, while a tight distribution will be tall and narrow.

Recall that this is a theoretical model with infinite tails. Thus, theoretically, it is possible for a score to exist anywhere. However, in the practical world it is highly unlikely, and, in the real world, it may even be impossible. This is a very important point for crime analysts to understand. For example, the average number of arrests for a given population may be 1.67 (which we know is impossible for an individual record) with a standard deviation of 0.85. If this is taken to three standard deviations, 99% of the population should fall between −0.88 arrests and 4.22 arrests. Obviously, no one caught up in any state correctional system has less than one arrest, and it is also impossible to be arrested a negative number of times. The variable "arrest" is not a continuous variable, and it cannot, in practical terms, carry on past zero on a continuum. In addition, the distribution of arrests across this population is most likely skewed, so applying the normal curve to this particular variable is problematic. It is important for analysts to know how the variables they are using are distributed. Therefore it is recommended that analysts make frequency tables or create **histogram** charts to help them understand how the data is distributed. A histogram is a specific type of bar chart used to display frequencies of class intervals or scores of a given variable. They use real limits rather than stated limits, so they are most appropriate for continuous, interval-level data. However, because this data rarely exists in crime analysis and mapping, histograms are commonly used for discrete, interval-level variables. **Figure 8–6** is an example of a histogram displaying state totals of violent crime in 2005.

The totals were grouped into numeric ranges of 10,000. Notice the modal category is "below 10,000" with a frequency of 19 states in this category. Also, notice the shape of the distribution. It looks nothing like a normal or bell-shaped curve. Analysis on this positively-skewed variable (the mean is greater than the median) is limited in its interpretation.

With criminal justice data, it is often difficult to make the assumption of normality because empirical distributions are often positively or negatively skewed. In their study of offender and victim movement in homicide cases, Groff and McEwen (2005) report that for homicides, both victims and offenders tend to stay close to home, creating a skewed distribution. A distribution can be *positively* skewed (where the mean is greater than the median) or *negatively* skewed (where the

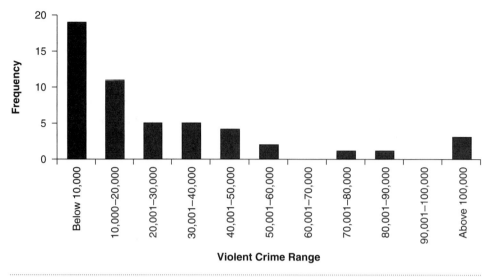

Figure 8–6 Histogram of State Violent Crime Totals
Source: Data from Bureau of Justice Statistics, http://www.ojp.us.doj.gov/bjs

mean is lower than the median). When dealing with skewed data, the assumption of normality cannot be made, which limits the interpretability of results produced from analyses that require this assumption. The data can still be described, but assumptions or inferences cannot be made directly from skewed data.

Many of the models produced by statistical analysis are based on assumptions of a *normal* distribution of a *continuous* variable. Thus, the results that are produced cannot always be strictly interpreted. In spatial analysis, crime (measured in raw counts) is a discrete variable that has a minimum value of zero. Statistical models based on the assumptions of normality and continuity will sometimes produce predictions of negative crime values. These results cannot be directly interpreted to mean that a negative number of crimes will occur in a given area, but that it is highly unlikely for any crime to occur in the given area. The limits of interpretation must be considered when crime maps and analyses are disseminated to others. If not, the results can be confusing and, worse, misleading.

There are far too many statistical tests and analyses that crime analysts have in their toolbox for us to provide a comprehensive discussion of each. As stated earlier in the chapter, it is beyond the scope of this book to provide those detailed discussions. Thus, the definitions and discussions of statistics is very much simplified. Our intention is to provide the beginning student with enough background in statistical operations to be able to understand what different tests in crime mapping do and how they can be interpreted. This is very important. Using the wrong test with the right data and vice versa can yield inaccurate and misleading results. Crime analysts must pay attention to the attributes of their data to avoid running the wrong tests.

The selection of the type of statistical analyses to run depends on some of the concepts and ideas we have already discussed in this chapter. Answers to the following list of questions can help the analyst choose which analyses to perform:

- What variables are available?
- What is the overall n?
- What is the unit of analysis?
- What do I want to know about the variable(s)?
- What is the level of measurement of the variable(s)?
- Are the variables discrete or continuous?
- How many groups will be compared in the analysis?
- Am I interested in just describing the data or finding inferences within it?

Throughout this book, the answers to these questions will be provided in the corresponding discussions for each statistical analysis. However, before discussing statistical tests, we need to examine the nature of relationships between independent and dependent variables.

The **dependent variable** (Y) is the variable that analysts are trying to explain (in crime mapping, the dependent variable is often some crime measure). **Independent variables** (X) are variables that produce a change in our dependent variable. For example, Catalano, Hill, and Long (2001), in an investigative analysis of a robbery series, created prediction models of where future robberies had the highest probability of occurring and where the offenders were likely to live and work in Phoenix, Arizona. In these models, freeway access and street density (independent variables) were used to predict which grocery stores were most likely to be targeted by the robbery gang (dependent variable).

Relationships can be modeled in many ways. The majority of mathematical formulas that are used in crime analysis require the assumption that X causes Y (**causal relationship**). However, in the real world, X and Y could be related in many ways. A third variable

(Z) might even be included as an **intervening variable** (X → Z → Y), or it could be factored in as an **antecedent variable** (Z → X → Y) or a **contingent variable** (X → Y → Z) in the relationship between X and Y. This is important because misspecified models in crime analysis can yield incorrect outcomes, which may result in inappropriate solutions designed to reduce crime problems. Problems of **multicollinearity** (when X, Y, and Z have overlapping measures of the same concept) and **spurious relationships** (when X and Y have no direct relationship but are both affected by Z) can also pose multiple problems for crime analysts. An example of multicollinearity would be median household income and median home value in a neighborhood. The value of a home is in part a function of how much money people make. Thus, these two variables overlap in what the analyst is trying to measure. When multicollinearity exists, special care must be taken in analysis and interpretation. Models that suffer from multicollinearity can produce false significance levels.

An example of a spurious relationship is poverty and crime. Spatial analyses consistently and frequently identify a strong link between areas with high levels of poverty and corresponding high crime rates. However, it may be the lack of opportunities that produce both. In this case, programs geared toward poverty instead of the lack of opportunities will probably yield minimal, at best, reductions in crime because deficiencies in opportunities produce both crime and poverty.

Multicollinearity   Spurious

Model specification of the relationship between variables is extremely important to crime analysis. Many common **tests of significance** depend on the assumption of a causal relationship between X and Y. Tests of significance, including *chi-square*, *z-tests* or *t-tests*, and *ANOVA*, should be common in data analysis. Essentially, they work by determining whether or not variable distributions or differences between groups or areas would be expected based on random chance. For example, if crime was distributed by random chance, we would not expect to see clusters or *hot spots* of crime. In crime mapping, tests of significance tell us if the distribution of crime across any given area or the difference between two or more areas is statistically significant. Analysts can also perform tests of significance to look at crime distributions across time. If crime was distributed randomly across seasons, we would not expect significant variations in crime from month to month. Tests of significance tell us how likely the distribution is to occur or how likely differences between groups are to occur by random chance. In crime analysis, if the significance level (often termed the "p" value) is 0.05 or smaller, analysts determine the results to be statistically significant. A significance level of 0.05 tells us that the probability of getting these results (a variable's distribution or differences between groups or areas) by random chance is expected only 5% of the time. A significance level of 0.01 tells us that we would expect the same results by random chance only 1% of the time.

Tests of significance cannot tell us how strongly related two variables are or in which direction they are related. To determine the strength and direction of a relationship between two variables, **measures of association** must be performed. Some commonly used measures of association in the analysis of criminal justice data (either in academics or in law enforcement) include *lambda*, *gamma*, *Kendall's tau statistics*, *Spearman's rho*, and *Pearson's correlation coefficient*. Each of these tests make certain assumptions about the variables used in their computations, including the level of measurement, so it is important when selecting which measure of association to perform that those important assumptions not be violated. Values for measures of association typically range from 0 to 1 (for tests that cannot determine the direction of a relationship) and −1 to 1 (for tests that can determine the direction of a relationship). Results close to zero indicate a weak relationship, and results closer to one indicate a strong relationship. Negative values designate an **inverse relationship** (as X increases, Y decreases; or as X decreases, Y increases), and positive values indicate a **positive relationship** (X and Y both increase or decrease together).

In mapping, there are several spatial correlation analyses. The two primary tests for **spatial autocorrelation** (using aggregate data) are *Moran's I* and *Geary's C*. These measures will be discussed in more detail later in this text. For the purposes of this chapter, these measures are interpreted similar to measures of association. In interpreting Moran's I, a value between 0 and +1 indicates positive spatial autocorrelation (or clustering). A value between −1 and 0 indicates negative spatial autocorrelation (random distribution). For Geary's C, values under 1 signify positive spatial autocorrelation, and values over 1 designate negative spatial autocorrelation.

In addition to running tests of significance and measures of association, an analyst may want to run a *regression* model. Regression analysis can be computed by employing one or multiple independent variables to determine their impact on a chosen dependent variable. The most common regression model, *ordinary least-squares (OLS)*, relies on multiple assumptions, including a **linear relationship** between X and Y; continuous, and thus interval-level or ratio-level variables; and that the **units of analysis** are the same. When using discrete variables or variables that are not measured at the interval or ratio level, direct interpretation of the results is difficult. When using variables that are discrete or measured at the nominal or ordinal level, other types of regression analyses should be employed. In crime mapping, regression models that employ spatial measures must still adhere to model assumptions, or the interpretation of results may be limited.

Regression models allow analysts to include multiple independent variables in the same model to identify how they affect the dependent variable. Depending on the regression model, analysts can make precise predictions about the conditions that will increase or decrease their dependent variable (typically crime). Interpreting OLS regression is fairly easy. First, the equation is stated as follows:

$$Y = a + b_1X_1 + b_2X_2 + b_3X_3 \ldots$$

In this equation, Y represents the dependent variable, X_1 and X_2 represent the independent variables, a is the value of Y when both X_1 and X_2 are zero (Y intercept), and b_1 and b_2 are the **coefficients** of X_1 and X_2, respectively. A typical output is contained in Tables 8–2 through 8–5. **TABLE 8-2** displays the descriptive statistics for the variables included in the model. (Assault rate per 100,000 is the dependent variable; the percentage of births to teen moms, median family income, and the percentage of persons older than 25 years who have not completed high school are the independent variables.) Note that the *n* is equal to 50 for all variables. This is because the unit of analysis for this data set is states, and there are 50 states included in this data set.

TABLE 8–2 Descriptive Statistics

	Mean	Std. Deviation	N
Assaults per 100,000 population	274.480	136.947	50
% of all births to teen moms	11.852	2.943	50
Median family income	48957.580	7314.200	50
% of population older than 25 years who have not completed high school	15.922	4.377	50

TABLE 8-3 provides the correlations between all the variables. This allows us to check for multicollinearity for variables that could potentially overlap in their measurements. Note that the Pearson's correlation for the percentage of persons older than 25 years without a high school degree is −0.513 and is significant at the 0.01 level. (Note that ** indicates a significant level at the 0.01 level or better. A separate t-test is conducted on values of association to test for significance.) This tells us that these two variables have a significant and inverse relationship. That is, as the number of persons older than 25 years who have not completed high school increases, the assault rate decreases. It is known that income and education overlap in their measurements, so we must be careful in our interpretations of these results.

TABLE 8-4 provides the model summary. Analysts are interested in the R-square and adjusted R-square. Because there may be issues of multicollinearity that might artificially inflate the findings, analysts need to use the adjusted R-square for their interpretations. The R-square and adjusted R-square will range from 0 to 1 and are interpreted as the percentage of variance in the dependent variables that is explained by the independent variables in the model. In our model, we can state that approximately 35% of the assault rate can be explained by the percentage of births to teen moms, median family income, and the percentage

TABLE 8-3 Correlations

		Assaults per 100,000 Population	% of All Births to Teen Moms	Median Family Income	% of Population Older Than 25 Years Who Have Not Completed High School
Pearson's Correlation	Assaults per 100,000 Population	1.000	**0.486	−0.127	**0.431
	% of All Births to Teen Moms	**0.486	1.000	**−0.753	**0.668
	Median Family Income	−0.127	−0.753	1.000	**−0.513
	% of Population Older Than 25 Years Who Have Not Completed High School	**0.431	**0.668	**−0.513	1.000

** Significant at the 0.01 level

TABLE 8-4 Model Summary

Model	R	R-square	Adjusted R-square	Std. Error of the Estimate
1	0.625	0.391	0.351	110.303

Predictors: (Constant), % of population older than 25 years who have not completed high school, median family income, % of all births to teen moms

of persons older than 25 years who have not completed high school. Not bad, but it also means that other variables not included in our model explain the remaining 65% of variance in the assault rate.

TABLE 8-5 displays the coefficients for our regression model. The first column "B" underneath the category "unstandardized coefficients" represents the values for "a" (constant) and $b_1, b_2,$ and b_3. Note that the value for "a" or the y intercept is -762.081. This means that when the percentage of births to teen moms, median family income, and percentage of persons older than 25 years who have not completed high school are all zero, the assault rate should be -762.081 per 100,000. Obviously, this value does not exist in the real world. An assault rate of zero is the absolute low. However, what this should tell you is that without the effects of these three independent variables, the assault rate is very low.

We can also use these coefficient values to make more precise predictions. When substituting our values into the equation, we have the following:

Assault Rate $(Y) = -762.081 + 35.934(X_1) + 0.0143(X_2) + 6.273(X_3)$

TABLE 8-5 Coefficients

Model		Unstandardized Coefficients		Standardized Coefficients	t	Sig.
		B	Std. Error	Beta		
1	(Constant)	−762.081	243.335		−3.132	0.003
	% of All Births to Teen Moms	35.934	9.395	0.772	3.825	0.000
	Median Family Income	1.043E-02	0.003	0.557	3.184	0.003
	% of Population Older Than 25 Years Who Have Not Completed High School	6.273	4.840	0.200	1.296	0.201

Dependent variable: Assaults per 100,000 population

If we know the values for our independent variables, we can use them to predict what the assault rate will be. Assume that a state has 5% of all births delivered to teen moms, a median family income of $40,000, and 7% of the population is older than 25 years and has not completed high school. Substituting these numbers, our formula should look like this:

$$\text{Predicted Assault Rate (Y)} = -762.081 + 35.934(5) + 0.0143(40,000) + 6.273(7)$$
$$\text{Predicted Assault Rate (Y)} = -762.081 + 179.67 + 572 + 43.911$$
$$\text{Predicted Assault Rate (Y)} = 33.5$$

Based on the model, if a state has 5% of all births delivered to teen moms, a median family income of $40,000, and 7% of the population is older than 25 and has not completed high school, we would expect about 34 assaults per 100,000 persons. Obviously, the model is based on probabilities (as all statistics are) and thus it is not perfect. But, depending on how good the model is (determined by a variety of factors, but a high adjusted R-square usually indicates a strong model), our estimates can be fairly precise.

As you can see, statistical analysis can be quite powerful in crime analysis for describing and making predictions about crime. However, statistical analysis requires analysts to obey rules and follow guidelines. When they do not, their results are inaccurate, misleading, and deceptive. But what, you may ask, does this have to do with making crime maps? The answer is that it has everything to do with crime mapping. Spatial analyses employ the same mathematical computations as nonspatial data analysis and thus rely on the same assumptions and require the analyst to follow the same guidelines and obey the same rules. If not, the crime maps produced can also be inaccurate, misleading, and deceptive. This by no means should keep analysts from making maps and using the various tools and tricks of the trade to identify problem areas when the commander requests their assistance, but it is imperative that they recognize the possible weaknesses in the data, and if they attempt to infer something about that data, that they do so with the utmost sense of responsibility.

■ Classification in Mapping

There are several methods to classify information in crime maps using some of the statistical concepts and ideas discussed in this chapter. Harries (1999) provides a good description of five common classification methods in *Mapping Crime: Principle and Practice*. Equal *interval* classification calculates the range (distance between the lowest and highest score) and divides the range into equal ranges or intervals. Typically, a map should contain between four and six intervals (or classes) but no more than seven in any case (Harries, 1999). A choropleth map is shown in **Figure 8–7**. Choropleth maps use shading variations according to their value on a chosen variable; in this case, burglary rates at the town level in the state of Massachusetts are shown.

Maps classified into quantiles place equal numbers of incidents or observations in each category. This method is strongly influenced by the number of categories selected. If five categories are selected (the default in ArcView), areas will be separated into 0–20th percentile, 21st–40th percentile, 41st–60th percentile, 61st–80th percentile, and 81st–100th percentile (see **Figure 8–8**).

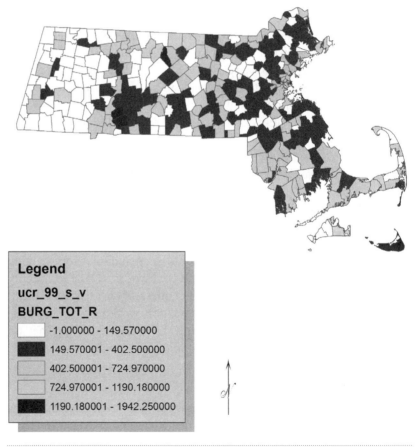

Figure 8–7 1999 Massachusetts Burglary Rate, Equal Interval Classification
Source: Map created from data files accessed at http://www.mass.gov/mgis/crime_statistics.htm

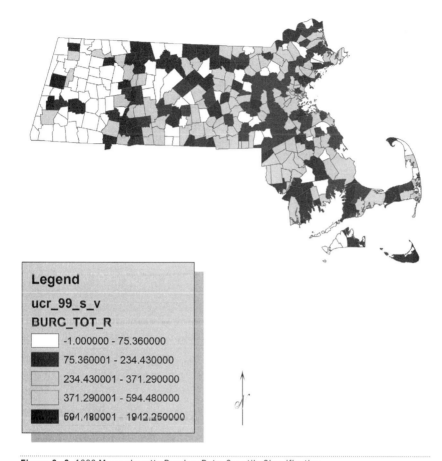

Figure 8–8 1999 Massachusetts Burglary Rate, Quantile Classification
Source: Map created from data files accessed at http://www.mass.gov/mgis/crime_statistics.htm

Equal area classification divides the distribution based on area rather than incidents or observations. When using the equal area classification, maps with areas that are roughly the same size will look similar to a quantile map. If the areas vary greatly in size, an equal area map will look substantially different from a quantile map.

Natural breaks classification method uses breaks or gaps in the data distribution to create categories. Typically, a procedure known as Jenks optimization (the default in ArcView) "ensures the internal homogeneity within classes while maintaining the heterogeneity among the classes" (Harries, 1999, p. 51; see **Figure 8–9**).

Standard deviation maps classify data based on the dispersion of data from the mean score. For example, if the average number of homicides for a city at the block group level is three, areas will be

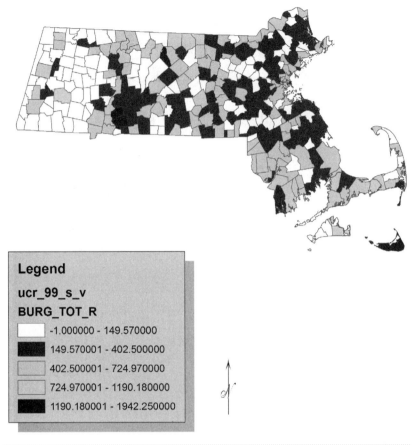

Figure 8–9 1999 Massachusetts Burglary Rate, Natural Breaks Classification
Source: Map created from data files accessed at http://www.mass.gov/mgis/crime_statistics.htm

categorized by how many standard deviations they fall above or below three homicides (see **Figure 8–10**).

As one can see, each of the four maps, using different classification methods, look very different and thus paint a different picture to the intended audience. The analyst must balance what classification method is best for the intended audience with rules of statistics. For example, standard deviation classification is a good tool for expressing extreme values. However, using this method assumes the variable is normally distributed. In this case, analysts cannot assume the burglary rate is normally distributed without looking at the frequency distribution first. This is easy enough to do in most spreadsheet or database programs (discussed in a little more detail in Chapter 6).

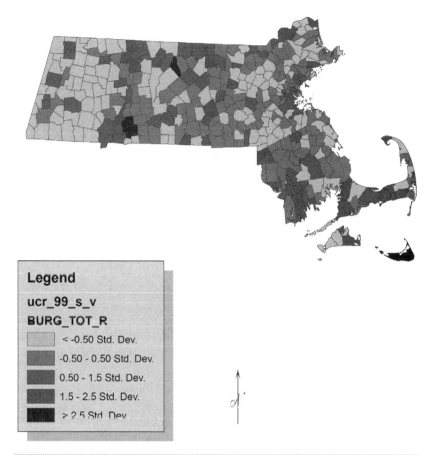

Figure 8–10 1999 Massachusetts Burglary Rate, Standard Deviation Classification
Source: Map created from data files accessed at http://www.mass.gov/mgis/crime_statistics.htm

■ **Online Resources**

- FedStats: http://www.fedstats.gov
- Dr. Tom O'Connor's comprehensive list and guide to modeling relationships: http://www.apsu.edu/oconnort/data.htm
- Modeling relationships between variables: http://www.apsu.edu/oconnort/crimtheo2.htm
- Statistics Every Writer Should Know: http://nilesonline.com/stats/

■ **Conclusion**

This chapter provided a brief discussion of statistics in crime analysis. Elias (1999), drawing on Huff's *How to Lie with Statistics* (1954), identifies six statistical "sins" for the analyst to avoid which forms a useful framework for summarizing the main points found in this chapter and integrating other relative points from previous chapters. First, analysts must acknowledge that even in randomly selected samples, statistical analysis, while computed without flaw, can yield inaccurate results. This can happen for a number of reasons. In self-report studies, people may report things that are untrue purposefully or accidentally. In survey research, potential respondents who decide not to participate in the study may be inherently different than persons who do decide to participate on key variables that matter to the study. Even if a sample is randomly drawn, there is still the possibility, although small, that it will not be representative of the larger population. In research that requires interviewing, characteristics of the researcher may influence respondents' answers to questions or their decision to participate in the research.

The second sin analysts engage in is reporting statistics that, while accurate, are deceptive. Reporting the mean, for example, with a skewed data distribution is misleading. Zedlewski (1987) used the arithmetic average in calculations to estimate the cost–benefit ratio for incarceration. He estimated that the typical offender committed 187 crimes per year, producing a "social cost" of $430,000 per offender, per year. DiIulio and Piehl (1991) in a similar study estimated that a typical felon only costs society $28,000 per year. Why are their conclusions so different? They likely used different measures of central tendency to estimate the number of crimes the typical offender commits and the average cost associated with these crimes. Because the distribution of both variables is likely to be skewed (recall our discussion in Chapter 6 that a small proportion of offenders cause a large percentage of the problem), using different measures of central tendency can yield vastly different conclusions.

The third statistical sin analysts commit is not presenting their audience with the whole picture. Focusing on only one month of crime, for example, without placing it in the larger picture of yearly crime can be misleading. Another problem is not including raw numbers along with percentages. Sometimes a large decrease in crime percentage may only amount to one or two crimes in raw numbers.

The fourth sin has to do with interpreting statistical significance. It is common in very large samples to achieve a finding of statistical significance for very small differences between groups or areas simply because there are a large number of cases or observations. It is important for analysts to run both tests of significance *and* measures of associations to determine whether or not their findings have *substantive* significance. For example, one group of offenders might self-report committing an average of 3.1 crimes per month, and the comparison group might report an average of 3.3 crimes per month. If the number of offenders included in the analysis is large enough, tests of significance may identify a statistically significant difference between the groups. However, from a practical standpoint, a difference of 0.2 crimes is insignificant to law enforcement and thus provides no meaningful difference.

The fifth sin relates to our earlier discussion of model specification. Most formulas for determining the nature of relationships assume a causal relationship between the independent and dependent relationship. Further analysis is needed to identify if a third factor, Z, is an

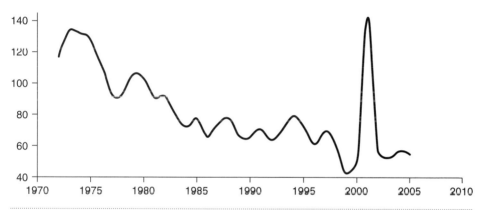

Figure 8–11 Officers Feloniously Killed, 1972–2005, Scale Example 1
Source: Chart created from data from Bureau of Justice Statistics, http://www.albany.edu/sourcebook/pdf/t31548006.pdf

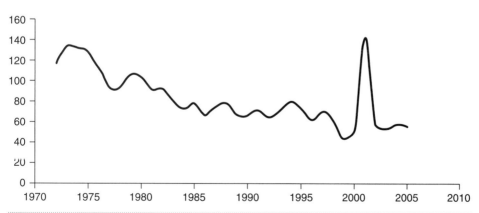

Figure 8–12 Officers Feloniously Killed, 1972–2005, Scale Example 2
Source: Chart created from data from Bureau of Justice Statistics, http://www.albany.edu/sourcebook/pdf/t31548006.pdf

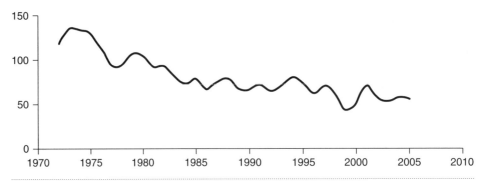

Figure 8–13 Officers Feloniously Killed, 1972–2005, Scale Example 3 [Does not include the 72 officers who lost their lives in the September 11, 2001 attacks.]
Source: Chart created from data from Bureau of Justice Statistics, http://www.albany.edu/sourcebook/pdf/t31548006.pdf

intervening, antecedent, or contingent variable that mitigates the relationship between X and Y. In addition, further analyses, such as partial correlations, must be performed to rule out spurious relationships.

The sixth sin an analyst can commit is choosing an improper scale to display his or her results. At one magnitude, the problem (or solution) may look much more impressive than at a different scale. The following figures illustrate the number of officers feloniously killed between 1972 and 2005. **Figures 8–11** and **8–12** use the same official figures but employ different scales.

Note that the changes from year to year look more dramatic in the first example, even though the figures used are the same. **Figure 8–13** shows the same information, except that the 72 officers killed as a result of the 9/11 attacks have been omitted.

This example reemphasizes the need to present the whole picture to the intended audience, making them aware of any known reasons for dramatic shifts in variable trends.

A seventh possible deadly sin would be for an analyst to fail to recognize the audience for whom the analysis is prepared. If an analyst wishes to simply describe crime in terms easy enough to let the audience understand the problems, many of these sins do not apply directly. However, it is incumbent on analysts to learn how to use these inferential statistics tools to better serve their law enforcement audience. Analysts may find that they need to describe their findings and analysis in an executive summary fashion and tell the police command staff what their findings mean in a laymen's terms rather than via statistical jargon. It may make the analysis more useful to them.

In conclusion, it is extremely easy to present inaccurate or misleading statistics (either intentionally or unintentionally), even if the calculations are performed perfectly. It is crucial for crime analysts and mappers to pay attention to the rules and guidelines of statistical analyses that have been addressed in this chapter to produce the most accurate and representative picture of what is occurring. Maps and analyses created by violating the assumptions and rules of various statistical analyses provide misleading information to policy makers and may result in wasted resources.

■ Recommended Reading

Elias, G. (1999). *How to collect and analyze data: A manual for sheriffs and jail administrators* (2nd ed.). Lafayette, CO: Voorhis Associates.

Walker, J. T., & Maddan, S. (2009). *Statistics in criminology and criminal justice: Analysis and Interpretation* (3rd ed.). Sudbury, MA: Jones and Bartlett.

■ Questions for Review

1. Explain *qualitative* and *quantitative* data. Provide an example of each.
2. What is a variable?
3. Explain the four levels of variable measurement. Provide an example of each in your discussion.
4. What is the difference between *discrete* and *continuous* variables?
5. What is a mean center and how is it used in crime mapping?
6. What is a standard deviation ellipse? Why do crime analysts only draw three or fewer standard deviation ellipses?
7. What is spatial autocorrelation? What are two commonly used tests of spatial autocorrelation?
8. Why do crime analysts need to have a solid understanding of statistics to create and explain crime maps?
9. How can data be classified in crime maps? How does an analyst know which classification method to use?
10. What are the statistical sins analysts should avoid?

■ Chapter Glossary

Antecedent Variable   In an antecedent relationship, a third variable, Z, is placed before X and Y.

Bimodal   A distribution is bimodal if it contains two modes.

Causal Relationship   In a causal relationship, X is assumed to have produced variation in Y.

Choropleth Map Choropleth maps use shade and color to denote different distributions of a variable. Done at the polygon level, they can be at different units of analysis including beats or precincts, census blocks or block groups, or towns.

Coefficient A coefficient is a value used in a mathematical equation that is a constant multiplicative factor of a certain object, such as in $9X^2$ where 9 is a coefficient of X^2. In simple terms, coefficients are the values and variables we use in statistics calculations.

Contingent Variable In a contingent relationship, a third variable, Z, is placed after X and Y.

Continuous Continuous variables *do* have values that fall between adjacent values on the same scale.

Dependent Variable The variation in the dependent variable (Y) is thought to be caused by the independent variable(s) (X).

Dichotomous Variables that are divided into two categories. Often called "dummy" variables, dichotomous variables are often used in parametric tests that require interval-level variables.

Discrete Discrete variables do not have values that fall between adjacent values on the same scale.

Exhaustive When a variable operationalization is exhaustive, there is a category for all values.

Frequency Distribution A frequency distribution is a visualization, in summary form, of how values are spread over categories or scales.

Histogram In a histogram, the categories of one variable are plotted on one axis, and the responses are plotted on the other axis. Histograms are best used on nominal and ordinal level variables but can be used on interval variables with collapsed categories (as in a frequency distribution).

Independent Variable In a causal relationship, the independent variable (X) is assumed to be at least a partial cause of the change in the dependent variable (Y).

Interquartile Range An interquartile range is the distance between the 25th and 75th percentile.

Interval Interval-level variables (data) can be ranked, such as ordinal variables, but the distance between each value is known.

Intervening Variable In an intervening relationship, a third variable, Z, is between X and Y.

Inverse Relationship In an inverse relationship, the values of one variable increases as the values in the other variable decrease.

Linear Relationship In a linear relationship, two variables (measured at the interval or ratio level) travel together in a seemingly straight line.

Mean   The average (usually the arithmetic mean) of all scores within a distribution.

Mean Center   In mapping, the mean center marks the location of the arithmetic average of all the incident locations on a map. The mean center is the approximate center of a series of events (typically crime incidents).

Measures of Association   Statistics that indicate the strength and nature of a relationship between variables

Median   The middle score in a distribution.

Mode   The most frequent score in a distribution.

Multicollinearity   Multicollinearity exists when variables in a model are highly correlated. This can mean that the variables are measuring the same thing.

Mutually Exclusive   When a variable operationalization is mutually exclusive, values can only fit into one category.

Nominal   Nominal-level variables (data) are variables that cannot be ranked. They are typically word categories that cannot be ranked. For example, gender and race are nominal-level variables.

Normal Curve   In a normal distribution, the values form a curved shape that is unimodal, symmetrical, and infinite in both directions. In normal distributions, the mode, median, and mean are all the same value.

Operationalize   This term denotes the process of measuring a variable by putting it into categories and assigning numbers to its characteristics.

Ordinal   Ordinal variables (data) can be ranked. In number or word form, the actual value of an ordinal measurement is not known; rather, it is included within a category that ranges in value.

Positive Relationship   In a positive relationship, the values of both variables travel in the same direction (either both are increasing at the same time or both are decreasing at the same time).

Qualitative   Qualitative research is usually narrative focused and is rich in information. Variables measured at the nominal level are considered to be qualitative data.

Quantitative   Typically number oriented, quantitative research typically utilizes statistical analysis to answer research questions.

Range   The simplest measure of dispersion, the range is the difference between the lowest and highest scores in a distribution.

Ratio   Ratio-level variables (data) are interval-level variables, except that ratio-level variables have a true zero.

Reliability   A variable is reliable if it consistently measures a variable from one study to the next.

Skewed   A distribution is skewed if the mode, mean, and median are not the same value.

Spatial Autocorrelation   Spatial autocorrelation tests indicate whether or not distributions of point locations are related to one another.

Spurious Relationship   A spurious relationship exists when X and Y are assumed to be related, but a third variable, Z, is actually strongly related to both X and Y. Once controlling for Z, the relationship between X and Y disappears.

Standard Deviation   The square root of the variance, the standard deviation represents the average deviation from the mean.

Standard Deviation Ellipses   Standard deviation ellipses are drawn on a map to represent typically one and two standard deviations from the mean center (the average distance from incident locations to the mean center). It also indicates the direction toward which the dispersion is oriented (north, south, east, west).

Tests of Significance   Tests of significance indicate whether or not the distribution of a variable or the relationship between two or more variables is statistically significant.

Unimodal   A distribution is unimodal if it only has one mode.

Unit of Analysis   A unit of analysis is the level at which variables are measured. In examining relationships using statistical models, it is important to keep within the same unit of analysis.

Validity   A variable has validity if it measures what it is supposed to measure.

Variables   Called "variable" because they vary in some way, variables are the characteristics of people, space (or even time), or behaviors that analysts study. People-related variables include age, race, gender, victimization, criminal behavior, etc. Place-related variables include distances from one place to another or the concentration of bars in an area.

Variance   The variance measures the average of the squared deviations from the mean.

■ Chapter Exercises

Exercise 14: Aggregating Data for Shaded Grid Maps

Data Needed for This Exercise

- _____
- _____
- _____
- _____
- _____
- _____

Lesson Objectives
- Aggregating data for various areas we use in law enforcement is a task that many of you will have to perform many times during your careers. In some cases we will have an RMS or CAD system that can do this aggregation for us and save us some work, other times we will need to perform spatial joins, summary tables, and regular joins in ArcGIS to get what we need, and at other times we will need to do some combination of these tasks. In this lesson you will learn a way to optimize some of this time using Excel to aggregate or finish aggregating several fields of data, and then use this data in ArcGIS to create several different kinds of hot spot maps. You will also learn about legend classifications and why these different methods make your data look different. We will also discuss some of the pros and cons to using different classification methods in an operational setting.

Task Description
- You have been assigned the task of figuring out what method of developing hot spots for the Glendale Police Department's Compstat program will be used to identify hot spots and hot areas for various crimes and disorder problems. The first focus of this effort will be to identify methods to develop hot spot maps for all Part I violent/person crimes for the Compstat process. The commanders have complained that the areas you show every week are always the same areas, and they would like you to use your skills to come up with a hot spot method that will make better use of rates (or crime per something) in the maps to better show hot spots rather than just by count. They have suggested rate maps for the following variables:
    - Part I violent crimes per 1000 population
    - Part I violent crimes per total crimes
    - Part I violent crimes per square mile
    - Part I violent crimes per all Part I crimes
    - Part I violent crimes per population between 18 and 25 years of age
    - Part I violent crimes per total calls for service in an area
    - Part I violent crimes per some other factor
- You need to decide which variables to try and what data you have available that could potentially answer these questions.

1. Start this project by opening the Excel spreadsheet called EX14.xls, which is located in the C:\CIA\Exercises\Exercise_14 folder.

When it has been opened, save a copy of it in your student folder and work with the copy.

2. You should see worksheets of data, one for census blocks and one for the grids.

3. One of the first things you need to consider is that when someone asks a question, such as "What is the rate per 1000 population?" or some similar question involving demographics, chances are you are going to have to use the census blocks, or tracts, to do the analysis rather than some political boundary that does not normally contain such data. This isn't always the case, and some engineering departments collect or interpolate population figures for various areas of the city based on some internal city regulation or need. You can also estimate the total population within census blocks to, for example, a quarter square mile grid polygon you might maintain, but this process can be somewhat tedious and time consuming. The basic concept is to find the smallest area for which you have population or other demographic data and try to distribute it evenly within the polygons you are trying to populate with the data. Unfortunately, they do not always align easily, so you make estimates. For example, we have 10 grids in a small area and 20 census blocks. The 10 grids are each exactly 0.25 square miles, but the census blocks vary in size over the area. For the census blocks that match up directly to the size of a grid, we can just carry the demographic data over from the census block to the grid. In the case of, for example, four census tracts that make up the same boundaries as one grid, we would need to summarize the totals from each census tract and carry that summary to the grid. In the case where the census tracts actually cover four grids, we would want to divide the demographics by four and assign that quarter of the demographic information to each grid within that census block's boundaries. Then there are census blocks that cover entire grids or only very small partial grids, or the census boundaries overlap several grids and do not share common boundaries in several places. In many cases the analyst would assign the ratio of demographics based on the size of the area. For example, a census area crosses two full grids and one-fourth of two other grids. A process for assigning that proportion of the demographics to each of the grid portions would be developed. This is not impossible, but it does cover some geoprocessing that is a bit further along than we want to get into here, but remember that chances are it will come up in your future as an analyst.

4. We have two worksheets with data that has been aggregated for the commanders request. We can use Excel to calculate the various rates instead of having to do them in ArcGIS. When we are done, we can add this data to the ArcGIS project and do a regular join between the grid name or the census STFID number, then join it back to the polygon, and we can classify all this aggregated data and rates into hot spot maps.

5. The first three rate formulas have been entered for you. Complete the rest of the rate formulas to complete each worksheet.

6. When you are doing your formulas, remember that dividing by 0 will give you an error message because this is not possible. You may have results like that, so the best thing to do is to make the values for those cells "null" or make nothing in them. Leaving the records blank will not affect the data import and will not make ArcGIS think the field is text instead of a number. You don't have to exclude null values in your legend in most cases, but you would have to deal with data not sorting correctly if you placed the word "null" in those records with no data and then imported it into ArcGIS. You might also have to calculate a new number field to do the mapping you want to do.

7. When you are done filling in the missing formulas for both tables and copying them all the way down, clean up the #DIV/0 errors by making them null, and save the copy of the spreadsheet again.

8. In the real world, when you are done, you would export each table from Excel to a DBF table or create an ODBC/OLE DB link to the spreadsheet and bring the data into ArcGIS to make the various maps you need. To speed the exercise along, however, this has already been done for you.

9. Close the Excel spreadsheet and open the EX14.mxd ArcGIS project in the C:\CIA\Exercises\Exercise_14\ folder.

10. After the project is open, save it as EX14.mxd in your student folder. You may want to make a new folder called "Compstat" or something else in your student folder. Anything you create for this specific exercise can be placed in this folder so you can easily delete it if it is no longer needed.

11. After you have saved the project, add the Grid.dbf and Census.dbf tables to the project from the C:\CIA\Exercises\Exercise_14\ folder.

12. Create a regular join between the GridName field in the Grids theme and the GridName field in the Grid.dbf table. Do another join between the STFID fields in the Census Block Groups theme and the Census.dbf table.

13. Open the attribute table of the Grids theme, and notice that the data that was in our table is now to the far right side of the polygon Grid theme's attributes. The same is true for the Census Data theme with all our counts and rates included.
14. For each of the two polygon themes (Grid and Census Blocks), change the legend so that each of the rate fields we calculated are displayed on the map. Again, the steps to accomplish this are as follows:
    a. Right click the **Grid** or **Census Block** theme.
    b. Go to **Properties** in the pop-up menu.
    c. Go to the **Symbology** tab.
    d. Choose a **Quantities** legend with graduated colors.
    e. Set the Fields box to each of the various fields we calculated one at a time (you may also want to try the Count Field option for reference).
    f. Apply the legend and compare the results.
    g. It would be a good idea to set the Grid layer and the Census Block layer to the same field value, then turn them on and off to see how the different rates compare.
    h. Decide which rate or count hot spot map best displays the data for the Compstat process.
15. You will likely find that many of the rate maps don't really show you anything useful when done in this fashion. The count maps or the crimes per square mile seem to do a better job of showing which areas are significantly high in Part I violent crimes, and the rate maps, although logical, do not always yield practical results for the day-to-day operations of a crime analysis unit. The problem isn't really with rate maps per se; it has more to do with the size and shapes of the regions we decide to use and how well they fit with the reality of the day-to-day policing functions for a city. The other problem is that the demographic data needed to do many of these rates maps is not always available for arbitrary political boundaries like beats, grids, and police districts, so estimates have to be used rather than good data. Switching to census boundaries gives you the underlying demographic information that may be a few years old or more (7 years in this case). Few cities actually forecast current population by small political boundaries on a regular basis, and if they do, they aren't the ones your police department wants to use. Your goal should be to learn more about these different rate maps that you could create and develop criteria that would be useful for your agency in the future.

16. Many people will disagree with this summation of rate maps, but sometimes what is statistically more significant or difficult is not always the best way to do things for a Compstat meeting or other analysis effort. Your maps and analysis products must be understood to be used. Any one of the maps you may have produced, using either geographic region (grids or census blocks), would work; you just need to explain what they mean, and they have to be understood by the command officers in your agency who need to act on them. You need to understand what the map is telling you so that you can explain it to them as well!

Exercise 15: Hot Spots, Hot Areas, Choropleth Maps, and Problem Areas

Data Needed for This Exercise
- All data needed is in the EX15.mxd file.

Lesson Objectives
- Learn the difference between ways to classify data for a choropleth map and how that data relates to an operational use for patrol deployment.
- Discuss the difference between raster and vector data and different ways to determine how to show a hot spot.
- Explore CrimeStat III's hot spot routines.

Task Description
- In this exercise we will explore the use of various types of hot spot techniques and classification methods for data that are available in the ArcGIS software. We will use stolen vehicle data to develop hot spot areas and discuss various concepts and processes to display data to show hot spots of activity, which can then be used to assist patrol or investigations to curb the upswing in stolen motor vehicles in Glendale.

1. Open EX15.mxd.
2. The MXD file should resemble DVD Figure 15–1.
3. In the table of contents (TOC), turn on the Total 2000 Census Population theme.
4. This data represents the total population from the 2000 census for each census block group in Glendale. As you can see from the legend, the red polygons contain between 3851 and 8678 persons, according to this data.
5. A *classification* is the manner in which a statistical method is used to display numeric data on the map. There are generally six

different ways ArcGIS can display numeric data (seven if you count manual methods the user can define on the fly).
   a. Natural breaks (Jenks)
   b. Standard deviation
   c. Quantile
   d. Equal interval
   e. Defined interval
   f. Geometric interval
6. The default in ArcGIS is almost always the natural breaks (Jenks), which is a process that puts a delineation between classifications where the data naturally breaks. For example, in the population data, there are 395 different census block polygons. The minimum population is 0 and the maximum population for any one census block group is 8676. A histogram of the data can be viewed by right clicking on the **Total 2000 Census Population** layer and choosing **Properties → Symbology → Classify** button (see DVD Figure 15–2).
7. The natural breaks classification method basically tries to find where there are breaks in the frequency table for the data. For our population data, the first break occurs at 628 where there is a gap between numbers, again at 1320, again at 2151, and so on. Although this method is statistically sound, you may find it difficult to explain this method and the resulting map to the command staff. Natural breaks just finds the places in the data where a break should be and displays it using those classifications.
8. It is important to note that whatever method you use, you should make sure that you limit the number of classifications to seven or less. Many studies indicate that the human mind cannot distinguish between more than seven colors or classifications at a time and make sense of them all.
9. Now let's change the classification method to standard deviation. Try to remember which polygons the natural breaks classification method showed as the highest.
10. To change the classification method, go to the **Properties → Symbology** dialog menu and click the **Classify** button on the right. Choose **Standard Deviation** from the drop-down list and apply the new classification method.
11. Notice that the same polygons are highest with this method and the natural breaks classification method, with one exception: The entire west side of Glendale is now also in the highest category.

The standard deviation classification method is good for showing areas that are significantly higher than other areas. In this case the west side of Glendale in the large census block group shows a population of 3679, and from looking at the histogram we can see that the data is slightly skewed, with most census blocks being between 800 and 1700 persons.

12. Now change the classification method to equal interval.
13. In this method, ArcGIS makes sure the interval within each class is exactly the same, in this case 1735. Take the lower number in the range for each class and subtract it from the higher number in the range, and you should get 1735 for each class. Notice also that the number of polygons that are in the highest category has decreased somewhat.
14. Now choose **Defined Interval**, and under the drop-down box, type in *1000* for the interval you wish to use. Notice how this affects the map. By reducing the number in the interval box, you will increase the total number of classes. You should find that about 1425 gives you seven classes for this data.
15. Now choose **Geometric Interval**. A geometric interval is a method of classification that works toward giving more intervals where the data is more abundant and fewer where the data is less abundant. The ArcGIS help file indicates that:

    > This is a classification scheme where the class breaks are based on class intervals that have a geometrical series. . . . The algorithm creates these geometrical intervals by minimizing the square sum of element per class. This ensures that each class range has approximately the same number of values with each class and that the change between intervals is fairly consistent. This algorithm was specifically designed to accommodate continuous data. It produces a result that is visually appealing and cartographically comprehensive.

16. All of the interval methods seek to optimize the display of data based on the distribution of the data across all of the polygons that are included by manipulating the ranges or intervals between values within each class.
17. Now switch to the quantile classification method.
18. This method is very similar to what many police commanders deal with, and it generally can be described as the percentage of polygons within a range. For example, if we had five classes using the quantile method, then each range would have approximately 20% of the total polygons ($5 \times 20 = 100\%$). You should also

notice that now a large number of polygons have been moved to the highest category for this data.

19. In reviewing all of these methods, remember that the data never changes. The change is only in the way the data has been classified, and thus the appearance of the map may have also changed. It is very common to use the natural breaks, standard deviation, and quantile classification types in crime mapping because, in general, they are a bit easier to explain. The standard deviation method most likely gives the analyst the best result for finding those polygons that are significantly higher than other polygons. Imagine using the quantile method of analysis and telling a commander he has 42 hot grids in his patrol sector for total volume of violent crime. A better tact might be to use a standard deviation method of classification and reduce that significant number down to 10 or less (of course this depends on the data).

20. It is a good idea to try several of these classification methods when doing any hot spot map to see which one tells the story that needs to be relayed to the decision makers in your police department.

21. We have been dealing with what is called choropleth mapping up to this point, which is sometimes referred to as shaded grid, hot spot, hot area, or graduated color analysis. When we think of hot spot maps, we generally think of a map that has different colors based on some numeric value within each of the areas we wish to study. In this case, we have been using the total population per census block group, and it is a count and not a rate. In the previous exercise we discussed how rates can be used and some of the drawbacks to them. So far, we have been dealing with what is called "vector" data. This means that the data is discrete spatially (and probably numerically as well, i.e., 2 instead of the continuous counterpart for nominal data of 2.35, etc.), or each polygon has a defined boundary that is shown by a line of some sort. Another form of hot spot data we often see is when the meteorologist from the news advises us whether to wear a raincoat or not. Spatial Analyst, an extension to ArcGIS, is often used to create weather map-like hot spot maps. Spatial Analyst takes point data and creates a raster grid of data that is spatially continuous and does not have specifically-designated boundaries for each area; instead the data is displayed using grids or pixels of color at different values that cover the entire surface of the area being analyzed.

22. Turn off the Total Population theme, and turn on the StolenVeh-Raster theme. Notice how the entire surface of the analysis area is covered with some color.

23. The red areas are the hot spots for stolen vehicles in Glendale for the past several years.
24. Turn on the Identify tool, and choose the **StolenVehRaster** theme from the drop-down box. Click on the map in a few dark blue areas and then several red areas, and notice that you get a pixel value between 0 and about 1800 or so. This tells us that there were 0–1800 stolen vehicles per square mile density from where you clicked. This does not mean there were 0–1800 stolen vehicles right where you clicked, they were within a square mile around that point.
25. Experiment with changing the classification methods for this raster grid layer and see how it affects the map. You will notice that the map does look quite a bit different as we change the classification types and the number of classes. It appears that natural breaks or standard deviation may give us the best view of the general hot spots we may find for auto theft. Again, it doesn't really matter which classification we use as long as we can properly interpret the data for the audience we are developing the map for.
26. Another point we need to consider is that creating one map for identifying or analyzing hot spots may not be the best practice. Any hot spot has other factors to consider that could affect the analysis.
    a. Are there any addresses within the hot spot that are actually causing the entire area to be hot?
        i. If so, we should find out by doing a graduated dot map as well (see exercise 19).
    b. What is the temporal displacement of crime within the area?
        i. Is crime consistently high there and it's always a hot spot, or does it change over time (seasonality, month to month, daytime versus nighttime)?
        ii. You may want to consider making maps for different hours of the day, month by month, or even hour to hour in some cases, and see if those areas are still hot at different times.
    c. Does the area show increased activity over time?
        i. What is the spatial trend of the data? Is it higher than last year, last month, etc.? How does it compare with other hot spots?
    d. When you have actually identified a hot spot, you will need to be able to track the actual crimes that have occurred there and be able to repeat the analysis at a later date.

e. The advantage of vector hot spot analysis is that the polygon boundaries for the hot spot do not change. The drawback is that they do not consider the surrounding polygons and potentially related crime in those surrounding areas that are either caused by activity within the hot spot (dispersion) or the effects of reductions in the hot spot over time (diffusion).

f. The advantage of raster hot spot analysis is that it is more sensitive to surrounding areas and crimes just outside the hot spot area, and it can be more specific when looking at individual crime types. The drawback is that enforcement can affect the shape of the hot spot, and it may be hard to exactly duplicate the analysis that exposed it in the first place. Another drawback is that you need the Spatial Analyst extension or similar software that does kernel density analysis.

27. What we often see is that polygons and aggregated data are often used to analyze counts within political or police boundaries, while raster data analyzes density of crime and ignores boundaries, for the most part. Another type of crime hot spot analysis comes from the free CrimeStat III software that was created by Ned Levine & Associates through a grant funded by the National Institute of Justice.

28. Turn off the Raster theme, and click the **plus sign** next to the Crime Stat Hotspot Layers group.

29. You should now see an additional 10 layers or themes, which are all shapefiles created by the CrimeStat III software.

30. Turn on the top layer, called K-Means Convex Hulls, and then turn on the K-Means Ellipses layer. By turning them on and off you can see that they are related because they cover approximately the same general areas. These layers are developed with the K-Means tool in CrimeStat III. They attempt to find areas where crime points are clustered significantly and then draw either the ellipses or convex hull polygons around the clustered crime points (in this case, auto thefts). Notice that they cover the entire city of Glendale for the most part. If you turn on the Stolen Vehicles point layer, you will see why.

31. Turn these themes off, and turn on each of the other themes to see what they show as clusters of crime, or hot spots.

32. The nearest neighbor clustering routine actually gives you three levels for most crime data. You will see large ellipses (or convex hulls—the outside extents of the points that are determined to be in a cluster are all connected into one large polygon), medium-

sized ellipses, and small ellipses. You can think of these as being on a scale level. For example, at the citywide level (level 3) we have crime clustering going on, at the beat level (level 2) we also have clustering, and at the local level (level 1) we also have localized hot spots or clusters of auto theft activity. The last two layers are hot spot clusters that are developed using the algorithms developed by Richard Block (Loyola University Chicago) and Becky Block (Illinois Criminal Justice Information Authority) from Chicago.

33. Within the CrimeStat III software, you can adjust the analysis settings to get smaller or larger ellipses as needed for the level of analysis you are doing. All you need is a DBF table from a shapefile that has the X and Y coordinates of each crime point in the DBF table.

34. If you turn on and off the stolen vehicle hot spot layers we have used in this exercise and compare them with one another, you can see that several of the same areas pop up no matter which method you use. This is a very good way to isolate significant areas that could use intervention. Find the areas where several methods overlap, and then use this new area as your focus area for enforcement activity for that crime type and track its progress over time.

35. The steps you should consider when doing a hot spot analysis should always include:
    a. Create a hot spot map through whichever method you are most comfortable (shaded grid, Spatial Analyst, CrimeStat III).
    b. Create a graduated dot map for the same analysis extents to see if any addresses significantly contribute to the hot spots identified in the previous step.
    c. You may want to remove the repeat address locations (e.g., police station and hospital) from the list and see if the hot spot still exists without that address, or at least understand the influence the address has on the identified hot spot.
    d. Zoom into the hot spot or hot spots and create additional hot spot maps of the more localized area (where within the hot spot are the hottest areas?).
    e. Create pin maps, graduated dot maps, and other maps that will help clarify where the problems are within the hot spot that can be influenced by enforcement or intervention strategies.
    f. Create charts, tables, or executive summaries to explain what you found and how it relates to enforcement or intervention strategies that are available to your police department.
    g. Be able to recreate this analysis to track the progress of any enforcement or intervention strategies that are put into place.

36. Turn on the graduated dot layer for stolen vehicles and one or two of the hot spot layers. Do you think specific addresses have some affect on hot spots in Glendale?
37. Besides the graduated dot, shaded grid, CrimeStat ellipses, convex hull hot spots, and raster grid hot spot maps, another way to analyze hot spots that is similar to the graduated point map is to assign data to a street segment and then show the various street segments in different sizes and colors based on the number of incidents that were found near that segment. An example of this is shown in the Stolen Vehicles per Street Segment layer. Turn the other hot spot map layers off, and turn on this layer.
38. The steps to create a layer like this are:
    a. Create a spatial join between the street layer and the stolen vehicles layer with the street segments giving up their data to the stolen vehicle points (or finding which street is closest to each point).
    b. Take the resulting spatially joined theme, find the unique identifier field name for the street layer that was added, and do a summary table on that field (count the total incidents per street segment).
    c. Take the resulting summary table and do a regular join back to the street segment theme.
    d. Create a graduated symbol legend for the street theme based on the count from the summary table.
39. We won't create this layer in this exercise because it would take too long with our data, but the purpose of this exercise is more to show you what is possible rather than how to do each method.
40. Each method of analysis has some issues that you need to be aware of, including the crimes per street segment method. In this method, incidents that happen within intersections will usually be assigned to street segments that come into the intersection, and the data may be displayed somewhat inaccurately at a large scale (zoomed into the city). Traffic crashes that happened within the intersection and along road networks can sometimes be inaccurately shown with this method. In this case, having a buffer of each intersection and hot spots at each intersection for crashes that happened within the intersection would be advisable, and only those crashes that happened 150 feet or more from the intersection should be displayed by the graduated street segment method. This method shows crashes that happened within the intersection and along

road networks most reliably at a large scale (1:8000); at a citywide or small scale (1:250,000) you may not be able to distinguish the various hotspots because they would be so small. The analyst should be careful when interpreting these results for law enforcement decision makers so that the map does not improperly help to make the wrong decision.

41. With any of these methods, the more data you have to analyze, the longer it will take ArcGIS to wade its way through it. The street segment analysis itself took at least 1.5 hours to complete due to the time it took to spatially join stolen vehicle points and the street segments.

42. If you compare the street segment hot spots with several of the other hot spot methods, you can also see some overlap and thus some confirmation that a few hot spots may be the most important to focus on in combination with repeat addresses and other analyses when the hot spot is identified.

43. This concludes this exercise. Close the exercise without saving it.

References

Catalano, P., Hill, B., & Long, B. (2001). Geographical analysis and serial crime investigation: A case study of armed robbery in Phoenix, Arizona. *Security Journal, 14*(3), 27–41.

DiIulio, J., Jr., & Piehl, A. M. (1991, Fall). Does prison pay? *Brookings Review,* 28–35.

Elias, G. (1999). *How to collect and analyze data: A manual for sheriffs and jail administrators* (2nd ed.). Lafayette, CO: Voorhis Associates.

Groff, E., & McEwen, J. T. (2005). Disaggregating the journey to homicide. In F. Wang (Ed.), *Geographic information systems and crime analysis* (pp. 60–83). Hershey, PA: IDEA Group.

Harries, K. (1999). *Mapping crime: Principle and practice*. Washington, DC: National Institute of Justice.

Kennedy, D. M., Braga, A. A., & Piehl, A. M. (1998). The unknown universe: Mapping gangs and gang violence in Boston. In D. Weisburd & T. McEwen (Eds.), *Crime mapping and crime prevention: Crime prevention studies* (Vol. 8; pp. 219–262). Monsey, NY: Criminal Justice Press.

Zedlewski, E. W. (1987). Making confinement decisions. *Research in brief*. Washington, DC: National Institute of Justice.

Distance Analysis

CHAPTER 9

▶ ▶ LEARNING OBJECTIVES

Chapter 9 begins with a discussion of several types of distance analysis. Each type of distance analysis has strengths and weaknesses that must be understood prior to interpretation. After studying this chapter, you should be able to:

- Identify the common types of distance analyses used in crime mapping.
- Explain the appropriate uses of distance analysis.
- Understand the strengths and limitations of distance analysis.

▶ ▶ KEY TERMS

Euclidean Distance
Manhattan Distance

■ Introduction

All analyses that rely on distance measures are distance analyses. Thus, the hot spot analyses that will be examined in Chapter 10 could also be viewed as distance analyses in that the determination of a clustering of events requires that individual events are located closer together than we would expect based on random chance. In this case, the distances measured are strictly those between criminal incidents (distances between homicides in a city, for example). This chapter discusses those analyses where distance is measured in relation to another point or in efforts to find another point (the mean center of a distribution, an offender's home, predicted next target, or some other point of importance such as a school, bar, or pawn shop). The distance analyses to be discussed in this chapter include mean center analysis, journey to crime analysis, spider distance analysis, proximity analysis, distance between hits analysis, and distance and time analysis.

There are several different methods of calculating distance. Each has its strengths and limitations in analysis. In crime analysis, the two most common methods of distance calculation are Euclidean and Manhattan. **Euclidean distance** is measured by measuring the distance between two points. Often referred to "as the crow flies" measurement, it is the shortest distance between two points on a map. The problem is that there rarely is a road that leads directly from point A to point B

335

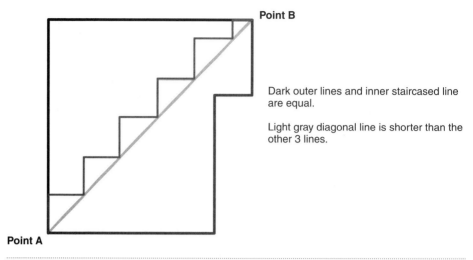

Figure 9–1 Distance Illustration

on a map (have you ever navigated Boston?), and travelers usually must take a series of twists and turns on their destination. Thus, Euclidean calculations of distance are typically smaller than the path actually traveled by an individual. **Manhattan distance** or "street" distance, as it is often called, is calculated by using right angles to get from one point to another. In practical terms, individuals do not always travel in right angles to get to their destination (again, have you ever driven in Boston?). Manhattan calculations are always larger than Euclidean calculations because the shortest distance between two points is a straight line (see **Figure 9–1**).

You may be wondering at this point which is the best calculation method to use. The answer depends on the type of analysis you are performing and the data you are using (including the street layout of the study area). Euclidean distance is much easier to calculate; however, Manhattan more closely approximates the distance traveled by individuals. Rossmo et al. (2005) observe that:

> Research has shown that Manhattan distance gives the most accurate result in the greatest number of cases, while not being significantly worse than other methods across the entire spectrum of cases—a finding true for both North America and Great Britain . . . As long as these specific exceptional cases can be identified and recognized through training and experience, the most reliable and practical method involves the use of Manhattan distance. (p. 111)

The best method of distance calculation is dependent on the type of analysis that is to be performed. Groff and McEwen (2005) found a strong correlation between Euclidean and street distances and thus argued that one could use the coefficients of Euclidean measures in a regression model to estimate street distances.

■ Distance Analyses

A variety of distance measures can be used in crime analysis and mapping, and in reality, distance and how features are related to each other on a map are what crime mapping is all about. The very technical term "Spatial Autocorrelation" has a fairly simple explanation to crime mappers, which is that those features that are closest to one another are likely more related to other features near it than to those further away. If we think about this within our own lives, we might often find our neighbors, friends, and co-workers nearer to us in geography than persons who live further away or have no similar occupation, interests, or residence. The first type of distance analysis we'll discuss is mean center analysis, which is used for tactical analysis purposes.

Mean Center Analysis

The mean center of any distribution is, very simply, the point at which the mean of X (latitude) and the mean of Y (longitude) of events meet on a map. It represents the average or the center of gravity of a spatial distribution (Levine, 2002). The problem with performing a mean center analysis is that it is sensitive to outliers (recall from Chapter 8 that the mean itself is sensitive to outliers and thus cannot be relied upon in skewed distributions). In addition, the mean center of distributions that are multimodal or odd shaped (such as in an L-shaped distribution) may be placed at a point where very few crimes actually occur. Thus, the utility of mean center analyses is not in finding a point on the map to throw more resources at or the apex of crime in an area. Rather, it is a reference point to be used in further analyses, such as in the comparison of two different distributions (same crime but different time, or different crime but same time), and as a starting place to begin prioritizing places and persons of interest (such as performing a standard deviation ellipses analysis; recall that standard deviation was discussed in Chapter 8). There are several different types of mean center analyses that an analyst can perform. CrimeStat performs several different mean center analyses, including the mean center, the harmonic mean, and the geometric mean. Note that when viewing **Figure 9–2** at a smaller scale, the three mean centers appear to be located at the same point.

However, in **Figure 9–3**, a larger scale reveals that the three mean center analyses are indeed three separate points.

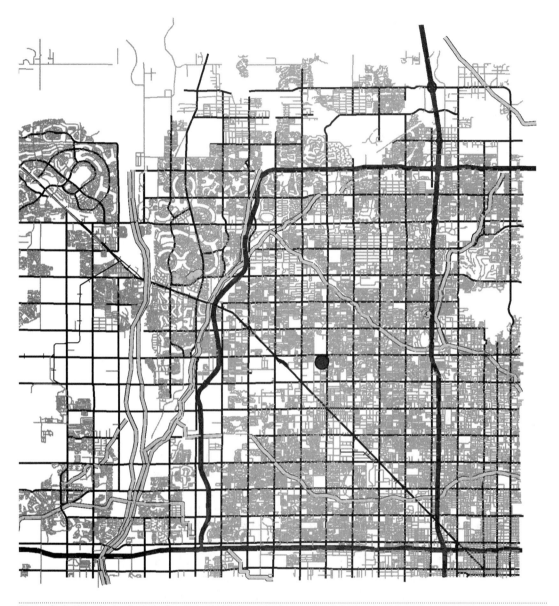

Figure 9-2 Mean Center Analysis, Burglaries 2003, Glendale, Arizona, Small Scale
Source: CrimeStat and ArcGIS

What does this all mean to the analyst? If we remember that maps are simply representations of the real world and our goal is to be as accurate as possible when doing analyses, then deciding which distance method to use is just part of that well-planned analytical process.

Figure 9–3 Mean Center Analysis, Burglaries 2003, Glendale, Arizona, Large Scale
Source: CrimeStat and ArcGIS

Journey to Crime Analysis

A journey to crime analysis is one type of distance analysis and is used primarily in investigations of a crime series that is thought to be attributed to an individual or group of persons acting together. It is conducted in hopes of prioritizing areas in which the offender or offenders are most likely to live or work. Recall from Chapters 3 and 4

TABLE 9-1 Average Distance to Crime in Glendale, Arizona

Crime Type	Number of Crimes	Mean Crime Trip (Miles)	Standard Deviation (Miles)	68% of Crime Trips Within (Miles)	Furthest Distance
Aggravated assault	8,526	1.18	2.86	4.04	6.91
Arson	62	1.32	2.68	4	6.67
Auto theft	1,058	2.88	3.55	6.43	9.97
Burglary	1,356	2.34	3.73	6.07	9.79
Curfew/loitering	164	1.59	1.85	3.44	5.28
Drug offenses	3,970	2.27	3.39	5.66	9.06
Murder	70	2.38	4.42	6.8	11.23
Other miscellaneous	18,237	2.42	3.9	6.32	10.22
Other sex offenses	404	1.61	2.66	4.27	6.92
Rape	85	1.35	2.79	4.14	6.94
Robbery	567	3.23	4.21	7.44	11.65
Runaway	5,958	0.19	1.29	1.48	2.77
Theft	5,139	3.18	4.28	7.46	11.75

that offenders are fairly routine in their travels for both criminal and noncriminal behavior (see **TABLE 9-1**).

In addition, offenders generally tend to travel greater distances for property crimes than they do for violent crimes, and the likelihood of offenders to commit any crime dissipates as they get farther away from their home. (Also recall that if offenders commit crime in relation to an anchor point that is not their home—for example, if the anchor point is their place of employment—distance decay measures from home to crime should not be used if more accurate work to home distances are available.)

The journey to crime research traditionally has examined the route undertaken by offenders between their home and place(s) where the crime was committed. However, this ignores the *victim's* journey to his or her victimization and also ignores other nodes where offenders maybe traveling from (work, school, parole office; Costello & Leipnik, 2003). Groff and McEwen (2005) found "clear differences in travel behavior between victims and offenders" in their study of homicides in Washington, DC (p. 60). They calculated both Euclidean and street distances for both offenders and victims and found that for homicides, victims traveled a median of 0.69 street miles (0.54 Euclidean miles). Median distances were chosen due to the skewed nature of the data set. (Both victims and offenders of homicide tended to stick very close to home.) The distance traveled for both victims and offenders varied according to the originating motivation for the homicide. For robbery-motivated homicides, victims were killed about 1 mile from their homes. For retaliatory, drug, and gang-related homicides, victims were killed a median of 0.67 miles from their homes. Male victims traveled farther to their murders than did female victims in every category of homicide (with the exception of domestic violence homicides).

You may be wondering why we included a discussion of *victims'* travel to crime. The answer is straightforward. Crime, in spatial analysis, must be viewed as a criminal event, and thus the behaviors and travel patterns of victims are equally important to the behaviors and traveling patterns of offenders. Although crime is largely opportunistic, and thus any number of targets may be equally desirable in an offender's eyes (although this not always the case in serial crimes), it is important to understand how victims and offenders interact with their environments to gain a better picture of how a crime unfolds from multiple perspectives. You will also find that law enforcement agencies do not often ask these questions of crime victims, and it may take considerable effort to change the vision of information collection to include collection of travel and behavior of victims in less serious crimes.

Spider Distance Analysis

Spider distance analysis draws lines from each point in a distribution to its centroid, or mean center. Spider analysis helps to answer several questions in crime analysis. First, in a crime series, is the offender likely to be a poacher or a hunter? Poachers or marauders exhibit fairly predictable patterns in their offending, usually committing crimes short distances around their central base (typically their homes, but not always). Hunters or commuters, on the other hand, are much less predictable in their offending patterns. Spider distance analyses can provide clues to analysts to whether or not a series is expanding outward or shrinking inward.

Buffer Analysis and Queries (Theme Selection)

There are two general types of proximity analysis: buffer analysis and queries. A buffer zone analysis is completed by drawing circles around points of reference, such as pawn shops or city parks, at distances determined by the analyst. Essentially, the points (locations of pawn shops or parks) serve as centers (or centroids) for the buffers with a radius set at a distance desired by the analyst (this can be done in miles, meters, feet, and so on). For example, an analyst might wish to draw 0.5-mile buffers around all pawn shops in his jurisdiction. He then may wish to add to a map layer of known burglaries within its jurisdiction and visually scan to see those burglaries that occur within a half mile of a pawn shop. At other times the analyst may want to analyze crime around city parks. A specific park may be the local hangout of transients, and the analyst may want to look at the crime around the park to determine if the crime rate in and around that park is higher than other parks in similar neighborhoods. This analysis may assist with planning efforts and enforcement of no camping ordinances and other issues that often surround high transient population areas. In this case a point is not the center of the buffer, but the limits of the park polygon would act in this manner, and then the buffer would be drawn, following the park outlines outward by 0.5 miles. Buffer zone analysis is a common type of distance analysis and is easily interpreted. See **Figure 9–4** as an example of a buffer analysis.

An analyst may also want to query incidents that are within a distance of some other type of point data, such as schools, liquor establishments, or pawn shops. For example, an analyst may be called upon to produce a map of crimes that are within a given distance of schools. Another example might be to query store robberies that are within a given distance of a freeway entrance. The benefit to using a query over a buffer analysis is that points that are not within the specified distance are removed from the map, lessening clutter and improving map clarity. Another related query is one that queries the number of points within a polygon, such as a police precinct or beat. In fact, any concept of space and relationships in space can be queried, such as within a distance of, totally within, intersected by, or adjacent to something else in a map layer in GIS. The analyst can even use graphic objects drawn on the map display to query points, lines, and polygons in relation to space and distance. In addition, one has the option with queries to create a new theme and save it as its own map layer. However, buffer analysis can be more visually appealing, such as in courtroom illustrations, because one can see the physical proximity between the incident and the point of reference. (For example, a map could have one point to represent a school and another to represent the home of a known sex offender.)

Distance Between Hits Analysis

A distance between hits analysis is the calculation of the distance between each crime in a crime series (crimes committed by the same

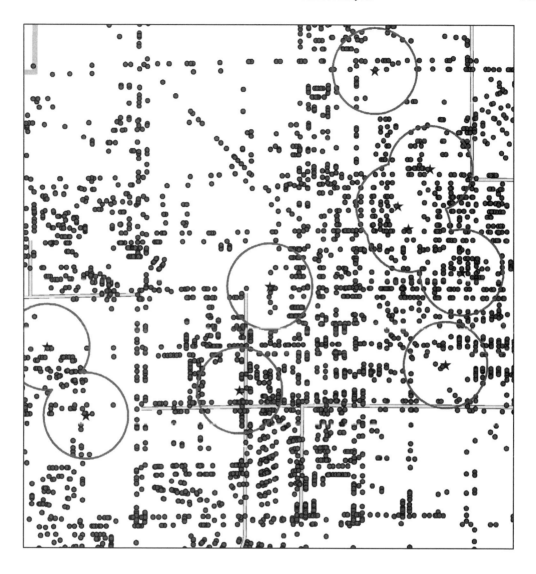

Figure 9–4 1000-foot Buffer of Glendale Schools
Source: ArcGIS

offender(s)) to determine the most likely distance the offender may choose from the most current hit. The concept here comes from the same mean and standard deviation calculations we learned in Chapter 8. We place the crimes in a series on the map and measure the distance from the first crime to the second, the second crime to the third, and so on. We then calculate the average and the standard deviation of those distances. Let's say the following were the distances we found between five hits in a crime series:

- First to second: 2.0 miles
- Second to third: 1.5 miles
- Third to fourth: 1.25 miles
- Fourth to fifth: 0.75 miles

By reviewing the distances between crimes in this series, we can determine a few things. First, the average distance this offender traveled was 1.38 miles with a standard deviation of 0.52 miles. This means that 68% of the time the offender traveled from 0.85 to 1.9 miles between hits, and 95% of the time the offender traveled from 0.33 to 2.42 miles between crimes. Second, by looking at the distances the offender traveled between crimes, we can see that a pattern is emerging between each hit in that the distance from one hit to the next decreased over time. Common sense would lead us to the conclusion that the offender will probably travel less than 0.75 miles from the fifth hit in this series to the next crime he will commit in the future. Depending on the circumstances of the crimes, the targets or victims that are chosen, and the modus operandi, we might be able to use this information efficiently to predict where the offender might go in the future.

Another side benefit of this analysis for tactical purposes is that a directional pattern might also be revealed. In **Figures 9–5** and **9–6** we can see the same crimes where there is no visible pattern and one crime where there is a specific directional pattern to the offender's activities.

Distance and Time Analysis

Distance and *time* are also very often related. When an analyst begins looking at the distribution of crimes based on how far away they are from another crime, a specific location, or a geographic region, he or she should also consider the tempo of the events. In the simple examples illustrated in Figures 9–5 and 9–6, if we looked at the time element, we might also see a pattern where the longer the offender waits between hits, the farther he or she travels between each crime in the series. This may relate to the amount of money the offender got

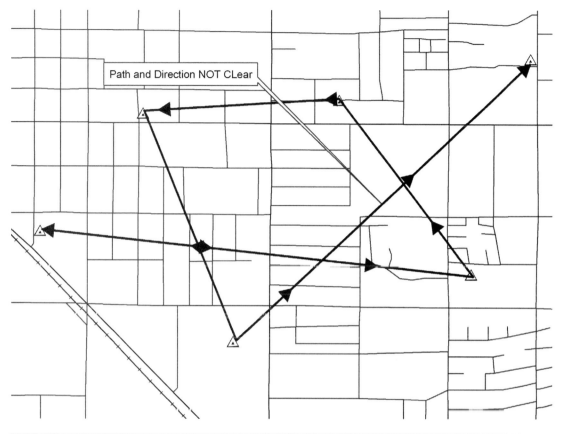

Figure 9–5 Path and Direction Not Clear
Source: ArcGIS

at each crime site and how much money that offender needs to satisfy a drug habit, for example.

Online Resources

- ArcGIS Spatial Analyst: Distance Analysis: http://www.esri.com/software/arcgis/extensions/spatialanalyst/about/distance.html
- CrimeStat III: http://www.icpsr.umich.edu/CRIMESTAT/
- CrimeStat III User Workbook. This entire workbook is extremely helpful: http:// www.icpsr.umich.edu/CRIMESTAT/workbook/CrimeStat_Workbook.pdf
- Spatial Predictive Analysis Crime Extension: http://www.bairsoftware.com/space/help.html

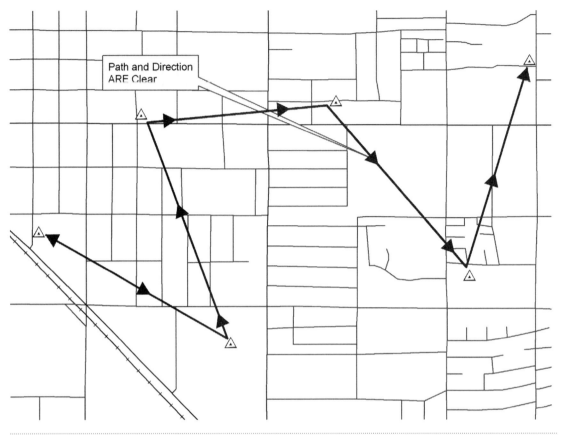

Figure 9-6 Path and Direction Clear
Source: ArcGIS

■ Conclusion

Distance analyses can be used to examine clusters of seemingly unrelated crimes (as in mean center analysis) or to investigate crimes thought to be committed by a single individual (as in journey to crime analysis, spider distance analysis, and distance between hits analysis). In addition, distance analyses can be used to provide a wide assortment of information based on the spatial distribution of crime with respect to how near, how far, within what distance, adjacent to what other feature, or simply how distant in space and time crimes have occurred. All of these forms of distance analyses can and should be used in conjunction with other analysis products to help decision makers make effective and productive decisions.

Recommended Reading

Lee, J. & Wong, D.W.S. (2001). *Statistical analysis with ArcView GIS*. Hoboken, NJ: John Wiley & Sons.

O'Sullivan, D. & Unwin, D. (2003). *Geographic information analysis*. Hoboken, NJ: John Wiley & Sons.

Wong, D.W.S. & Lee, J. (2005). *Statistical analysis of geographic information with ArcView GIS and ArcGIS*. Hoboken, NJ: John Wiley & Sons.

Questions for Review

1. What are the different methods for measuring distance? How does the crime analyst know which method to employ?
2. Explain the different types of distance analyses discussed in this chapter.

Chapter Glossary

Euclidean Distance Euclidean distance is measured by measuring the distance between two points.

Manhattan Distance Manhattan or "street" distance is calculated by using right angles to get from one point to another.

Chapter Exercises

Exercise 16: Count or Density

Data Needed for This Exercise

- _____
- _____
- _____
- _____
- _____
- _____

Lesson Objectives

- Perform simple spatial autocorrelation routines in ArcGIS.
- Review operational applications of distance decay concepts to restrict search locations for possible offenders in a crime series.

Task Description

- Spatial autocorrelation can be described as follows:

    Spatial autocorrelation statistics measure and analyze the possible degree of dependency among observations in a geography. Classic spatial autocorrelation statistics include Moran's I and Geary's C. These statistics tell us if the data is clustered, dispersed, or random, and to what degree. If data appears random or dispersed, then we know that typical crime mapping procedures will tell us very little. If the data is clustered, then crime mapping procedures will provide us some useful information and predictions in most cases. Classic spatial autocorrelation statistics seek to find out if features closer to one another are more related than those further away and analyze the value of this relationship. The concept which makes it useful is that if features that are closer together (clustered) are related, then crime analysis and mapping techniques will provide good results.

Moran's I and Geary's C statistics are also available in Crime Stat III.

- "Spatial autocorrelation" assumes that there can be three basic types of spatial data distributions:
  - Random: There does not seem to be any relationship between incidents, and they occurred by chance.
  - Clustered: There does appear to be some relationship between incidents spatially that we would not expect to occur by chance.
  - Dispersed: Incidents are not clustered or random.
- Distance decay is described as follows:

  Distance decay is a geographical term which describes the effect of distance on features within a specific geography. The distance decay effect states that the interaction between two features declines as the distance between them increases, or in other words, as the distance between crimes increases, the less likely a relationship exists between them.

  Distance decay is graphically represented by a curving line that swoops concavely downward as distance along the x-axis increases. Distance decay can be mathematically represented by the expression $I = 1/d^2$, where I is interaction and d is distance, among other forms. It also weighs into the decision to migrate, leading many migrants to move less far away from their current location than they originally contemplated.

- Distance decay has application in crime analysis to related incidents and also to how far a suspect may travel to commit a crime. Chances are the offender will go far enough away from his home to avoid identification but not so far that he feels uncomfortable with his surroundings. Therefore, a graph with the distances traveled by an offender to his crime locations would show a curve that has many more incidents near his home than incidents farther away from his home.

- In this exercise we will learn to use a simple spatial autocorrelation tool in ArcGIS to find out if your data appears to be clustered, random, or dispersed and use what we know about distance decay to estimate the most probable location where the offender may live.

1. Open EX16.mxd.
2. The top layer is a series of 16 Honda thefts that occurred in late 2003 and early 2004 in this area, which surrounds a hot spot we identified in exercise 15.
3. Click on the **red toolbox** to open the ArcGIS tools.
4. Find the Spatial Statistics tool and open it.
5. Open the tools section called Analyzing Patterns.
6. Double click on the **Average Nearest Neighbor** tool to open it.
7. Complete the tool dialog as follows (see DVD Figure 16–1):
   Input feature class: *Hondaseries*
   Distance method: *Manhattan Distance*
   Check the **Display out Graphically** box.
8. Click **OK** to run the tool.
9. You should see a dialog menu that comes up indicating that the data is slightly dispersed (see DVD Figure 16–2) but that it could be by random chance.
10. This tells us that the Honda thefts may not have any spatial relationship that can be determined from the distance they are from one another, and on the map they do look like they are somewhat randomly distributed across the area being analyzed. If you open the attribute table, you can see that two addresses had the majority of incidents, 4730 W Northern Av and 4949 W Northern Av, which are both apartment complexes.
11. Although they do not show any clustering based on this simple nearest neighbor spatial autocorrelation routine, they are specific to at least two addresses in a big way.
12. To apply some concept of distance decay to these stolen Honda incidents to find the best location for where the offender may live, we need to first find the center of these incidents. We can do this by adding up all of the XY coordinates and dividing by 16 to get the average XY coordinates as shown in DVD Table 16–1.
13. Click on the **Drawing** toolbar and find the tool that adds a graphic point to the map. Move your cursor onto the map display, and watching the XY coordinates in the lower right of the project

window, move your mouse around until it comes close to the 624313.3940 by 928582.8966 measurement, and then click.

14. After adding the point, right click it, and then in the pop-up dialog menu, click the **Size and Position** tab. Enter the correct X and Y coordinates in the appropriate boxes and Apply it. The point will move to the exact XY coordinates we entered.

15. Notice that this point is very close to the two Northern Av addresses we mentioned and does not appear to be at the actual center of the incidents by visual inspection. We would expect to see this point a bit farther to the west based on just what we can see on the map.

16. This is because we are dealing with an average of the X and Y coordinates and because those two addresses make up 11 of the 16 stolen vehicle reports we are working with. The mean center is therefore nearer to the two Northern addresses.

17. Delete the point after you have memorized where it was located.

18. Another way of finding this mean center is to use a tool in ArcGIS. Turn on the ArcToolbox if it is not already on. Find **Spatial Statistics Tools → Measuring Geography → Mean Center** and double click it to launch the tool.

19. The input feature class is the HondaSeries again. Name the output feature class MeanCenter.shp and place it in your student folder.

20. Click **OK** to run this tool, find and add the new layer the tool created in the project, and turn the theme on.

21. How about that? The new mean center is in the same exact place.

22. Another tool related to this is the Central Feature tool. Double click on this tool in the toolbox, call the output feature Centralfeature.shp, and save it to your student folder.

23. Use the Manhattan distance (instead of the straight line, how the bird flies, or Euclidean distance) because it is more representative of the street network the suspect most likely has to use to travel from crime to crime.

24. Add this new theme to the project when it is completed, and notice that this tool creates a new point for the incident that is the most central to the other features, which in this case is 4949 W Northern Av.

25. DVD Table 16–2 indicates that the average suspect travels 8.12 miles from his known home location to the location of a stolen motor vehicle crime. Because we know this distribution is skewed (recall distance decay, which is more close trips than far away trips) we need to use the median distance of 6.65 miles as our central measurement or average. So an auto theft suspect will travel on

average 6.65 miles from his home to a place where he steals a vehicle. We know just from looking that these distances are too large to be of much use in this analysis.

26. Another way we can look at our data is to determine the percentage of trips that were within specified distances from a suspect's home or what is often called a cumulative frequency table.

27. Creating a cumulative frequency table for every trip taken by known auto thieves reveals that 31% of those trips were 1 mile or less, 55% were 3 miles or less, and 65% were 5 miles or less. Considering that we do not verify every suspect's address when they are arrested, and that the addresses the suspects give are questionable in 30–50% of the arrests, we cannot rely on this information with great confidence (but it is all we will likely get.) However, we can say that the suspect could likely live within 1 mile of the mean center of the Honda auto thefts we are analyzing, especially because no incident is greater than 1 mile from the mean center, or the central feature, in this imaginary series.

28. We can then look for possible auto thieves who live within a 1-mile radius of the mean center or central feature who meet our criteria for the suspect description. We may need to compare lifted prints, etc. to them. In general, we can think about the distance decay and the suspect's thought patterns and try to determine why he may be targeting Hondas in this neighborhood. Could he be targeting Hondas outside this hot spot area we don't know about? _____. Would knowing where these Hondas were recovered help us? _____. If 11 out of the 16 Hondas were recovered in the same general location, would this be more or less indicative that the offender probably lives within this area?_____.

29. You are now an expert in spatial autocorrelation and distance decay theory as it applies to crime analysis and your daily duties!

30. Close the project and save it in your student folder as EX16.mxd.

Exercise 17: Distance Between Hits and the Mean Center

Data Needed for This Exercise
- All data needed for this exercise is in EX17.mxd.

Lesson Objectives
- Learn how to create predictive layers in a tactical analysis for the distance from last hit and distance from mean center.
- These two processes assist in giving us some ability to predict high-risk areas that are commonly outside of the area around the mean center where most standard deviation methods highlight.

Task Description
- In this exercise we will take a robbery series, a burglary series, and an indecent exposure series and analyze them based on how far each crime is from the others. We will use this information to develop a mean center buffer and a distance from last hit buffer that we can use to help us narrow down our analysis and prediction.

1. Open EX17.mxd.
2. When the project opens we see our typical robbery series from exercise 9, a data frame for burglary, and another one for indecent exposure.
3. We will use each of these imaginary crime series to analyze the distance between hits and create a buffer around the last hit in the series based on the mean and standard deviation of the distances between hits. The concept here is that if the offender has already traveled n miles on average, then one could assume that he will travel at least that far to the next hit.
4. We will also do a similar analysis, but instead of using the last hit as the start of our buffer, we will use the mean center to draw a buffer around it. The assumption with this analysis is that if all of the crimes were within an average of n miles of the mean center, then it could be assumed that the next hit may also be that far away from the mean center.
5. Let's start out by turning on the mean center layer.
6. Notice that the next hit would be expected to be in the empty area in the middle of the crime series.
7. Now turn on the Robbery Central feature layer. This point shows us the crime that is closest to the mean center or the central feature of the points on the map. Either of these could be used as the "center" of our crime series and the decision to use one or another really depends on the logic we derive from what we know about our suspect's behavior. In this robbery series, we know that the offender is starting to move east a little, so the central feature could be a better "center" point for that buffer than the mean center. It probably makes little difference; however, it is something to consider based on the modus operandi (MO) information you know about this offender and this series.
8. Turn off these layers and turn on the Robbery Path layer. Open the attribute table of the Robbery Path layer.

9. Notice that the only fields in this layer are FID, Shape, and ID, and each of the ID numbers are zero. These are the default fields that ArcGIS creates when you create a new shapefile.
10. To find the mean and standard deviation of the distance between these hits, we need to know the length of each line segment (distances between each hit). A simple tool in the Table view, called Calculate Geometry, allows us to calculate this.
11. First we need to add a field to hold the value we are going to calculate.
12. In the Table view, click on the **Options** button → **Add Field**.
13. We will call this new field Length and make it a Double data type.
14. Click **OK** to create this new field.
15. Click on the **Length** field name, and then right click it.
16. Click on **Calculate Geometry**, and in the dialog box complete it as follows (see DVD Figure 17–1):

    Property: *Length*
    Coordinate System: *Use coordinate system of the data source*
    Units: *Feet US (ft)*

17. Now we have the length in feet for each line segment that separates all of the crimes on the map and the sequence between them.
18. Right click the **Length** field name again, and then pick **Statistics** from the drop-down menu.
19. We find that the mean is approximately 8391, and the standard deviation is approximately 3124. Converting these figures to miles, we get _____ and _____.
20. From this quick calculation we find that the average distance between each hit was 1.6 miles, 68% of the hits were between 1 mile and 2.2 miles, and 95% of the hits were between 0.7 miles and 2.8 miles. From several years of analysis with different crime series types, I have found that the suspect very seldom hits again within the lower range of the 68% percentile and not generally greater than the maximum range of the 95% percentile. Our sweet spot is actually greater than 1 mile from the last hit and less than 2.8 miles from the last hit. To portray this on the map, we can create some simple graphics.
21. Using the Draw toolbar, find the Circle drawing tool.
22. Make sure to zoom out twice from the series extents before starting the next step.

23. Find the last hit in the series (labeled 8), and starting at that point, drag a circle out the distance of 1 mile (or as close as you can get it).
24. Right click on the new graphic, make the fill no color, and make the line red and 2 points.
25. Repeat this by drawing another circle out 2.8 miles (14,784-foot radius) around the last hit in the series.
26. Make this circle graphic hollow with a red outline as well.
27. What we really want is a donut-shaped polygon where the area that is less than 1 mile from the last hit is empty. To create this we need to select both circle graphics and then right click the map display screen. In the pop-up menu, choose **Graphic**. Click **Operations → Remove Overlap**. This will create the donut shape we desire and make the two graphics into one.
28. Change the fill color to yellow to see that our donut does exist now.
29. To see through this graphic donut we need to use a different type of fill pattern than a solid color. Right click the donut and choose **Properties** from the pop-up menu. Click the **Change Symbol** button, and in the symbol choices on the left, scroll down to the bottom and find symbols that are crosshatch and line patterns.
30. Choose the type that says **10% Simple Hatch**. Change the outline box to red and 2 pts, and change the fill color to red and apply the new properties.
31. Where the standard deviation ellipses and this donut overlap is our most likely next hit location. If we add in the directional analysis we discussed in exercise 9, we then limit the area down a little bit more.
32. In some series you may only want to create a buffer graphic of just the mean distance traveled (if they are all very regular). If the offender's pattern seems to show he is traveling less and less between hits, you may want to create a buffer graphic of just the low ranges of the 68% and 95% distances instead of using the low side of the 68% and the high side of the 95% ranges. The great thing about this method is that you can adjust your analysis based on the suspected behavior of the offender, and it is not as mean centric as the other methods out there. I suggest using this method in conjunction with the other methods, however, and not by itself.
33. Turn on the mean center layer and the mean center distances layer.
34. The length field has already been calculated for you, so check the statistics of the length field in the attribute table from the mean center distances layer.

35. The average distance from the mean center to each crime in the series is 1.1 miles, and the standard deviation is 0.26 miles.
36. There are no robberies within a half a mile of the mean center, and the closest robbery is the central feature (labeled 6) at around 0.64 miles from the mean center.
37. Our minimum distance should probably be 0.5–0.6 miles.
38. Our maximum distance should be the mean (1.1) + 1 standard deviation (0.26), so the maximum distance is 1.36 miles. We therefore need to create two buffers: one at 0.55 miles and one at 1.36 miles (or we could go out to 1.1 + (2 × 0.26) or 1.62 miles).
39. Instead of creating graphic buffers this time, let's use the buffer tool to create our buffers.
40. Open the ArcToolbox and find the Analysis tools.
41. Open the **Proximity** section and double click on the **Multiple Ring Buffer** tool.
42. The input feature for this theme is the robbery mean center.
43. Save the output to your student folder and call it MeanCtrBuff1.
44. Enter *0.5* in the distances box, and then click **Add**. Enter *1.36* in the distances box and click **Add** again.
45. Change the buffer unit field to *Miles*.
46. Make sure the Dissolve option is *All*, and click **OK** to run this tool.
47. Add the buffer layer to the project and change the legend so that you can see each buffer separately. You can see that they appear to overlap. Change the legend so that only the 1.36 mile buffer has a class in the legend (click on the **0.5 mile** class and click **Remove**) and uncheck the **<all other values>** box. This results in the same donut we created using graphics.
48. Now we have a mean center area and a last hit buffer area. Where these two overlap is the high-risk area for a new hit in the series.
49. Combine these two methods with the standard deviation ellipses or rectangles and our directional analysis, and we have a pretty specific prediction for a high-risk area.
50. If we combine the mean center buffer, the last hit buffer, and the two standard deviation ellipse and keep only those areas where those layers overlap, we would get a high-risk area prediction that looks similar to DVD Figure 17–2.
51. In this robbery series it appears that the offender would most likely hit where crimes 1, 2, 3, and 5 were located, or the offender might shift to new territory east of where he has been hitting. Finding

and then adding all of the similar businesses this offender has hit in the past could give you a very small number of targets to ask surveillance units to watch, which will make your analysis more useful.

Burglary

52. Activate the Burglary data frame and save your project as EX17.mxd in your student folder.
53. In this series we are dealing with smash and grab business burglaries. The offenders hit the businesses between midnight and 5:00 a.m., and so far on Thursdays and Tuesdays. They generally find some large rock, brick, or other locally found landscape item and smash out the front window, enter the business, and take product and any cash they may find in the store register. The only similarity among the businesses in this series is that they are all located within small strip malls and they sell some product within the store and have store displays in the window (RadioShack, Cigarettes Cheaper, Kline's Hobby Shop, Sally Beauty Supply, etc.).
54. Review the data in the attribute table and look for any patterns.
55. Review the sequence of the burglaries and look for any patterns.
56. There is not too much there. Thursday and Tuesday and the time frame (from midpoint) all look like a pattern, but the sequence of events does not look very promising. In addition there are huge gaps in hits, so we are likely missing some related cases.
57. Let's start by using the ArcToolbox to create the mean center point and the standard deviation ellipses for this series.
58. These tools are both located in the Spatial Statistics toolbox. Perform these analyses now and add the new layers to your project and save the project.
59. We want to create a line path layer between hits so that we can determine the distances between each hit (we can also just use the Measure tool). To create a new shapefile, we can also use the ArcToolbox. In the ArcToolbox, find the tool section called **Data Management → Feature Class**.
60. Double click on the tool called **Create Feature Class** (see DVD Figure 17–3).
61. For the output location use the directory folder address for your student folder. (*We are looking for the folder name, not a file name.*)
62. In the output feature class name, type in *BurgPath*.
63. The geometry type will be *Polyline*.

64. At the very bottom we need to enter a projection for this data. This will make sure that the lengths and distances are accurate.
65. Click the **Browse** button and then the **Select** button in the pop-up menu.
66. In the Browse for Coordinate System window, double click on **Projected Coordinate Systems**.
67. In the next window, choose **State Plane**, then choose **Nad 1983 (feet)**.
68. In the next window, choose **Nad 1983 StatePlane Arizona Central FIPS 0202 (feet). prj**.
69. Click the **Add** button.
70. Click **Apply** and then **OK**.
71. Click **OK** to run the tool.
72. Now add the new, empty shapefile you created into the project.
73. If you turn on this layer, nothing will show on the screen, and if you check the attribute table, you will find no records. This is because you have not created any graphics or added any attributes to this new layer. We just created the shell to hold that information, and now we need to actually add some data.
74. Click on **View → Toolbars → Editor** to turn on the Editing toolbar.
75. Click on **Editor** in the toolbar, and choose **Start Editing**.
76. In the Editing pop-up menu, choose the folder where you placed the BurgPath file you created.
77. Click **OK**.
78. In the Editor toolbar, make sure that the Target box says *BurgPath* for the name of the layer you wish to start editing.
79. Also make sure the Task says *Create New Feature*.
80. We need to create each line between the crimes in the series individually. Click on the little **pencil** in the Editor toolbar, and then click once on the first crime in the series (**label 0**).
81. Move your mouse to the next crime in the series (label 1), and then double click to save the line segment while on top of the next crime (label 1). Click once on **label 1** again to create a new line segment, and then move your mouse over the next crime in the series. Double click on that point to create the new line segment. Repeat this action until you have created six line segments that connect the crimes by sequence of occurrence.

82. Now we need to stop editing and save our changes. Click on **Editor** on the Editor toolbar, and choose **Stop Editing**. Indicate **Yes** to save the changes.
83. You've just created a sequential crime path for your series.
84. Now open the attribute table for this path, and add a new field called *Length* (use double as the type). Right click the name of this new field in the table and in the drop down menu. Click **Calculate Geometry** and add the **length** of each line segment in **miles** to the table.
85. Find the statistics for the paths.
86. Create a multiple buffer ring around the last hit in the series (make sure to select it first from the attribute table for the crimes in the series) at 0.82 miles and 2.54 miles, and remember to set the dissolve to All.
87. Add the new theme and see what it looks like compared to the standard deviation ellipses and what you think would be the most likely area for a new hit in the series.
88. Use the Measure tool to measure the distances between the mean center and each crime in the series and figure out what the average distance from the mean center is. (Click on the mean center with the Measure tool, drag it out to each point, and record the distance. Add the distances and divide by the total number of distances measured.)
89. Using the Buffer tool in the ArcToolbox, create a single buffer from the mean center that average distance.
90. Now look at where the mean center buffer, the last hit buffer, and the standard deviation ellipses all overlap. This small area is where the next hit is most likely to occur. You can use the Draw tool to generally follow this overlapping area and use this for your prediction or highlight the overlap in some manner (see DVD Figure 17–4). Alternatively, you can get rid of the layers that you used to create the final analysis area and keep the drawing of the overlapping area alone (see DVD Figure 17–5).
91. Save your project in your student folder as EX17.mxd, and then activate the indecent exposure data frame (right click on the data frame name and choose **Activate** from the pop-up menu).
92. Open the attribute table for this crime series and see if there are any temporal patterns that are useful in helping you figure out what this offender is doing and why.
    a. Are the hits on any specific day of the week?
    b. Do the hits occur during any specific time period?

c. Is the tempo of the hits increasing or decreasing over time (days between hits)?
93. Is there any travel pattern than can be seen by following the crime sequence on the map display?
   a. With this many cases, you may find it more difficult to see a travel pattern than with the robbery series or the burglaries.
   b. Generally, the offender does seem to be crossing back and forth over the center area where there are no crimes, which indicates the offender may _____.
   c. If we use the Measure tool, we find that the offender traveled a little over 63 miles between hits. Because there are 20 hits in the series, this means that there were 19 trips, and the average distance traveled between hits was _____ miles.
94. Create the mean center point and standard deviation ellipses for this series using the tools in the Spatial Statistics toolbox.
95. Using the Measure tool, what is the average distance between the mean center and each hit in the series? _____.
96. Using the distance measures for the last hit and mean center, create at least one buffer around the last hit in the series and the mean center for the average distance traveled, and add them to the map.
97. Using the Draw tool, create your best estimate of where the next hit in this series might be using the standard deviation ellipses, mean center buffer, last hit buffer, and any directional analysis you may have come up with.
98. See DVD Figure 17–6 for the final result map.
99. Save and close your project.

Exercise 18: Spatial Autocorrelation and Distance Decay

   Data Needed for This Exercise
   - All data needed for this exercise is in EX18.mxd.

   Lesson Objectives
   - Take a crime series from start to finish for a tactical analysis.
   - Use the CrimeStat III tools to create:
     - Mean centers
     - Centers of minimum distance
     - Spatial deviation ellipses
     - Correlated walk analysis to get path between hits
   - Experiment with interpolation in CrimeStat III.

Task Description
- This exercise will take you through the typical tactical robbery series analysis and cover several topics of information that can aid you in recreating a tactical analysis on your own data.
- In this exercise, we will take data from records we have entered into a robbery matrix in Excel and convert them to tabular format for manipulation and adding additional fields. We will then learn how to export this data for use in ArcGIS to get the best results and geocode the data for our analysis.
- We will also use this data in CrimeStat III and do several spatial statistics calculations to provide us with ArcGIS output we can use to assist us in developing a final analysis result, which will help us predict the next location in a crime series.

1. The robbery detail has come to you with six cases and indicated they think the same suspect is responsible for each robbery. The report numbers are 7-2345, 7-3680, 7-5015, 7-6350, 7-7685, 7-11690, and 7-14360. The basic information you get is that a white or Hispanic male enters a video store, acts like he is going to rent a movie, then pulls out a handgun, threatens the clerk, and takes money and sometimes DVD movies during the robberies. You then build a robbery matrix and search the records for other cases that have similar characteristics. You find a total of 15 cases in 2007 that could be the same offender.

2. You read each report and create a robbery matrix and list of the offenses in Excel. The results of your in-depth analysis can be found in the C:\CIA\Exercises\exercise_18\RobberyMatrix.xls file. Open this file.

3. There are two tabs in the worksheet; one is the matrix and the other contains a tabular list of each offense.

4. Review the matrix and decide which cases besides the six the investigator gave you should also be included in this robbery series. Be sure to think about the witness reliability factor and MO characteristics the robber used in each robbery. When you have decided which robberies to include, make the robbery list spreadsheet active in the Excel workbook (you can also use the robbery list spreadsheet if you like reviewing the cases in a tabular form rather than in a matrix; see DVD Figure 18–1).

5. You will need to think about what justifications you may have for including all of the cases, or some of the cases, or only the original six cases provided by the investigator. You may also want to notify

the investigator about the additional cases so surveillance video can be pulled and compared to the other six robberies in real life. Whatever rational thought process you use to exclude or include reports, you must be able to defend it when you do your analysis and provide a tactical prediction. In most cases, when you begin reviewing the cases, you will become the expert in this series, and your opinion will be important in the analysis.

6. You may find that in a real situation, the detective may even argue with you on which cases should be included. This is a good thing and should be encouraged if the logic is sound, and you should consider his opinion and work toward making the relationship a cooperative one; however, it is still your judgment and opinion that decides which cases are included because you are responsible for the reliability of the prediction.

7. For this exercise we will include all of the robberies in the matrix and continue.

8. In the robbery list spreadsheet, we see all of the details of the offenses with the primary date, time, and location information to the left and the other MO information to the right (see DVD Figure 18–2). This is a fairly standard way to set up this information.

9. Some analysts use the matrix and use a paper copy to compare the robberies side by side, some do it electronically, and some just compare everything in a list like the robbery list and never use the matrix. What you use is not as important as your logic and how you justify including or excluding one case or another. How many similar MO factors do you need to say a case is related? How much do you trust the witness accounts of the incidents and their descriptions? Do you have video or other physical evidence to verify that the same suspect is involved? Do you need that?

10. In the robbery list we see that I have sorted the crimes by date and time of occurrence, and I added a datetime field, DBH (days between hits) field, a CDBH (cumulative days between hits) field, and a DOLPDAY (dollars per day) field. These extra fields can be used to help you sort the data and also do some quick temporal predictions in Excel and then later in CrimeStat III.

11. Click on the temporal analysis spreadsheet and open it (see DVD Figure 18–3).

12. Review the spreadsheet and notice how the temporal analysis can also help you decide which cases should probably be included. The hour of day this offender has hit has been very consistent until

the last hit, which was at 0721 hours instead of late afternoon. This case could be questionable unless other MO information strongly indicates it should be included. If it is included (which it is for our analysis), it could be an indication that the offender is changing his MO or the temporal pattern for some reason. The DBH field kind of shows us that there has been a different tempo to the offender's behavior throughout the robberies. At first he always took 30 or more days before the next hit, and toward the end he hit as frequently as 2 days apart. This could be due to the amount of money he was getting for his efforts, or he could have been in jail or out of town for a while during recent months. Granted, much of this is pure speculation by the analyst, but this is something you should discuss with the investigator. He may have some insight and ideas of his own that can help you make sense of this or ignore it. Perhaps a news announcement ran on television that showed video of the offender. This often causes some change in MO patterns in a robbery series.

13. Review the formulas used to make all of these temporal predictions and calculations. You can save the Excel spreadsheet in your student folder or thumb drive if you want to refer to it later.

14. The process from this point would be to take our temporal data and place it in a Word document as a bulletin and get our tactical analysis started. Because this class is about GIS, not Excel or Word, we won't spend much time on actually creating a bulletin in these software products.

15. After we have accomplished all of this, we would probably create some text fields for the datetime field because Excel generally truncates such datetime fields to just a date when we export the data to a DBF table. ArcGIS does not always handle dates and times as we would like either, so often we need to create a text equivalent of the datetime and time fields by using the Text function in Excel. For example, the format to make the datetime value a text value but retain the datetime look and feel is =TEXT(a2,"mm/dd/yyyy HH:mm").

16. After we have set up all our fields the way we like them, we need to select all of the data and then go to **Insert → Name → Define** and call this range of data *GIS*. We can then either export it as a DBF table or do a direct OLE DB link from inside ArcGIS 9.2 to that named range. In this case, the DBF table has already been created for you to save time and a lot of questions about Excel.

17. Close the Excel spreadsheet and open EX18.mxd.

18. Add the C:\CIA\Exercises\Exercise_18\RobberyList.dbf table to the ArcGIS project, and save it as EX18.mxd in your student folder.
19. Open the RobberyList.dbf table and view the tabular data.
20. Notice in the table that despite all our best plans, the Bethany Home Rd addresses only have an "R" and not "RD" for the street type. Let's see how well this geocodes for us.
21. Right click on the robbery list table and choose **Geocode Addresses** from the pop-up menu.
22. Use the only geocoding locator currently in the project, and save your geocoded data in your student folder as rob.shp.
23. All 15 robberies should geocode in a few seconds. Accept the geocoding results and close the geocoding dialog. (If you have problems, please bring them to your instructor's attention.)
24. Look at the spatial distribution of your robberies because this is another factor we want to look at when deciding if our cases are related. Do they appear clustered (bunched together), random (all over the map and no pattern), or dispersed (spread out but a possible pattern)?
25. Label the points using the SEQNUM field.
26. Follow the travel between hits and see if there appears to be any pattern in travel behavior. Do you see one or maybe two?
27. At this point we could use the Spatial Statistics toolbox to create our spatial deviational ellipses and mean center as we did in the last lesson, but this time we will take a different approach.
28. If you remember correctly, we had the XY coordinates in our table, and those extra fields were called DBH, CDBH, etc.
29. Let's open CrimeStat III and run some routines to help us analyze this series.
30. Run the CrimeStat III executable file at C:\CIA\Resources\CrimeStatIII\crimestat.exe.
31. After a few seconds the welcome screen will disappear, and you will have the data input menu in front of you (see DVD Figure 18–4).
32. Click on **Select Files** and find the rob.dbf (part of a shapefile along with the .SHX, and .SHP files that are stored when you create a new shapefile) you created in steps 22–23. If you cannot find your file, you can find a version in the exercise_18 folder.
33. Click on the column item for the X variable and choose the field named "**X**". Choose the field named "**Y**" for the Y variable,

and choose **CDBH** as the time variable, as shown in DVD Figure 18–4.

34. Change the type of coordinate system to projected (Euclidean), and the data units should be *feet*. The time unit should be *days*.
35. Click on the **Spatial Description** tab to get the next menu, and make sure the Spatial Distribution tab is selected (see DVD Figure 18–5).
36. We will create the mean center, standard ellipses, center of minimum distance (closest point to all crimes), and the convex hull polygon (join the outer points of the series). Click on each of these items, and then click the **Save Result To** button for each item (see DVD Figure 18–6). In the Save Output To box, choose **ArcView SHP**, and save each layer to your student folder.
37. I placed a 1 in each section, and CrimeStat III will add other information to the 1 in the final file name.
38. Click on the **Compute** button when you have given each file a name.
39. You will see a dialog menu, and when it says Finished at the bottom left, you can then close this dialog box only (see DVD Figure 18–7).
40. Do *not* close the CrimeStat III window where we set all this up because we can use it again. Minimize the CrimeStat III window.
41. Go back to the ArcGIS project and load the new files that you named.
42. If you did everything correctly, you should have nine new shapefiles. Load each one.
43. There will be four point files and five polygon files altogether. The four point files are:
    - GM: Geometric mean
    - HM: Harmonic mean
    - MC: Mean center (mathematical center of crimes)
    - MCMD: Center of minimum distance (closest point to all crimes)
44. See Chapter 2 (quick start guide) or Chapter 4 in the CrimeStat III manual for a description the spatial description routines and the science behind each of them. These chapters can be found in the folder called "Manual" in the CrimeStat folder.
45. Turn on each of the points and see that they are very close together. This generally indicates that the offender is very consistent in his or

her crimes and that you may have a very good chance of predicting the next hit area accurately. It also gives us a pretty good indicator that the offender could live very near these points.

46. You can remove the harmonic and geometric mean points from the project and turn the other two off.
47. Turn on the first polygon layer that starts with 2SDE. This is the two standard deviation ellipse like we created using the Spatial Statistics toolbox in ArcGIS in the previous lesson. Its companion starts with SDE and is the one standard deviation ellipse. Turn these on and see how your crimes are distributed. Change the symbology as needed so you can see through each layer.
48. Turn on the layer that starts with Chull. This layer connects all of the outer points in the crime series, and it basically tells us that all crimes were within this odd-shaped polygon. The convex hull is another method that crime analysts have used to predict the next hit in a crime series. Studies in Glendale have shown that in robbery series, the convex hull contains the next hit 83% of the time. But as with the standard deviation ellipses, it can be very large.
49. Turn on the layer that begins with SDD. This is very similar to the standard deviation rectangle because it is calculated much the same; however, the result is a circle rather than a rectangle.
50. Turn these layers off, and turn on the last polygon layer we added that begins with XYD. This is the 68% standard deviation rectangle.
51. Turning all these layers on gives us something that looks like DVD Figure 18–8. All of these layers are called "centrographic" measures because they involve finding the middle of the crime series spatially and then create some buffer around the center based on a mathematical formula that uses standard deviation.
52. You can probably see why the following Glendale statistics for 24 crime series were possible:
    a. 54% of the predicted next hits were within the one standard deviation rectangle.
    b. 91.7% of the predicted next hits were within the two standard deviation rectangle.
    c. 71% of the predicted next hits were within the one standard deviation ellipse.
    d. 95.8% of the predicted next hits were within the two standard deviation ellipse.
    e. 50% of the predicted next hits fell within the average distance between hits buffer from the last hit.

f. 83.3% fell in the mean plus two standard deviations buffer.
 g. 83% of the predicted next hits fell within the convex hull polygon area.
53. Now let's go back to CrimeStat III and create some more mystery.
54. Restore the CrimeStat III window we minimized.
55. Uncheck all the spatial description items we selected the last time.
56. Click on the **Data Setup** tab again, and then click on the **Reference File** tab.
57. In the section called Reference Grid, enter the following coordinates:

   Lower left X: *591037*   Lower left Y: *900088*
   Upper right X: *641107*   Upper right Y: *943763*

58. Now click on the **Measurement Parameters** tab under Data Setup.
59. Enter the following data:

   Square miles: *78*
   Street miles: *1532.6*

   Check the box next to Indirect (Manhattan) for the type of distance measurement.
60. Now click on the **Spatial Modeling** tab.
61. On the Interpolation tab, check the **Single** check box to run a density analysis for our points using a single input file.
62. Accept all of the default values in the rest of the dialog, and then click on the **Save Results To** box. Save this like you did with the other layers in your student folder. Again, I just called this file "1" because CrimeStat will add a prefix to the file name. Make sure to choose and ArcView SHP as the type.
63. Click on the **Space Time Analysis** tab.
64. Under the Correlated Walk Analysis section, click on the **Prediction** check box.
65. Change the time, distance, and bearing method boxes to *Median*.
66. Click on the **Save Results To** button again, and save this file to your student folder (again, I used "1").
67. Click **Compute** to run your new advanced spatial statistics.
68. In the Results window, click on the **CWA-P** tab and find a section that says:

   • Predicted time: 314.56500
   • Predicted X coordinate : 629498.82389
   • Predicted Y coordinate : 922686.82626

69. Write down the predicted time value because we will use it later.
70. Close the results window, and minimize the CrimeStat menu, but do not close it.
71. Restore the ArcGIS project and add your new layers.
72. There will be two point layers, one polygon layer, and three line layers.
73. The point layer that starts with POOrigL is the location where CrimeStat detected as the best point to predict the next hit based on analysis of time, distance, and bearing between hits.
74. The point layer that starts with PredDestL is the actual location where CrimeStat predicts the next hit to be. I have found this to be very reliable at predicting the next date and the distance from the last hit but horrible at predicting the right direction from the last hit. Therefore, this point is not likely to show you the actual hit location, and in fact it is probably 45–180 degrees away from where the hit will be.
75. You can remove these two points from the project.
76. Now open the table for the rob.shp and look at the last value in the CDBH column. It is 297.51. Subtract 297.51 from 314.57 and you get 14.06, or 14.06 days from the last hit on October 25, or November 8 or 9.
77. Notice how all of the methods for predicting the next date of hit are all within the same approximate range. This is a very good indicator that your next hit date prediction is very good. I generally always use the Correlated Walk Analysis date and then add three days before and after it. In this case November 5, 2007 through November 12, 2007 would be a great date range to advise the investigator when surveillance would be optimum for catching the suspect committing a new offense.
78. Close the attribute table.
79. The Line theme that begins with Path shows the probable path to the next hit. The theme that starts with PW shows the general direction of the next hit or bearing line.
80. The Events theme represents our lines between each hit in the series starting with case 1 and ending with case 15.
81. Unfortunately the Events theme does not give us each individual segment distance (open attribute table to see that the events theme has only one record), but we can change the legend to include multiple arrowheads to give us a general idea. We can also add a new field called Length and use the Calculate Geometry tool (by

right clicking **field name** and choosing **Calculate Geometry** from the drop down menu), which eventually gives us 2.8 miles average distance between hits (e.g.; (209774.177186 feet ÷ 5280) ÷ 14 trips = 2.83785 miles).

82. We can create a buffer of 2.8 miles around the last hit in the series for our final prediction and see if it helps us narrow down the area.
83. Perform this buffer analysis as we did in the previous exercise.
84. If you have not been able to follow this lesson or had difficulties, sample layers have been created for you in the exercise_18 folder.
85. The final layer we will look at is the result of our interpolation density run in CrimeStat. This layer starts with a "K" and then whatever you named it.
86. Turn this layer on.
87. Right click the layer and then choose **Properties → Symbology**.
88. Choose a **Quantities → Graduated Color** legend type, and use the field called "Z" as the classification value.
89. You should see something that looks like DVD Figure 18–9.
90. By adjusting the settings in CrimeStat for the interpolation function, you can predict the next location, or probabilities, or create density maps that look very similar to what you can get out of the Spatial Analyst extension. By using a quartic method, a bandwidth of fixed interval or 1 mile, and relative densities under output units, the result more closely resembles a spatial analyst density map.
91. Some analysts use this as a predictor for the next hit. The highest density areas where the most hits have been is the high-risk area where the next hit *could* be.
92. Save and close your project.

References

Costello, A., & Leipnik, M. R. (2003). Journeys to crime: GIS analysis of offender and victim journeys in Sheffield, England. In M. R. Leipnik & D. P. Albert (Eds.), *GIS in law enforcement: Implementation issues and case studies* (pp. 229–231). New York: Taylor & Francis.

Groff, E., & McEwen, J. T. (2005). Disaggregating the journey to homicide. In F. Wang (Ed.), *Geographic information systems and crime analysis* (pp. 60–83). Hershey, PA: IDEA Group.

Levine, N. (2002). *CrimeStat: A spatial statistics program for the analysis of crime incident locations* (Vol. 2.0). Ned Levine & Associates, Houston, TX, & the National Institute of Justice, Washington, DC. Retrieved March 2009, from www.icpsr.umich.edu/NACJD/crimestat.html

Rossmo, D. K., Laverty, I., & Moore, B. (2005). Geographic profiling for serial crime investigation. In F. Wang (Ed.), *Geographic information systems and crime analysis* (pp. 102–117). Hershey, PA: IDEA Group.

Hot Spot Analysis

CHAPTER 10

> ▶ ▶ **LEARNING OBJECTIVES**
>
> Chapter 10 explains several types of hot spot analyses and discusses the strengths and weaknesses of each type. The type of hot spot analysis used depends largely upon two factors: the type of data available and the questions about various crime clusters that need to be answered. After studying this chapter, you should be able to:
>
> - Provide a working definition of a "hot spot."
> - Be able to explain different types of hot spot analyses.
> - Identify the strengths and weaknesses of each hot spot analysis.
> - Interpret the results of any hot spot analysis in a manner that is useful for practical applications.

■ Introduction

The available research provides several definitions for the term "hot spot," which can be slightly confusing for the beginning student. For example, a hot spot may be loosely defined as "a single address, a cluster of addresses close to one another, a segment of a streetblock, an entire streetblock or two, or an intersection" (Taylor, 1998, p. 3). Conversely, a hot spot may have to adhere to strict requirements, such as being no longer than a standard street block, not being within a half of a block from an intersection, and being at least one block away from another hot spot (Buerger, Cohn, & Petrosino, 1995). Boba (2005) utilizes the definition put forward by Sherman, Gartin, and Buerger (1989) whereby a hot spot is a specific location or small area that experiences large amounts of crime. She distinguishes hot spots from hot dots, hot products, and hot targets. Hot dots represent people who are repeatedly victimized, hot products are items or property that are repeatedly victimized, and hot targets are places that share similarities (fast food stores, convenience stores) that are frequently victimized.

Eck et al. (2005) suggest that although the definitions of hot spots vary within the literature, the "common understanding is that a hot spot is an area that has a greater than average number of criminal or disorder invents, or an area where people have a higher than average risk of victimization" (p. 2). They suggest that hot spots can be in the form of a hot spot place, hot spot street, hot spot neighborhood, hot spot city, or hot spot region. In addition, recall from Chapter 8 that selecting the appropriate level of analysis is far more important than adhering to a rigid definition that does not fit the situation.

Another point to make is that hot spots are dynamic; they change over periods of time. They can move, change shape, expand, contract, or disappear and reappear depending on two variables: the types of crimes being analyzed and the duration of time over which those crimes occurred. This chapter reviews some of the common types of analyses used to identify hot spots and clusters of crime. During our examination, we will refer to several concepts and ideas presented in the crash course on statistics in Chapter 8. If you find yourself having difficulty understanding some of these analyses, you may need to revisit the relevant sections found in Chapter 8 for additional help.

Hot spots, hot areas, hot places, or hot street segments are classifications designed to take a large amount of data and identify places where crime is at a higher level than other places and to help identify those areas for problem solving efforts through enforcement. A hot spot analysis is just the start of any good crime analysis effort and can be used to find single event hot spots at which to address enforcement or to track progress of a tactical action plan over time. When the analyst has identified a problem area by time of crime and temporal distribution, the work has usually just begun. The analyst will need to drill down into the hot spot and often may need several methods of hot spot analysis to pick up all of the issues causing that particular hot spot. For example, we might look at residential burglary within our city. We want to first look for the hot spots across the entire city, perhaps for the entire year. Our scale at this point is a citywide analysis. This will yield us some results and identify generalized hot spots for residential burglary. Once we have identified several hot spots, we want to choose the ones that have the most number of burglaries and then look at each individual hot spot. We will likely find that apartment complexes make up a large number of our residential burglaries in some of these hot spots. To find out if this is the case, we may want to look at repeat burglary locations or addresses and create a graduated point or "hot place" map. If these hot places are inside our hot spots, we may want to exclude them and retry our hot spot analysis and see if our hot spot remains a hot spot. When this analysis is done, we might even consider changing our unit of measurement from individual crimes to rates such as burglaries per 1000 residents to level out the playing field between the single family homes and the apartment complexes. Our next step will likely include analysis efforts toward identifying what temporal patterns we may see in our hot spots, what types of specific modus operandi are present, or if the residents are just failing to lock their garage doors. Although a hot spot analysis is a valuable tool with which to analyze crime, as with any tool, it is made for only one job, and you will need other tools and analytical processes in addition to hot spot analysis to make your analysis useful to those decision makers.

■ Types of Hot Spot Analyses

There are essentially five broad categories of analyses used to identify and examine clusters of crime (Boba, 2005; Eck, et al. 2005; Paulsen & Robinson, 2004). Various software applications (such as the free CrimeStat software) have the ability to perform many of these analyses. CrimeStat even comes with an extensive manual for the beginning analyst to use and learn from.

Manual or Eyeball Analysis

The first method for identifying hot spots, the *manual* or *eyeball* technique, requires the analyst to produce a simple point or pin map of criminal events. The analyst visually scans the distribution in search of points that are clustered together. This is the modern version of the classic paper pin map approach. This technique, while simple and seemingly easy to complete, is as limited in its usefulness as it is in its complexity. For example, analysts who are unfamiliar with the areas they are mapping (recall from our earlier chapters that this is an offense) may incorrectly identify or ignore hot spots. The basic idea here is to look for an abnormally large number of crimes clustered within small areas relative to the overall study area. However, in areas that experience large amounts of crimes, even after filtering for crime type and time, the map may be covered with dots. In addition, addresses that have more than one crime incident will only have one visible dot. This is because in the geocoding process, repeat addresses receive a dot in the same place every time (see **Figure 10–1**). (Please note that the maps are not formatted in a way that would be used for presentation. This will be discussed in Chapter 11.) However, for our purposes in this chapter, it is more important that you understand how they differ visually depending on the type of analysis you decide to employ.

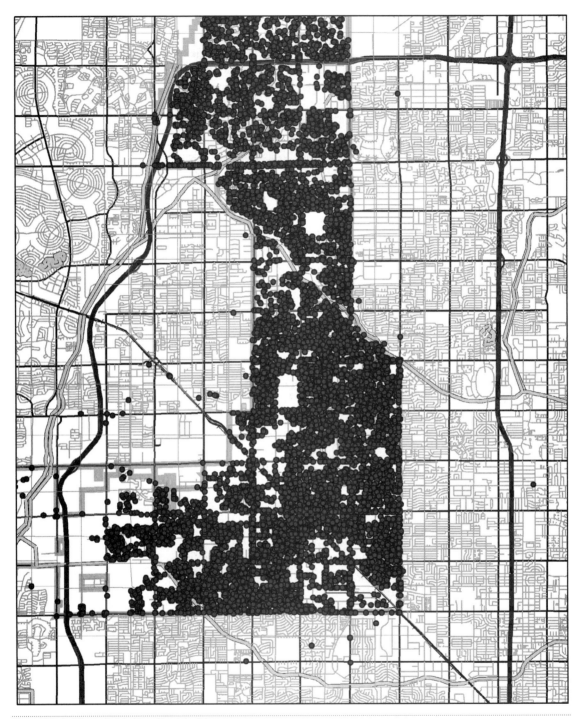

Figure 10–1 Offense Points Map, Glendale, Arizona
Source: ArcGIS

Another technique employs the use of graduated symbols (usually circles) so that addresses with repeat occurrences are represented by larger dots. Those addresses with the largest number of incidents will be represented by the largest dots. This can make a cluttered map slightly easier to read, but again the process is imprecise because the size of the circle is determined by the analyst's ability to filter the data correctly (see **Figure 10–2**).

A second technique, termed repeat address mapping (RAM), allows the analyst to choose the number of incidents an address must be involved in before a dot becomes visible on the map. For example, an analyst can make visible only those addresses representing the highest 10% of repeat incidents. Thus, addresses that do not make the top 10% will not be visible on the map.

Graduated Color Map or Choropleth Map Analysis

A second approach, using *graduated color maps* or *choropleth* maps (sometimes referred to as shaded grid maps), uses color intensity to shade areas on a map according to the number of criminal events those areas experience. This approach is dependent upon the analyst choosing an appropriate level of analysis (census blocks, block groups, beats, cities, etc.). Caution is advised when choosing the unit of analysis. Selecting a unit of analysis that is too large will result in obscuring smaller hot spots placed in larger areas with limited crime (see **Figure 10–3**). In the map in Figure 10–3, the unit of analysis is a uniform grid layer that is overlaid on top of the city of Glendale. However, this process can be done (and may be more useful) with polygons that represent beats or precincts.

Grid Cell Mapping Analysis

The third type of hot spot analysis utilizes *grid cell mapping*. This type of analysis is sometimes called a *density* analysis. There are several individual methods to this approach, and results vary depending upon the assumptions made by the analyst. Essentially, a grid is placed atop a map that contains address-level information for criminal events (similar to Figure 10–3). The grids contain equally-sized cells (much like a piece of graph paper), and grids can range in size according to the analyst's needs. However, the typical size is between 50 and 500 ft because this makes for a detailed and visually pleasing map, depending on the scale of the map being developed (Boba, 2005). The analyst determines a *search radius* (also called a bandwidth), which calculates the number of incidents within the radius and divides that number by the size of the search area. In simple density analysis, the

376　CHAPTER 10　Hot Spot Analysis

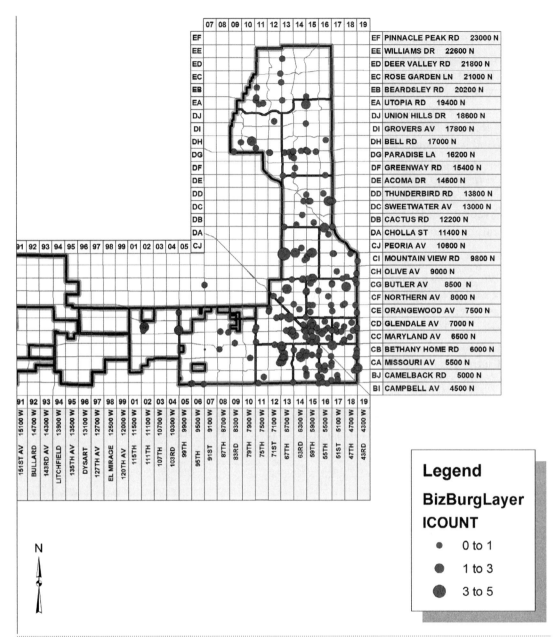

Figure 10–2 Graduated Symbol Map, Glendale, Arizona
Source: ArcGIS

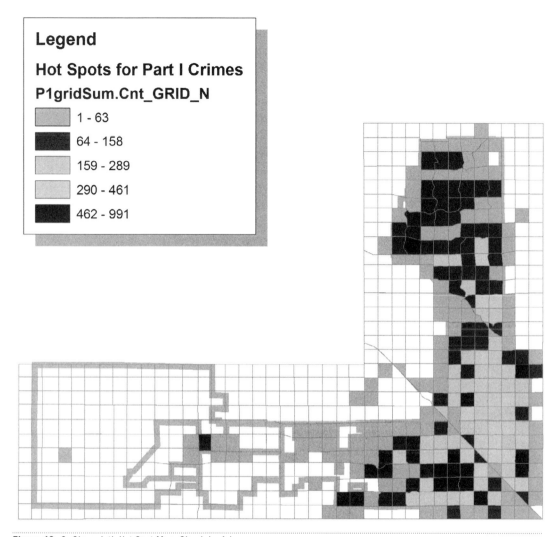

Figure 10–3 Choropleth Hot Spot Map, Glendale, Arizona
Source: ArcGIS

score computed for each cell does not represent the number of crimes within that cell; it represents the number of crimes that are near the cell divided by the area around the cell. Thus, the score is not a count of crimes but a ratio of crimes in an area divided by the size of the area (also referred to as a density calculation). The cells are assigned colors based on their scores. Typically the darkest colors represent cells with the largest scores. The smaller the cells, the greater the resolution (and the smoother the picture). Map results can vary significantly as the cell size and search radius are adjusted. See **Figure 10–4** for an illustration of a grid cell analysis.

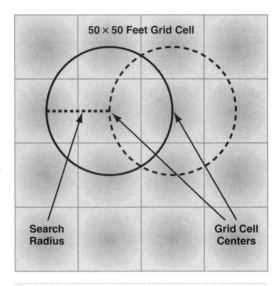

Figure 10–4 Grid Cell Analysis

A more precise type of grid cell analysis using kernel density calculations is also common. Kernel density interpolation employs a grid analysis methodology to estimate crime density across an entire study area by assigning greater weight to those incidents that occur closer to the center of the search radius and lesser weight to incidents that occur farther out. The results then provide information on where crime is clustered together, but it also provides a density value relative to the entire study area. In addition, because hot spots are not perfectly shaped circles or ovals, kernel density interpolation projects a more realistic image of the shape of the hot spot distribution (Levine, 2002). **Figures 10–5** and **10–6** represent single kernel density maps set at different bandwidths (search radius). Figure 10–5 employs an adaptive approach where the bandwidth is smaller in crime-dense areas and larger in less crime-dense areas, which allows for a minimum number of points to be found. Figure 10–6 depicts a fixed interval search radius of 1 mile. Note the difference in the shape of the outputs between the two maps.

Another type of analysis, dual kernel density interpolation, allows the analyst to produce a risk value associated with crime density. This is an important advantage, and these maps can be very useful to law enforcement. For example, in dual kernel density interpolation, analysts can compute victimization risk relative to the population. This allows for a more accurate analysis of areas that are sparsely or densely populated relative to other areas being studied. In addition, this technique can

Types of Hot Spot Analyses   379

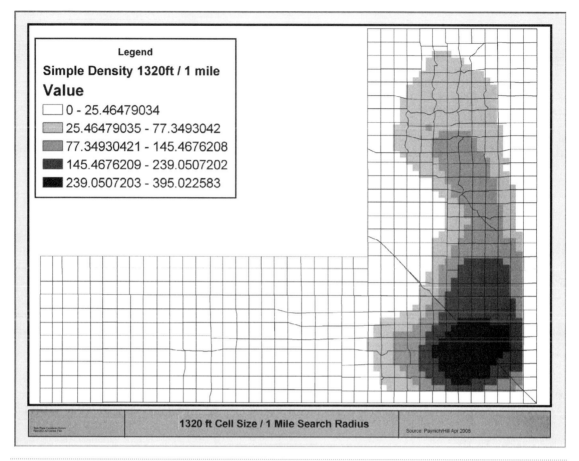

Figure 10–5 Single Kernel Density Adaptive
Source: CrimeStat and ArcGIS

be used to compare crime density for two different crime types and to compare crime densities for two different time periods (Levine, 2002).

Point Pattern Analysis or Cluster Analysis

The fourth type of hot spot analysis, *point pattern analysis* or *cluster analysis*, as it is sometimes called, uses an arbitrary starting point (called a "seed") to calculate whether or not incident points are closer in proximity than we would expect by random chance.

> Typically, an arbitrary starting point ("seed") is established. This seed point could be the center of the map. The program then finds the data point statistically farthest from there and makes the point the second seed, thus dividing the data points into two groups. Then distances from each seed to other points are repeatedly calculated, and clusters based on new seeds are developed so that the sums of within-cluster distances are minimized. (Harries, 1999, p. 117)

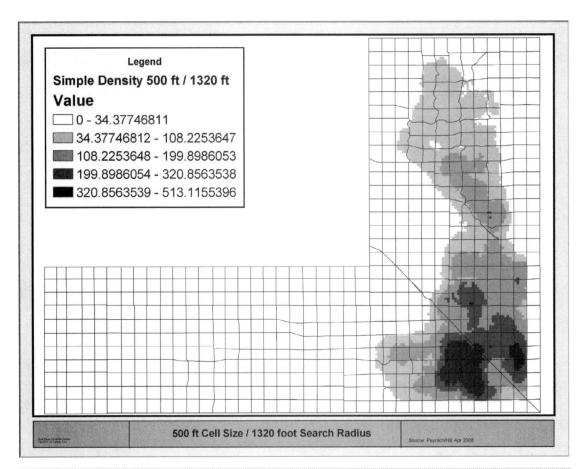

Figure 10–6 Single Kernel Density Fixed 1 Mile
Source: CrimeStat and ArcGIS

There are several common point pattern analysis techniques. The first, called *fuzzy mode* analysis, works by creating a search radius to use at each point (as performed in a grid analysis) and counting all incidents that occurred at a specific address and in the defined search radius around the location. The key here is to set an appropriate search radius. Using settings of a mile or larger will create false hot spots. Setting the radius too low will ignore true hot spots (Levine, 2002). See **Figure 10–7** for an example of a fuzzy mode map. Note that hot spots are represented by graduated symbols.

Again, it is important that the crime analyst intimately knows the study area he or she is working with. The second type of point pattern analysis, *nearest neighbor hierarchical clustering* (Nnh), uses a confidence interval and a set minimum number of incidents in determining

Types of Hot Spot Analyses 381

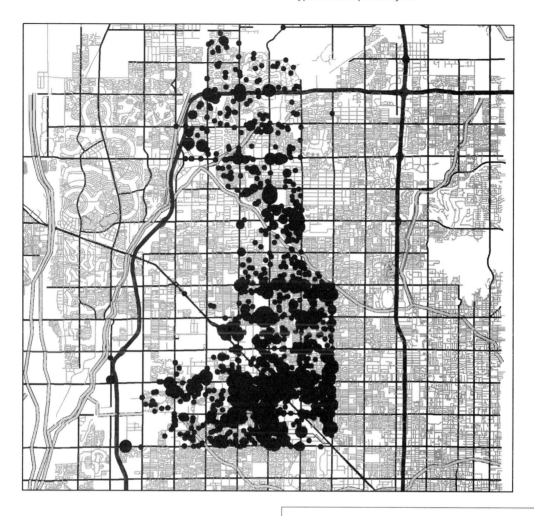

Figure 10–7 Fuzzy Mode, Burglaries 2003, Glendale, Arizona
Source: CrimeStat and ArcGIS

hot spots. The confidence interval (similar to the concept of statistical significance) is set in an effort to more accurately identify clusters we would not expect by random chance (Levine, 2002). The bigger the confidence interval (usually 0.10 is the largest acceptable confidence interval), the bigger the search area. A confidence interval of 0.10 allows for a 10% margin of error. That is, we are 90% confident that this is a true hot spot in the case of a 0.10 confidence interval. Another component to Nnh is its ability to identify second and third order clusters of crime. Essentially, you may have several hot spots that are clustered in close proximity in a larger area that has high concentrations of crime. Recall our discussion of milieu from the examination of environmental changes and crime in Chapter 3. Milieu is the placement of defensible space within a larger area that is also characterized as defensible space. This notion also works in reverse. In second and third order groupings, Nnh identifies local hot spots that are placed within larger crime clusters. In other words, smaller hot spots of crime are situated in larger hot spots of crime. Another type of point pattern analysis, *Risk adjusted Nnh* (Rnnh), allows analysts to control for a third variable (often population) to identify hot spots of crime based on the relative risk of victimization based on the third variable (such as population) rather than just using simple counts and proximity (Levine, 2002).

Spatial autocorrelation analyses assume that criminal events that occur in different locations (yet close in proximity) are related; said in another way, events that occur in time and place closer to one another are more likely to be related than those events that occur farther away or later in time. The computations and interpretations are similar to some of the measures of association discussed in Chapter 8. The two most commonly employed tests for spatial autocorrelation are *Moran's I* and *Geary's C* (Levine, 2002). Both tests require aggregate data (point data aggregated to some polygon feature, such as a census block). The variable that is being computed must be in a continuous format (such as a crime rate). In Moran's I, the value is computed by comparing each area's crime rate to all other areas. The value ranges from -1.0 to $+1.0$. Positive results indicate that spatial autocorrelation exists: Areas with high crime rates are clustered together, and areas with low crime rates are clustered together. A negative value indicates no spatial autocorrelation and thus a random distribution of crime (Levine, 2002). Geary's C is a little more precise and is used for small neighborhoods. Because the computations are based on the squared deviations of incidents (similar to computing a variance), the resulting value for Geary's C ranges between 0 and +2.0. Results between 0 and 1 indicate that spatial autocorrelation exists (high crime areas are clustered with high crime areas, and low crime areas are clustered with low crime areas). No spatial autocorrelation exists if the value is greater than 1. Because Geary's C is more sensitive in smaller areas, it is possible to get conflicting results when running both a Moran's I and a Geary's C on the same data. If this occurs, analysts should visually scan the graduated color maps to determine the appropriate interpretation to use (Levine, 2002).

Standard Deviation Analysis

A fifth type of hot spot analysis is called *standard deviation analysis* (recall the definition and computation for standard deviation from Chapter 8). The outputs are in the shape of a rectangle or an *ellipse*, depending on the specific type of analysis performed, and they are drawn around criminal incidents that are clustered together and that we would not expect based on random chance. Ellipses are oval shapes that vary in size in accordance with the incident distribution. Ellipses and rectangles can be short or tall, narrow or wide, depending on the north to south distribution of incidents and the east to west distribution of incidents. Of course, visually, these analyses cannot illustrate the true shape of a hot spot, but they are useful in comparing changes across crime types and across time. In addition, some standard deviation analyses, such as *standard deviational ellipse analysis* (SDE), are better suited for skewed data distributions (recall from Chapter 8 that this is common with crime data; Levine, 2002). See **Figures 10–8** and **10–9** for illustrations of different types of standard deviation analyses.

A slight variation of these standard deviation ellipses that uses a similar development method is the convex hull representation of hot spots. A convex hull hot spot analysis technique results in an odd polygon shape that follows the outermost points within the cluster of points rather than a spatial deviation ellipse (see **Figure 10–10**).

The concept here is that if you have a group of points that has been determined to be included in the hot spot, the software application (CrimeStat III) will then draw a polygon that joins the outermost points so that all of the other points fall within the convex hull polygon. Because this process differs only in the output of a convex hull rather than a standard deviation ellipse, it is not considered to be a different hot spot analysis technique.

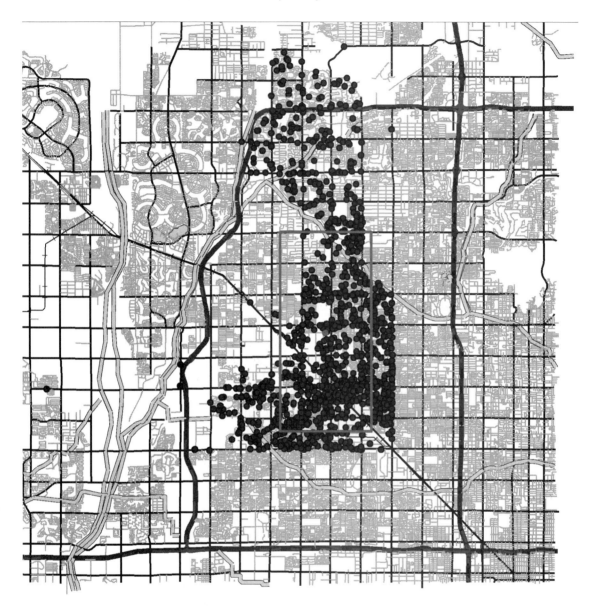

Figure 10–8 Standard Deviation of X and Y Coordinates, Burglaries 2003, Glendale, Arizona
Source: CrimeStat and ArcGIS

■ Online Resources

- "Crime Hot Spot Analysis and Dynamic Pin Map" paper: http://proceedings.esri.com/library/userconf/proc97/proc97/to600/pap575/p575.htm

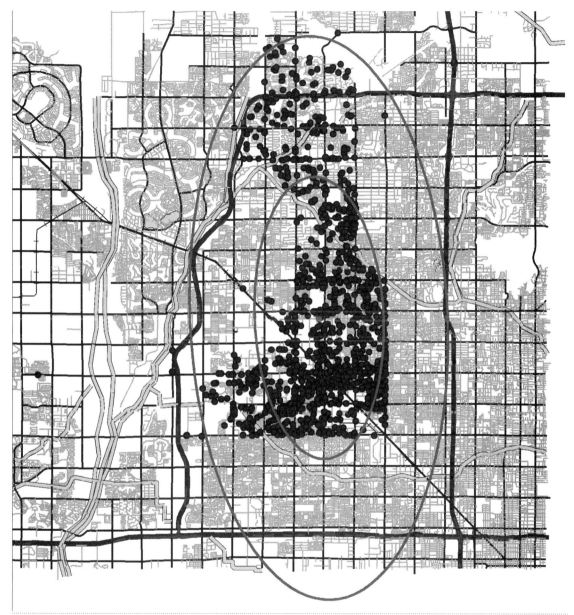

Figure 10-9 One and Two Standard Deviations of X and Y Coordinates, Burglaries 2003, Glendale, Arizona
Source: CrimeStat and ArcGIS

- ESRI Mapping Center's Hot Spot Analysis of 911 Calls: http://mappingcenter.esri.com/index.cfm?fa=maps.hotSpot911
- Crime Reduction Toolkit: http://www.crimereduction.homeoffice.gov.uk/toolkits/p031306.htm

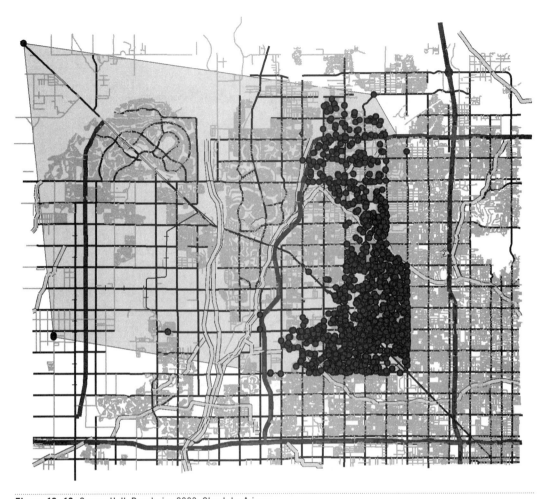

Figure 10–10 Convex Hull, Burglaries 2003, Glendale, Arizona
Source: CrimeStat and ArcGIS

■ Conclusion

Various analyses exist for identifying hot spots. Some of these are very simple (the eyeball method), and some are much more complex (dual kernel density interpolation). Because the choice of which analysis to employ is dependent upon the type of data (Do we have point or aggregate data? Is the distribution skewed?) and the types of questions that need to be answered (Do we need to know the shape of the distribution? Are we comparing changes across crime types or different periods of time?), the type of hot spot analysis we choose makes a difference. Of course, the analyst's skill level is important in selecting the appropriate method, but most software applications come with a manual that will school the analyst on the finer details and practical steps involved in completing the analysis. Each type of hot spot analysis (as with all statistical-based analyses) comes with varying strengths and weaknesses that must be understood prior to both performing and interpreting the results. It is vitally important that the analyst remembers who the audience is, what the purpose is for developing the hot spot, and remembering the offenses of analysis from Chapter 8. Creating a beautiful hot spot map is a great thing, but if the audience you made it for already knew the areas on the map were hot based on their enforcement experience, then what did you accomplish, except for maybe validating their opinions? The analyst should consider that the hot spot analysis chosen may not be the only method that should be experimented with for a given analytical project; in fact, you may find that you will need to perform a hot area, hot place, and hot street segment analysis to really get down to the identified problem, and then perform temporal, modus operandi, and other analyses to really provide a useful product. A hot spot map is never the end of an analysis; it is typically just the beginning. This chapter provided an overview of the most common types of hot spot analyses. However, students are encouraged to seek alternative sources for more advanced methods.

Recommended Reading

Anselin, L., Cohen, J., Cook, D., Gorr, W., & Tita, G. (2000). Spatial analyses of crime. In *Measurement and analysis of crime and justice. Criminal justice 2000* (Vol. 4, pp. 213–262, NCJ 182411). Washington, DC: US Department of Justice, Office of Justice Programs.

Holden, K. (1995). Vector autoregression modeling and forecasting. *Journal of Forecasting, 14,* 159–166.

Questions for Review

1. Explain the different methods an analyst can use to identify hot spots.
2. What are the strengths and weaknesses associated with each method of identifying hot spots?

Chapter Exercises

Exercise 19: Graduated Points

Data Needed for This Exercise
- All data for this exercise is contained in EX19.mxd.

Lesson Objectives
- In this exercise you will learn how to use the spatial statistics tools in ArcGIS to create a graduated point map, or what is commonly referred to as a "repeat calls" or "repeat location" map. These maps are good for identifying businesses and other locations that may be crime attractors or high crime locations where focused enforcement action could be useful.

Task Description
- The police chief has requested that you identify the top 10 locations in the city of Glendale for the crime of business burglary. He needs a map that shows these high locations and does not want a pin map or a hot spot map. He wants only those top 10 business addresses that have been burglarized multiple times in 2004.

1. Open EX19.mxd.
2. When you open the project you should see the standard Glendale map we have used numerous times throughout these exercises and the following themes:
   a. Business_Burglaries_2004
   b. Grids
   c. Beats
   d. City_Boundaries
   e. Street_Labels
   f. Street_Outline_Boxes
3. Click on the little **red toolbox** in the menu bar to open the ArcGIS tools.
4. Find the Spatial Statistics toolbox.
5. Find the Utilities section and the Collect Events tool.
6. Now go back up to your Business Burglary theme and export a new copy of it into your student folder with a name of your choice (I'll name mine BizBurg.shp). You do not need to add this exported version to your map document.
7. Double click the **Collect Events** tool to launch it.
8. In the Input Incident Features box, click the **Browse** button and navigate to your new theme.
9. In the Output Weighted Point Feature Class box, navigate to your student folder and save this new file as BizBurgGraduated.shp.
10. Click **OK** to run this tool.
11. When it is done, you should see that nothing was added to the project. The tool did count the number of incidents

that occurred at the same XY coordinates and produced a table called BizburgGraduated.dbf and a shapefile with a count by location.

12. Click the **Add Data** button, navigate to your student folder, and load in the shapefile BizburgGraduated.shp.
13. Open the attribute table and sort the ICOUNT field in descending order (right click the field name and choose **Sort Descending** from the pop-up menu).
14. Notice that there are 23 records or locations with two or more burglaries reported in 2004. The top one had four burglaries reported at these business locations.
15. Close the table.
16. Find the Spatial Statistics toolbox, then the Utilities toolbox under it. Click on the **Count Rendering** tool to launch it.
17. Fill in the dialog menu as follows:

    Input feature class: *BizburgGraduated*
    Field to render: *ICount*
    OutputLayer file: *Student folder/BizBurgLayer.lyr*
    Number of classes: *3*
    Symbol color: *Bright_red*
    Maximum field value: *5*

18. Click **OK** to run the tool.
19. When it is done, close the toolbox and add data.
20. Find the LYR file you created (BizBurgLayer.lyr) in your student folder and load it into the map.
21. Turn off the original point file. You can remove it from the project and TOC at this time.
22. Now you have graduated symbols that show the locations with multiple business burglaries—or do you? No. Do you have all the locations correct?
23. The chief only wants the top 10, but we have a problem, don't we? We have one location with four burglaries and 22 locations with two burglaries and then everything else. We need to limit the legend to show only those locations where the ICOUNT ≥ 2.
24. We can accomplish this in a few ways:
    a. We can create a theme definition.
    b. We can edit the table and delete any data for records with less than two burglaries.
    c. We can change the legend to show only those locations with two or four burglaries.

25. We'll manipulate the legend this time. Right click the layer in the TOC and choose **Properties → Symbology**.
26. Click on the **Classify** button at the far right.
27. Click on the **Exclusion** button in the pop-up menu.
28. Under the Query tab, enter *"ICOUNT" = 1*.
29. Click **OK** and **OK** again to get back to the Symbology menu.
30. In the Symbol window, click **Show Class Ranges Using Feature Values**.
31. Right click on the second 2 value in the symbol section of the menu, and choose **Remove Classes(s)** to get rid of it.
32. You should now only have a symbol for 2 and a symbol for 4, with 4 being larger.
33. Apply the legend and click **OK**.
34. Create a nice looking layout for this map. Enter any disclaimer text you feel is necessary, titles, a north arrow, a legend, logos, pictures, etc. that you think the map needs to look professional and useful for the chief.
35. Save your project in your student folder.

Exercise 20: CrimeStat III Hot Spot Analysis Techniques I

Data Needed for This Exercise
- Start this exercise by opening the C:\CIA\Exercises\EX_CrimeStat folder and opening the CSHotspots.mxd file.

Lesson Objectives
- This analysis will help you learn how to create nearest neighbor, k-means, STAC, and interpolation hot spots using CrimeStat III. The exercise will discuss the many options for using all of these hot spot techniques and why you may wish to use all of these different processes for the same analysis effort in some cases.
- You will also learn that the CrimeStat III manual is a wealth of information, knowledge, and examples of how these techniques are used.

Task Description
- You have been assigned the task of determining what areas within the city of Glendale would be the most in need of enforcement and intervention to prevent Part I violent crimes.
- To establish which areas are the hottest with respect to these crimes, we will use a standard deviation ellipse analysis to determine how the crimes are dispersed throughout the beat.
- We will then use several hot spot analysis techniques in CrimeStat III to identify the most significant hot spots and to

do some different kinds of outputs to help us fully understand the violent crime in the city of Glendale in 2003.

1. Open the CSHotspot.mxd project located in the C:\CIA\Exercises\Exercise_20 folder.
2. The map should resemble DVD Figure 20–1.
3. A good analyst will always explore the attributes of the data he or she is using for the analysis. You might want to look at the date range and see if it includes data from the entire year, if the count of 1016 Part I violent crimes is within the number normally received in a year, and check for null fields. We also need to verify if the X and Y coordinates of the data exist in the table. Open the attribute table for the Part1Violent theme in the table of contents (TOC).
4. In this table the X and Y coordinates are the sixth and seventh fields from the left. However, we obtained this information from another source and have reprojected it upon receiving it, so we are not sure if these coordinates are correct.
5. We can easily modify these existing X and Y fields using the Calculate Geometry tools that are available when viewing tables. If the X and Y fields did not exist at all, we could use the ArcToolbox → Data Management Tools → Features → Add XY Coordinates tool to add Point-X and Point-Y fields.
6. Make sure that *no* records are selected in the table or else the following steps will only be done on the selected records.
7. Because we have X and Y fields and we just need to calculate them again, right click on the X field name to get the pop-up box. Choose **Calculate Geometry** from the pop-up dialog menu (see DVD Figure 20–2).
8. Answer *Yes* that you do want to perform the calculation function outside of an edit function.
9. The property we wish to choose is **X Coordinate of a Point**.
10. We also want to select **Feet US (ft)** as the units.
11. Now click the **OK** button to have all of the X coordinates recalculate.
12. Repeat steps 6 through 11 for the Y field (of course use the **"Y Coordinate of a Point"** as the property we wish to calculate).
13. Several of these coordinates do not appear to have changed, but unless we know for sure, it is always a good idea to recalculate them in this manner.

14. You should have reviewed the table and found that the midpoint date range goes from 01/01/2003 through 12/16/2003 and not through 12/31/2003. Normally this should trigger a notion that the data may not be complete for the entire year. In this case, the data is adequate for what we are going to be doing, and I will advise you that the data is complete for the entire year. In most cases, you should question the completeness of this data with simple things like this.
15. Close the attribute table.
16. Look over the data on the map with the Part1Violent theme turned on.
17. Closely observe the points and the distribution of the points on the map and try to find at least three different clusters of points on the map that you think could be a hot spot. Print the map and draw these clusters, or use the draw tool in ArcGIS to draw the clusters as you see fit.
18. I came up with some rather large clusters (see DVD Figure 20–3). Your hot spots may be slightly different than mine, but with this data you can see several potential clusters of points that could indicate where a great deal of Part I violent crime occurs simply by eyeballing the data. One of the dangers in doing this is that each point on the map could actually be several points at the exact same location or address. There could be several locations where there are multiple crimes that occurred rather than just one, so seeing only one dot could be deceiving to the naked eye. The scale at which you are viewing this data could also change the way the clusters appear to you. Zoom in very close on my cluster at the lower right of the map. When you zoom in on this, you might see smaller clusters within the larger one, or in some cases a cluster might disappear as the scale changes.
19. Now that we have mastered the visual perception skills we need to do a proper hot spot map, we can explore the world of spatial analysis using the statistical methods and functions inside CrimeStat III.
20. Right click on the **Part1Violent** theme in the TOC and choose **Properties → Source**.
21. Note where this shapefile is located (see DVD Figure 20–4).
22. We need to remember where this file is when we are using CrimeStat III. While we are here in ArcGIS, let's collect some more information that we will need with CrimeStat to do our hot spot analyses.

23. First we need to know what the extents of our Part I violent crime points are. Extents are the minimum bounding rectangle we can draw around our points so that every point is within the box. Use the rectangle draw tool and draw a box around the points so that all of the points are within the rectangle.
24. Ignore the point that is way over in the upper left-hand side of the screen at 19032 NW Grand Av.
25. Make sure that all of the points are within the rectangle you draw and that the rectangle is as small as it can be to just contain all the points. You do not have to be exactly within a certain tolerance; close will do.
26. Now right click on the rectangle and choose **Properties** from the pop-up dialog box.
27. Go to the Size and Position tab, and make sure the blue box at the bottom left of the nine boxes is pushed in. This is the lower left corner coordinates of the box we drew. Write down on a piece of paper or copy and paste into WordPad or Excel the X and Y coordinates of this bottom left corner. When copying and pasting, *do not* include the "ft" at the end.
28. When done, click the **blue box** at the upper right corner of the nine boxes.
29. The coordinates have changed to provide the X and Y coordinates of the upper right corner of the box we drew. Copy these as with the lower left coordinates.
30. Click on the **Area** tab now, and change the units to Square Miles.
31. Copy this number, which is the area in square miles of our rectangle. We have now determined the output parameters for our work in CrimeStat III.
32. In addition to the extents of the points or the minimum bounding rectangle of the points, we need to know the approximate footage or mileage of the streets in this area.
33. Make sure the rectangle is selected in the map display (use the black arrow **Select** tool to click on the edge of the rectangle until you see the drawing's handles).
34. Go to **Selection** → **Set Selectable Features** in the menu bar, and make sure that Gln_Surrounding_Streets is the only layer chosen. Turn it on so it draws.
35. Click on **Gln_Surrounding_Streets** to make it the active theme, and then go to **Selection** → **Select by Graphics** in the menu bar.

You should now see the selected streets in the map display showing as blue.

36. Open the attribute table of the Gln_Surrounding_Streets theme.
37. Find the field called Length, which is the length of the street segments in feet.
38. Right click this field and choose **Statistics**. You will get statistics for the selected street segments.
39. Copy the **"Sum"** value from the Statistics results window.
40. Your data may not be exactly the same as shown on DVD Table 20–1, but it should be fairly close (see DVD Table 20–1 to compare your data to mine).
41. Save these numbers because we will need them and the directory location you wrote down for the Part1Violent shapefile when we open CrimeStat III.
42. Open CrimeStat III, which is located at C:\CIA\Documents\Tutorials_Programs\CrimeStat_III\CrimeStat.exe.
43. When the introduction screen opens, click it anywhere to get to the program. After about 1 minute it will automatically take you to the program, but you can bypass this welcome screen just by clicking on it.
44. In the Data Setup Tab, under the Primary File tab, choose **Select Files**.
45. In the Select Files dialog menu, choose **Dbase Files (.DBF)** as the type and then navigate out to the location where the Part1Violent .dbf file is located: C:\CIA\Exercises\EX_CrimeStat.
46. After adding this file, modify the X and Y coordinate settings under Variables and check **Projected (Euclidean)** as the type of coordinate system (see DVD Figure 20–5).
47. The data units should also be set to feet. Note: If you are using decimal degree or latitude and longitude data, you may need to reproject the data to a state plane system appropriate for your data and location because it is much easier to obtain all of these settings and values. If you do use decimal degree data, be aware that the square miles of the minimum bounding rectangle need to be converted from square degrees to square miles manually based on the instructions in Chapters 3 and 6 of the CrimeStat III manual. ArcGIS does not do this calculation for you, although you can reproject on the fly for this one value and then process everything else using the decimal degree data.

48. Now that we have told CrimeStat III where our data file is located, we need to tell it where our minimum bounding rectangle (MBR) is located.
49. Click on the **Reference File** tab.
50. Enter the X and Y coordinates you copied down for the lower left and upper right corners of your MBR. Make sure you notice where each value goes, and do not use the tab key inside CrimeStat because it does not perform as expected.
51. For this analysis, leave the Cell Specification → By Number of Columns setting at 100. This means that there will be 100 columns in the grid we are going to create within the MBR we drew.
52. If we increase the number of columns, we can get a much finer grid and a much smoother product for mapping, but for now we will just use this 100 setting.
53. Check the **Use Lower-Left Corner for Origin** box (located on the right side of the dialog menu.)
54. You've just told CrimeStat where the analysis will take place and what area the analysis will cover. Because CrimeStat does many of these hot spot analyses based on map distances and distances between points, we now need to tell CrimeStat what the overall square mileage is and what the street mileage is in the MBR.
55. Click on the **Measurement Parameters** tab.
56. In the Coverage → Area field, enter your square mileage of the MBR and change the units to square miles.
57. In the Length of Street Network section, enter the total street length you calculated back in ArcGIS.
58. Click the check box next to **Indirect (Manhattan) Distance** at the bottom.
    - Euclidean distance is the shortest distance between two points or "as the crow flies" distance.
    - Manhattan distance is the estimated street route distance a person would have to take to get from point A to point B. This is measured in right or 90 degree angles.
    - There are many discussions and arguments in published materials about which method is better. You should explore these publications and choose which method best fits your data.
59. In our case, Manhattan distance calculations for Glendale data seems to be logical and common sense, and it typically performs the best. Now that we have accomplished all of these tasks, we can start creating hot spots.

60. Click on the **Green Spatial Description** tab at the top and then **'Hot Spot' Analysis I**.
61. Check the **Nearest Neighbor Hierarchical Spatial Clustering (nnh)** box and then click the **Save Ellipses To** button. Save this as a shapefile called "1" in your student folder. The directory string should appear as C:\CIA\StudentFolder\1 in the file location box.
62. Click **Save Convex Hulls To**, and also choose a shapefile as the output type and "1" as the file name in your student folder. CrimeStat III will automatically add information to the file name when it is exported.
63. For our analysis, we will use the default of Random NN Distance. We will also keep all of the other default settings at this time.
64. Now click on the **'Hot Spot' Analysis II** tab.
65. Check the boxes next to **STAC** and **K-Means** as well, save the ellipses and convex hulls for these analysis methods as "1" in your student folder, and output them as an ArcGIS shapefile. Keep the defaults for each of these methods.
66. Notice that you can change many different variables for each routine to include the search radius, degree of ellipse to return, and how many points need to be in a cluster. All of these settings are explained in Chapter 6 of the CrimeStat III manual in great detail, so for speed I will only use the defaults for these methods in this exercise. Refer to the manual, and test various settings to see what the results are with your own data to learn more.
67. Click on the **Spatial Modeling** tab → **Interpolation**.
68. Choose a **Single Density Estimate** and check the box.
69. Under method of interpolation, choose **Quartic**.
70. Under Choice of Bandwidth, choose **Adaptive**.
71. Under output units, choose **Relative Densities**.
72. Click **Save Result To** and save this as "1" again in your student folder.
73. Click the **Computer** button at the bottom of the CrimeStat III window.
74. Under the NNH tab in the Computer Results dialog, we can scroll down to see that the program found 14 first order clusters and two second order clusters in the data. Under the STAC tab we can see that it found five clusters. Under the K-Means tab we can also see that five clusters were found. Finally, with the KernelDen-

sity (Interpolation) routine we have some random numbers that don't tell us too much other than the routine did run and produce something.

75. Close the computer dialog, but leave the CrimeStat III program open and minimize it because we will use it again in a few minutes.
76. Return to ArcGIS and close any tables you may have open, turn off any unnecessary layers, and go to **Selection → Clear Selected** features to clear any selections we may have done.
77. You can delete your rectangle at this point if desired. If you want to keep it, you can right click it, go to **Properties**, and choose a no fill color and a 2 point red outline.
78. Click on the **Add Data** button and add in the following layers from your student folder. If you cannot find them, go to the C:\CIA\Exercises\EX_CrimeStat\ folder and add those layers the instructor created when developing this exercise.
    - CKM1.shp (convex hulls for the K-Means routine)
    - CNNH11.shp (convex hulls for the Nearest Neighbor routine—first order)
    - CNNH21.shp (convex hulls for the Nearest Neighbor routine—second order)
    - CST1.shp (STAC convex hulls)
    - K1.shp (interpolation result)
    - KM1.shp (K-Means ellipses)
    - NNH11.shp (Nearest Neighbor ellipses—first order)
    - NNH21. shp (Nearest Neighbor ellipses—second order)
    - ST1.shp (STAC ellipses)
79. Turn all of these layers off initially, and let's go over them one by one.
80. Turn on the ST1.shp layer. You might notice that these ellipses seem to be surrounding the same general areas you probably visually circled at the beginning of this exercise. STAC is a reliable tool for finding citywide hot spots and can be useful for identifying them within smaller areas (just reduce your MBR and other measurements and limit the points to those inside the MBR by creating a new layer of just those points).
81. Turn off the ST1.shp layer and turn on the CST1.shp layer. As you can see, the program draws convex hull polygons around the same general hot spot areas and includes all of the points the program

used to qualify the area as a hot spot. Many analysts find the convex hull result to be more useful because it actually contains all the points in the cluster rather than estimating where the hot spot is by an ellipse.

82. Turn off the CST1.shp layer and turn on the KM1.shp and CKM1.shp layers.
83. With the settings we chose, we are basically getting clustering for the entire city. To get smaller clusters we may want to increase the total number of clusters, the separation distance between clusters, or the standard deviation setting for the ellipses. With some work and experimentation, k-means will provide similar results as STAC with a lot more clusters.
84. Now turn these layers off, and turn on the NNH11.shp and NNH21.shp layers.
85. Move the NNH21.shp layer under the NNH11.shp layer in the TOC so you can see both layers.
86. The NNH routine generally finds more localized hot spots and also has the added benefit of finding first, second, and third order hot spot clusters, which means it finds hot spots and different levels or scales in size. This could be very useful for determining focus areas for enforcement and then finding the specific events that are causing a hot spot at each order or scale.
87. Turn those layers off and turn on the CNNH11.shp and CNNH21.shp layers. As with the other methods discussed, the convex hull provides the analyst with the actual points that make up the hot spot and may be more useful for further analysis.
88. The final hot spot map we will look at is the results of the interpolation routine called K1.shp. Turn off any other hot spot layers, and turn on the K1.shp layer.
89. You should see a bunch of very small grids. We need to change the symbology of this grid. Right click the theme and choose **Properties**.
90. Choose the **Symbology** tab in the Properties dialog menu.
91. Choose **Quantities** → **Graduated Colors** as the type of classification.
92. Choose the field called "Z" as the Value field.
93. If you get a message that indicates you need to use a larger sample size, click **OK**. To avoid having this message pop up we can easily change the sample size limits for this ArcGIS session.

94. Click the **Classify** button.
95. In the Classification dialog menu, click the **Sampling** button. Change 10000 to 1000000 by simply adding two zeros to the end.
96. You should notice that the chart changed slightly. If you change this number back and forth, you can see that there are likely many Z score values for grids that are null or 0 and that the graph simply gets moved more to the left when we change the sampling. This does not make too much of a difference in the overall output, but the instructor generally changes it to 1000000 and keeps moving forward.
97. Change the outline of each classification range to no color and apply the symbology and classification. (DVD Figure 20–6 shows what the result will resemble.) As you can see, the interpolation routine highlighted one area as the most dense for Part I violent crimes. This is because of the search radius and type of hot spot and interpolation we chose. To get a different result, we need to modify how far out the kernel density model searches, and to make the entire result slightly smoother in appearance, we also need to change how many columns we will have, which is often called the cell size.
98. Turn off all of these hot spot layers, and save your ArcGIS project in your student folder as CrimeStat.mxd.
99. Maximize the CrimeStat III program again and go back to the 'Hotspot' Analysis I and 'Hotspot' Analysis II sections and uncheck all of the k-means, nearest neighbor, and STAC choices so we do not recreate those layers.
100. Click on the **Spatial Modeling** tab, and make the following changes:
    Method of interpolation: *Quartic*
    Choice of bandwidth: *Fixed Interval*
    Interval: *1320*
    Interval unit: *Feet*
    Area unit:points per: *Square Miles*
    Output units: *Relative Densities*
    Save result to: *Save this as "2" in your student folder*
101. Press **Compute**, and when it is done, close the results window, minimize CrimeStat III, and open your ArcGIS project.
102. Add K2.shp to the project.

103. Change the classification and legend as we did with K1.shp and apply it.
104. You should now see a much more focused hot spot map that has many smaller hot spots rather than the one large one we saw before (see DVD Figure 20–7).
105. We still see that the hot spots are about in the same places that we saw them in the other hot spot routines, but this is more of a weather maplike result and highlights smaller hot spots. Let's change the settings in CrimeStat just a bit more and do this again.
106. Turn off K2.shp and any other hot spot layers, and then maximize CrimeStat again.
107. Go to the **Data Setup** tab and the **Reference File** tab within it.
108. Make the following change in the Reference Tab: Check the box that says **By Cell Spacing** and change the value to 1320 (meaning 1320 feet because our data units (XY) are in feet).
109. Do not make any changes to the Spatial Modeling tab, except change the output file name to 3.
110. Compute the new interpolation routine and add K3.shp to the map and classify it as you did before. Notice that by changing the cell size we have much larger grids that are 1320 feet by 1320 feet. This gives us the ability to create a grid layer of any sized grids we wish, which can be used for a lot of other things.
111. Now let's try to get the best result we can with the interpolation routine that gives us the densest areas for Part I violent crimes and only a few hot spots to deal with instead of 20 or more as with our attempt in K2.shp.
112. Maximize CrimeStat again, and then go to the **Data Setup** tab → **Reference File** and change the By Cell Spacing setting to 50.
113. Go to the **Spatial Modeling** → **Interpolation** tab and change the interval to 2640 and the output units to Relative Densities.
114. Name this as 4 and compute the hot spot interpolation routine. Notice how much closer it runs because you are now using 50 feet by 50 feet cells for the output grid, and this increases the number of grids by huge margins, but it will make the final map look much smoother.
115. When it is done, close the output dialog, and change the method of interpolation to **Triangular in Spatial Modeling** → **Interpolation**.

116. Change the cell size under **Data Setup → Reference File** to 150.
117. Name this new file 5 and run it again.
118. Because of the amount of time waiting for CrimeStat to run and to classify the Z Field with 50 feet by 50 feet cells, we can assume that this is just overkill, and using a cell that small doesn't add much to our analysis.
119. If you open the attribute tables for K1.shp, K4.shp, and K5.shp, you will notice that the total number of records dramatically increases from K1.shp to K5.shp, and even more so with our 50 foot cell size in K4.shp. With 1,154,252 or so cells (because each of us drew our rectangle slightly differently) the results are very smooth, and there is no choppiness in the image with K4.shp. With K5.shp we still get the lack of choppiness or pixelization effect, and we only have 128,656 cells, which also draw faster. For a citywide analysis, chances are that 150 foot cells are about right, and we could even probably increase this to 250 or 300 feet as needed depending on the square mileage we are analyzing. As the size of our area decreases, so should the cell size and search radius.
120. There are many methods for determining the appropriate cell size and search radius; however, after some practice, most analysts can run the CrimeStat routine several times and come up with what is about right. With experience you keep using those same settings at a citywide analysis level for consistency. With Glendale this is about a 150–300 foot cell size and 2640 foot search radius for nearest neighbors. This works well for most citywide hot spot mapping that uses the interpolation routine.
121. There are several differences with the method of interpolation in the final results, but Quartic and Triangular appear to work well for citywide hot spot analysis. Experiment with these different methods and change only one variable to see how it affects the final map result, and over time you will learn which works best for your agency. Another way to test to see if the interpolation variables are set correctly is to overlay the k-means, nearest neighbor, and STAC results, and in some cases you can overlay the points as well. If it looks like the hot spots are in agreement in a few places, they are the most serious hot spots and should be the ones about which you notify your command staff (see DVD Figure 20–8).

122. There is one thing you need to watch for when doing any kernel density estimation, such as CrimeStat's interpolation routine. If you use an incorrect search radius, you can get an effect where the hot spot actually appears between where clusters of points exist. This phenomenon happens because the search radius for the scale you were analyzing was too large and oversimplified the hot spot because it found too many points during its cell search and summary. We found that this occurred in one small hot spot in the Foothills patrol division in our K5.shp map (see DVD Figure 20–9).

123. In this case the red dot to the east of the hot spot or hot area is actually an apartment complex with seven Part I violent crimes under the one dot.

124. You should be *aware* of this problem, but don't be *frightened* about using this density estimation method in CrimeStat. This is the only area on the map I saw that seemed to be related to this issue, and we probably would not have advised our supervisor that this location was a hot spot when compared to the other ones we saw.

Exercise 21: CrimeStat III Hot Spot Analysis Techniques II

Data Needed for This Exercise
- Start this exercise by opening the C:\CIA\Exercises\EX_CrimeStat folder and opening the CSHotspots2.mxd file.

Lesson Objectives
- This lesson will cover using weights and dual kernel density estimations and how they can help you find significant areas that need enforcement actions.

Task Description
- You have been assigned the task of determining where hot spots of Part I violent crime are occurring based on an intensity value (like a hot spot rate) assigned to each crime type.
- To perform these analyses, we need to use CrimeStat III and add a variable, called a weight or intensity value, to our analysis to do dual kernel density estimations using interpolation in CrimeStat III.
- The second analysis will be a dual kernel density estimation, which uses a primary file and a secondary file of robberies to show the relative increase or decrease of density between 2002 and 2003.

1. Open the CSHotspots2.mxd project located in the C:\CIA\Exercises\Exercise_20 folder.
2. If you closed CrimeStat III after exercise 20, reopen the program and collect the minimum bounding rectangle (MBR) lower left and upper right XY coordinates, MBR square area, and street footage inside the MBR as we did for exercise 20 to speed this process along. If you still have the data from exercise 20, use it for this exercise.
3. Under the Reference File tab, set Cell Spacing to 150.
4. After you have done the data setup in CrimeStat III for the primary file, reference file, and measurement parameters, go back to the primary file, and under the Z (Intensity) section under the XY Coordinate field choices, choose the field called **Intensity** as the column. This field is calculated as 100 for murder, 50 for rape, 25 for robbery, and 1 for aggravated assault offenses listed in the Part1Violent crime shapefile (Part1Violent.shp). We will use these intensity values to provide a different interpolation result based on only the Part 1 Violent crimes but the intensity of the various crimes. Areas where a murder took place will be a given a higher value than aggravated assault and so on. In this case, the commanders thought that aggravated assault was the crime they could have the least impact on, so they would rather have hot spots that reflect the greater focus on murder, rape, and robbery.
5. Go to the **Spatial Modeling → Interpolation** tabs, check the single kernel density estimate box, and set the variables as follows:
    Method of interpolation: *Triangular*
    Choice of bandwidth: *Fixed interval*
    Interval: *2640*
    Interval unit: *Feet*
    Area units: points per: *Square miles*
    Use intensity variable: *Check this box*
    Output units: *Relative densities*
    Save output as: *EX21_1* in your student folder
6. Click the **Compute** button to run this interpolation.
7. When it is done, minimize the CrimeStat III program, but close the output results.
8. Maximize the ArcGIS project, and save it as EX21.mxd in your student folder.
9. Add the KEX21_1.shp file to the project, and change the legend classification to be a quantities classification, using a natural breaks

classification with seven classes, and green to red dichromatic color scheme.

10. You may have to set the sampling higher to see all of the data as we did in exercise 20.

11. If we compare the result of this interpolation run with K5.shp (exercise 20), which had the same settings, except we used an intensity value this time and in K5 we did not, we can see that a new hot spot has emerged. This new hot spot (see DVD Figure 21–1) is mainly there because there were 15 or 16 robberies, which made up the majority of offenses in this small area.

12. In addition to making hot spot maps that are based on the density of points (crimes) in your city, you can also find out what hot spots exist for specific crime types by providing a higher intensity value for the crimes you want to focus on more. All you need to do is add a number field to your attribute table for the point data and then calculate different values for the crime types you wish to emphasize. Perhaps your agency has decided that Part I violent crimes are its main focus; however, the agency wants hot spot maps of all crimes, with Part I violent crimes having greater weight. You now know how to accomplish this quickly and efficiently using CrimeStat III.

13. Chapter 8 of the CrimeStat manual discusses kernel density estimation or interpolation in great detail. I urge you to review it as time permits during your career for a better understanding of how these tools work.

14. Now you are an expert at interpolation in CrimeStat, and we can now move on to dual kernel estimations.

15. Our goal with this part of the exercise is to determine where Part I violent crimes have changed in Glendale between 2002 and 2003. We will use both the P1Violent2002.shp and the Part1Violent2003 layers in the ArcGIS project for this analysis. *(Note: The Part1Violent2003 theme in the table of contents (TOC) is actually the shapefile called Part1Violent.shp and has been simply renamed in the TOC for this project.)*

16. For our purposes this time, we are not going to use the intensity value for either layer.

17. Open CrimeStat again or maximize it. If you just maximized it, then all your previous settings should still be there. If you closed it, you will need to reset all of the settings for the primary file in addition to adding in a secondary file.

18. In our case we want the primary file to be the 2003 Part I violent crime points (C:\CIA\Exercises\EX_CrimeStat\Part1Violent.shp) and the 2002 Part I violent crime points (C:\CIA\Exercises\EX_CrimeStat\P1Violent2002.shp) to be the secondary file for this analysis
19. In CrimeStat III, click on the **Data Setup** tab and the **Primary File** tab. Make sure that you have the part1violent.dbf file as the only file in the primary file section and that the X and Y fields are entered under the Variables section. If you also have the intensity field chosen, you can change it back to <none>.
20. Now click on the **Secondary File** tab and find the P1Violent2002.dbf file and add it here.
21. Choose the X and Y fields for the variables, and you are done.
22. Go to the reference file, and make sure that By Cell Specification is set to 150.
23. Make sure you have data entered in the Measurement Parameters tab for the total square mileage, and street mileage/footage as needed. Also be sure to check the **Manhattan Distance** box.
24. Click on the **Spatial Modeling** tab.
25. Check the **Dual Interpolation** box, and uncheck the **Single Interpolation** box if it is checked.
26. Set the following:
    File to be interpolated: *Primary*
    Method of interpolation: *Triangular*
    Choice of bandwidth: *Fixed interval*
    Interval: *2640*
    Interval unit: *Feet*
    Area units: *Square miles*
    Output units: *Absolute differences in density*
    Save results to: *EX21_2.shp* in your student folder
27. Click **Compute** and wait for the computation to complete.
28. When the computation is done, close the Compute dialog window, and leave CrimeStat open but minimized.
29. Add the DKEX21_2.shp file to the ArcGIS project (note that DK was added by CrimeStat III to denote Dual Kernel estimate).
30. Change the classification and legend to a Quantities legend using seven classes in natural breaks classification type with the color band going from green to red as in previous exercises.

31. I used a gray scale (see DVD Figure 21–2) to show the absolute differences between the 2003 and 2002 data. The light or white (red on your map) areas are those places where Part I violent crime went up, and the darker areas (green on your map) show where the Part I violent crimes showed a decrease.

32. Practice with comparing the robberies between 2003 and 2002. You will need to do a Select query on each layer, create new shapefiles of just the robberies for both years, and repeat the analysis. If you are using only robberies, you may need to redraw the extents or MBR of the points if the rectangle covers too much area that doesn't have any points.

33. You can use this analysis technique to compare time periods or check to see if robbery, auto theft, or other crimes are higher in certain areas based on population by using a census block theme as the secondary file (this typically needs to be a block centroid point file with the total population as one of the fields).

34. There are really no limitations as to how you can make use of the interpolation routine in CrimeStat III, and you can accomplish everything for which another analyst currently uses the Spatial Analyst extension. CrimeStat III has an added benefit in that it is free!

35. You may close CrimeStat and ArcGIS after saving your project.

Exercise 22: Density as a Predictor

   Data Needed for This Exercise
   - All data is contained in EX22.mxd.

   Lesson Objectives
   - In this exercise you will use the interpolation routine in CrimeStat III to create a density layer that will try to predict the next likely hit location for a new crime in a series.

   Task Description
   - We will use the interpolation routine in CrimeStat III to create a hot spot map for a strategic analysis project as well as a prediction in a tactical analysis for a crime series. Although the concept behind these tasks are a little different, the setup and usage is basically the same in CrimeStat, so we can learn to create hot spot maps and predictive maps with this easy-to-use tool.

  1. Create a new project and save in it your student folder as EX22.mxd.

2. Open ArcCatalog and drag and drop the following layer files into the project:
   a. C:\CIA\Geodatabase\layers\
      i. Beats.lyr
      ii. City_Streets.lyr
      iii. Grids.lyr
      iv. Street_Labels.lyr
      v. Street_Outline_Boxes.lyr
   b. C:\CIA\Exercises\Exercise_22\
      i. AllRezBurgs.shp
      ii. RezburgSeries.shp
      iii. Extent.shp
3. Close ArcCatalog.
4. Save your project again.
5. We need to create a density layer (hot spot) for the AllRezBurgs theme, and we'll use CrimeStat III to accomplish this.
6. To do this type of analysis in CrimeStat III, we need a primary file that has the X and Y coordinates in the attribute table, and we need to know the extents of the area we wish to analyze.
7. To find out what the extents are of our analysis area, we need to select all of Glendale. Place your mouse on the bottom left corner of the grid layer where the grid boxes of BI and 81 come together. In the bottom right hand side of the ArcView project window there is an X and a Y coordinate value. Write them down here _____, _____.
8. You do not need to record the decimal values, but you can keep them if desired.
9. Now click in the very upper right of the grid layer, where the grid boxes of 19 and EF come together, and write those coordinates down here: _____, _____.
10. Turn on the Measure tool, and this time we need to know the area of this extent.
11. Click the second icon from the left that looks like a small pointed box (see DVD Figure 22–1).
12. Now draw a box starting at the lower left and click once to lock it; go to the lower right and click once to anchor it; then go to the upper right and click once; and then go to the upper left and double click it to calculate the total area of _____ square miles.

13. Now click on the **plus sign** icon in the Measure tool (measure a feature) and click on the **Extents** theme somewhere outside of the city limits.
14. You should get about the same square mileage area, around 266 square miles, through either process.
15. The final tidbit of information we need for CrimeStat III is the total street mileage within this extent. We'll use a Select By Location query to select all the streets from the City_Streets.lyr theme that intersects the Extent.shp theme. When all of these streets are selected, open the attribute table of the City_Streets.shp theme, and using the Statistics option, find the sum of the street lengths and write it down here: _____.
16. Check your measurements with mine:
    Lower left XY: 530991, 910446
    Upper right XY: 631169, 984625
    Total square miles: 266
    Total street miles: 3023.4 (the length field is in feet)
17. Now open CrimeStat III.
18. In the Data Setup section, we will choose the AllRezBurgs.dbf table as our primary file (see DVD Figure 22–2).
19. The X coordinate will be the field called X, and the Y coordinate will be the field called Y.
20. Set the type of coordinate system to projected (Euclidean), and set the data units to feet.
21. Now click on the **Reference File** tab within data setup.
22. Enter your X and Y coordinates (from step 16) in the appropriate boxes (see DVD Figure 22–3).
23. Click on the **Measurement Parameters** tab within data setup, and enter your total area and total street mileage (see DVD Figure 22–4).
24. Check the **Indirect (Manhattan)** check box at the bottom.
25. Now click on the **Spatial Modeling** tab (blue), and choose the **Interpolation** tab under it.
26. Check the single check box, and complete the remaining portions of the dialog menu as follows (see DVD Figure 22–5):
    Method of interpolation: *Normal*
    Choice of bandwidth: *Fixed interval*
    Interval: *0.5*
    Interval unit: *Miles*

Area units: points per: *Square miles*
Output units: *Absolute densities*
27. Make sure to save the result to an ArcView SHP file, and save it in your student folder with a name you can remember (see DVD Figure 22–6).
28. Click on **Compute** to run this interpolation routine.
29. Close the CrimeStat results screen when it is done running, but leave the main portion of CrimeStat III open and just minimize it.
30. Add the layer you just created into your project. Change the legend to a graduated color legend based on the field called "Z."
31. Your map should look something like DVD Figure 22–7.
32. Now turn this theme off and maximize CrimeStat III again, but this time change the dialog parameters to the following and rerun the interpolation routine (see DVD Figure 22–8):

Method of interpolation: *Quartic*
Choice of bandwidth: *Fixed interval*
Interval: *0.25*
Interval unit: *Miles*
Area units: points per: *Square miles*
Output units: *Relative densities*

Save the result to another file name so you can keep the original for comparison.
33. Your new result should be more specific than the first with much smaller hot spots (see DVD Figure 22–9).
34. By adjusting the number of cells in the data setup section where you entered the coordinates of the lower left and upper right extents of the analysis area, or what is sometimes referred to as the minimum bounding rectangle (MBR), you can further enhance the look of the final result. In the following example, I went back and changed the default of 100 columns and rows to 250 columns and rows and reran this analysis (see DVD Figure 22–10).
35. By changing the size of the little polygons (the number of columns in the data setup) and the size of the search radius (the bandwidth interval), you have an infinite number of results you can obtain. The ones used in this exercise are fairly standard, but you may wish to experiment with several of the options, including setting the bandwidth to "adaptive," which means that the search radius changes based on how dense your points are. Each one of these routines only takes a few seconds to run in CrimeStat, so you can

easily run 3–10 different routines in just a few minutes. Just make sure you change the name of the output file each time if you wish to preserve each version, or else CrimeStat will just write over the older version.

36. Remove your interpolation layers and save your project again.
37. Now we will use this routine to help us predict where a new hit in a crime series is possible.
38. Open the attribute table for the RezburgSeries.shp theme.
39. There are no X or Y coordinates in this table, so we will need to add them before we can use this data in CrimeStat III.
40. There is an ArcToolbox tool that can do this for us.
41. Click on the red toolbox and open the **ArcToolbox**. Find the tools called **Data Management Tools → Features → Add XY Coordinates**, and double click it.
42. Choose the RezburgSeries theme as the input layer, and click **OK** to run it.
43. If you closed the attribute table, open it again to see that at the end of our table we now have Point_X and Point_Y fields, which hold our coordinate values.
44. We will use these fields to set up CrimeStat in the Data Setup → Primary file section.
45. Close CrimeStat and then reopen it to clear all the entries for the last section.
46. Under the **Data Setup → Primary file** tab, choose the RezburgSeries .dbf table as the source, and choose the Point_X and Point_Y fields as the source for the X and Y Coordinate values.
47. Set the Type of Measurement as Projected (Euclidean) and the data units to feet as we did before.
48. Go to the Reference File tab, and enter the following information into the empty cells:
    Lower left: *608672.35069, 952777.509997*
    Upper right: *619322.725081, 961387.464331*
    By number of columns: *200*
49. Go to the Measurement parameters tab and enter the following settings:
    Area: *3.289258 sq miles*
    Length of street network: *299157.59887 feet*
    Check the box next to **Indirect (Manhattan)**.

50. Go to the Spatial Modeling tab, choose the single interpolation function, and set the rest of the menu as follows (see DVD Figure 22–11):

    Method of interpolation: *Quartic*
    Choice of bandwidth: *Adaptive*
    Minimal sample size: *100*
    Area units: points per: *Square miles*
    Output units: *Probabilities*

51. Add the new shapefile you created (I called mine Prob1) to the project.
52. Change the legend to a graduated color legend and apply it. Your map should resemble DVD Figure 22–12.
53. Notice how it shows that the actual highest probability area is near the center of the crimes (red) and where none have occurred before. We know that this is more likely the place where the offender resides, so this particular analysis is not as helpful as we may have expected.
54. Go back to CrimeStat III, change the following settings, and rerun the analysis:

    Method of interpolation: *Negative exponential*
    Choice of bandwidth: *Fixed interval*
    Interval: *0.25*
    Interval unit: *Miles*

55. Add the new layer into your project and create a graduated color legend for this new theme. Your map should now resemble DVD Figure 22–13.
56. As you can see, the lower right section (red) of the analysis area now has the highest probability, and the middle and upper sections have a secondary probability (orangish). I have found that the triangular or negative exponential method of interpolation comes the closest to predicting the same or similar areas as does the standard deviation ellipses and other methods we have already discussed. Much more research into this method needs to be done before we can assume that it is a reliable method, but it is another tool that is cheap, fast, and useful for crime analysis in GIS.
57. Save and close your project.

Exercise 23: A Probability Grid Concept

Data Needed for This Exercise
- All data is contained in EX23.mxd.

Lesson Objectives
- Discuss what a probability grid is, delve slightly into how one is created and used to analyze a tactical crime series, and discover what things to consider when using multiple statistical layers to predict a next crime area for a series.

Task Description
- The probability grid method, or PGM, is a tactical prediction method that really resembles what many geographers do all of the time: They use geoprocessing to create one final output. Geoprocessing is simply the concept of using multiple layers derived from some basis in fact or theory and using where these layers overlap to develop a final output layer that has the best of all layers used in the process.
- We can compare the PGM to the method that many businesses use nowadays to determine where to build their next store. They take known information from the customers who currently shop at their stores and enter other parameters—such as distance from a freeway, distance from another shopping center or store of the same type, total population in an area, and other factors—to come up with the best location for a new store. All of these individual elements can be queried and developed inside of GIS, and then where they overlap the most (or don't overlap at all), they have the new store location.
- Doing this analysis for a crime series generally has the same steps. First we figure out what logic can be used to help us figure out what the suspect is doing. We consider the following factors:
  - Victimology: Who are our victims, and where are similar victims located?
  - Spatial statistics: Where will the standard deviation rectangles be located?
  - Distance analysis: How far between each hit has each crime been, and how far will the offender likely travel the next time?
  - Directional analysis: Can we see the pattern between hits and make a logical estimate of which direction the offender may travel to next?

- Motivation and modus operandi (MO): Do we know what is motivating our offender(s) (thrill, money, reward?), and is the MO in each case we are including sufficient enough to tell us if the same offender(s) is committing these crimes?
- How can we create analysis layers for all of these logical items and use them together to create one operationally useful analysis and map?

1. Open EX23.mxd.
2. The standard deviation ellipses and other layers have already been created for you using CrimeStat or the spatial statistics tools in ArcGIS.
3. The scenario is that this is a sexual assault series for which you have been assigned to create a tactical prediction. The suspect has been targeting young women between the ages of 18 and 27 years while they are walking or jogging for exercise. These women are all small in stature but athletic, none have been taller than 5′4″, and the heaviest victim weighed approximately 105 lbs. The women each left their homes, which are within the same area, and ran to the nearby park, where they ran laps around the park and then ran or walked home. During their walks home, the suspect, described as a black male with very dark skin, approximately 17 to 20 years of age, 5′11″, 135 lbs, with short black hair and brown eyes and wearing a black hoodie and dark pants, started running or walking alongside the women, scaring them. The women have stated that the suspect "came out of nowhere" and began walking or running beside them and trying to make small talk about exercise. After a few seconds of this conversation, the suspect said "goodbye" and then acted like he was dropping back to leave or go in a different direction, then he ran up behind the women, dragged them to the ground, and in each case attempted to or successfully removed their pants while the women fought with him. The suspect was successful in making vaginal entry in only the last four cases, but he has been getting increasingly violent with each attempt. In the beginning the women were able to fight their way free, and the suspect only got their pants or top off and grabbed their breasts or vagina, but the women were able to kick and fight their way free and run. The suspect did not follow the women when they escaped but instead ran the other way. In the last four cases, the offender punched one victim unconscious, beat another victim, and threatened the last two victims with a knife to their throats and then sexually assaulted them in a secluded location in a neighborhood from which the

victim was coming, a short distance from the park. The victims all believed that they had seen the suspect in the neighborhood before and felt as if he must have been watching them for several days before the attempted or successful sexual assault.

4. In addition to the standard layers, a PGM Grid has been created. This is basically just a fishnet of polygons cast across the surface of where the rapes have occurred. Open the attribute table for the PGM Grid theme.

5. You will see fields for the Stdev2, Stdev1, ConvHull, Last Hit Buffer (LHBuffer), and the probability result from CrimeStat's interpolation routine. I have *selected by location* where each of these layers intersected the PGM Grid and then scored each grid in the PGM Grid with a score between 1 and 3 where they overlapped. The Stdev2 and ConvHull intersecting grids get the highest score at 2 or 2.25 because they are the best predictors of a new hit based on research mentioned in previous exercises. The Stdev1 and Last Hit buffers get a value of about 1 or 1.5 because they are good predictors about 50% of the time, etc.

6. When I score the PGM, I don't want to use values from 1–10 or some other number values because a score of 10 will far outweigh a score of 1, and all of these variables perform about the same, so we want to keep the ability to score one aspect or another higher so that it cannot unduly influence the final PGM Grid result.

7. The last item we need to add is the buffer from the city parks. In this rape series, the suspect always confronted the victims after they left the city park where they went to exercise, so we want to find out what the average distance is from each park where the offenses occurred and create a buffer around the available parks at that distance. This buffer zone will then be scored in the PGM, and the final result will be totaled and shown on the map. First, we need to find out what the average distance is.

8. If we just use the Measure tool and measure a few of the distances between parks and the known incidents, we can see that most are within 0.25 miles, or 1320 feet, from the park. There is one that is about 0.29 miles, but that could be a simple measuring error on my part. However, if I choose a distance of 0.25–0.35 miles as my buffer distance, chances are I would be very safe in that estimate.

9. Now we need to create a buffer around each of the parks within 0.25–0.35 miles or so of selected parks in this analysis area or minimum bounding rectangle (MBR).

10. Make sure that the City-Parks layer is the only selectable theme (**Selection → Set Selectable Layers**).
11. Now use the Select tool to select all of the parks in and around the rape series points.
12. Use the ArcToolbox and create a single buffer around all of the selected parks at a distance of 0.25 to 0.35 miles as you deem appropriate. Save this new theme in your student folder as Buffer_of_City_Parks.shp.
13. Add the new theme into the project, if it was not automatically added, and turn it on.
14. You should now have a map that resembles DVD Figure 23–1.
15. Now we need to score the PGM Prediction theme where the buffer of the parks intersects the PGM polygons. We will use the Select By Location query tool for this.
16. Click on **Selection → Select By Location**. We need to select the features of the PGM Prediction theme that intersect the Buffer_of_City_Parks theme.
17. After selecting the small polygons in the PGM layer with this method, we need to score the PGM Grid with a value of 2.5 (we are very positive and use a score very close to 3 because all of the crimes in the series had this MO factor).
18. Open the attribute table for the PGM Prediction theme.
19. Right click on the field named "ParkBuffer" and choose **Field Calculator** from the pop-up menu.
20. Type in *2.5* in the window, and make sure the Calculate for Selected Records Only check box *is* checked and press **OK**.
21. Now 8543 selected grids that intersected the buffer of the city parks we created have been scored at 2.5.
22. Click the **Options** button and choose **Clear Selection** so that none of the grids are selected.
23. Now we need to add up all the scores in the Total field, so right click the **Total** field name and choose the field calculator. In the calculate window, enter the following information:

    [IntResult] + [Stdev2] + [Stdev1] + [LHBuffer] + [ConvHUll] + [ParkBuffer]

24. Click **OK** to execute the calculation.
25. Close the attribute table for the PGM theme.
26. Right click the theme in the TOC and choose **Properties → Symbology**.

27. We need to refresh the graduated color legend using the total field. Click on the **Value** field, and choose the **Total** field again to refresh the view.
28. Change the legend so that the highest values are bright red, then orange, yellow, white, and blue (to show hot and cold), and apply the new legend.
29. I generally use a standard deviation classification method legend at this stage because I only want the areas with the highest scores to be red.
30. Clear all selected features (**Selection** → **Clear Selected Features**).
31. Turn off any layers that you do not need anymore (like the buffer of parks, etc.).
32. Your final prediction map should look something like DVD Figure 23–2.
33. Notice that we now have only four small areas that are in the peak risk for a new crime in our series. It will now be much easier for the surveillance teams to watch these areas, and a site visit might be appropriate as well to see what is at these locations and decide which one of the areas could be the most fruitful.
34. Although I personally feel that the PGM process is a much better overall prediction method than any single method currently being used, it does not mean that I always use it to make predictions. If the area is very small, I may use only a standard deviation ellipse and last hit buffer analysis to give the detectives a good target area.
35. Common sense has to take over here.
36. If it takes you 14 hours to create a good PGM map, and the detectives have already made an action plan without your map, chances are you are spending too much time and should use a simpler method, but be sure it is your "best analysis" effort and not a "safe" effort. I know many analysts who will just provide the points where the crimes have occurred and will not put their reputation on the line at all to make a prediction. There could be many causes of this, which include lack of faith in their own abilities with GIS software or lack of assurance that the detectives will actually use their analysis. I have often had the latter opinion, but you should do the analysis anyway because it is what you are being paid for, even if the detectives do not follow your prediction advice.
37. CrimeStat III is a good tool to create the overall polygon grid theme. You simply create an interpolation layer, like we have in the

past, then add fields to that grid for the various layers used for decision making in our geoprocessing effort, and calculate them.
38. Other extensions that are available are Hawth's tools and several other scripts and extensions, which you can find at http://arcscripts.esri.com.
39. Hawth's tools is a free extension that works right in ArcGIS. It can make paths between points and polygon grid layers, as well as density kernel estimations (interpolation routine like CrimeStat III), and it can be very useful for analysts.
40. Go to the http://arcscripts.esri.com Web site and see if you can find this extension.
41. You may download this extension as needed, but you will also find the latest version in your C:\CIA\Documents\Tutorials_Programs\ folder as a ZIP file that you can unzip, install, and begin using back at your office or home.
42. Save and close your project.

Extra Practice

43. If you wish to work on another crime series and use the Hawth's tools or CrimeStat III to create a PGM Grid prediction for a new hit in a crime series, open EX20A.mxd and do the analysis for that crime series.

■ References

Boba, R. (2005). *Crime analysis and crime mapping.* Thousand Oaks, CA: Sage.

Buerger, M. E., Cohn, E. G., & Petrosino, A. J. (1995). Defining the "hot spots" of crime: Operationalizing theoretical concepts for field research. In J. E. Eck & D. Weisburd (Eds.), *Crime and place* (pp. 237–258). Monsey, NY: Criminal Justice Press.

Eck, J., Chainey, S., Cameron, J. G., Letner, M., & Wilson, R. E. (2005). *Mapping crime: Understanding hot spots.* Washington, DC: National Institute of Justice.

Harries, K. (1999). *Mapping crime: Principles and practice.* Washington, DC: National Institute of Justice.

Levine, N. (2002). *CrimeStat: A spatial statistics program for the analysis of crime incident locations* (Vol. 2.0). Ned Levine & Associates, Houston, TX, & the National Institute of Justice, Washington, DC. Retrieved March 2009, from www.icpsr.umich.edu/NACJD/crimestat.html

Paulsen, D. & Robinson, M. (2004). *Spatial aspects of crime: Theory and practice.* Boston: Pearson.

Sherman, L. W., Gartin, P. R., & Buerger, M. E. (1989). Hot spots of predatory crime: Routine activities and the criminology of place. *Criminology, 27*, 27–55.

Taylor, R. B. (1998). *Crime and small-scale places: What we know, what we can prevent, and what else we need to know* (pp. 1–22). Crime and Place: Plenary Papers of the 1997 Conference on Criminal Justice Research and Evaluation, US Department of Justice, Washington, DC.

Mapping for Your Audience and Future Issues

SECTION IV

CHAPTER 11   Mapping for Your Audience
CHAPTER 12   Future Issues in Crime Mapping

Mapping for Your Audience

CHAPTER 11

▶ ▶ **LEARNING OBJECTIVES**

Chapter 11 discusses the importance of creating appropriate maps for your audience. Elements of content, style, and color are addressed. Because crime maps must also be accompanied by explanations for those who will be reading and using the maps, it is important to tailor maps and their discussions to the intended audience. After studying this chapter, you should be able to:

- Identify the different uses of crime maps by different criminal justice practitioners.
- Create different maps and write tailored descriptions for the intended audience.

■ Introduction

The belief that "one map fits all" is severely flawed. Each audience has a unique set of objectives it is trying to achieve (Harries, 1999). For example, patrol officers need maps that can help them make day-to-day decisions about the people and places where they are assigned. Investigators need maps to help them locate suspects in the crimes they are investigating as well as help them predict where and when perpetrators are likely to strike again. Administrators need maps to help them improve the effectiveness and efficiency of their operations. Community members want maps to inform them about problem areas in their communities that need attention. Court officers need maps to tell a story to the jury about a criminal event. And finally, corrections officers need maps to help them track and assist offenders who are under their supervision. Given this wide array of goals, one can see how maps and analysis must be tailored for each unique audience. This chapter provides suggestions about how the beginning analyst might create different maps for different audiences. The audiences discussed in this chapter by no means represent a complete list. However, they are the most common audiences a crime analyst must please and thus are included in the discussion. GIS is a powerful desktop application. It allows you to create maps and analyze data that has some spatial or geographic relationship to something else (on the Earth's surface). We could also say that GIS is any data a police department collects as a normal part of its activities. Most data collected by a police department can be related to, or used with, location data (addresses) that deal with crime, disorder, terrorism, or any other issue assigned to a police department to solve. Police departments routinely collect geographic data on where a crime happens or even where an apartment complex is within their jurisdiction. They track where the "bad guys" live and where they arrested them. They keep track of citation data and the related racial-profiling data that many departments have begun to collect or analyze in the past several years.

All of these datasets can be used with a GIS to enable quick access to data and information and put it in the hands of the decision makers. With this increase in the amount of data available to decision makers, we have to be able to make sense of all of it. In the business community this is often described as "business intelligence" or the process of taking the bits and pieces of *data* we collect everyday, ordering it and organizing it so that it makes sense and provides *information*, which administrators, detectives, and patrol officers can then use as *knowledge* to prevent crimes, catch criminals, or enhance public safety.

■ GIS in Criminal Justice

There are a variety of GIS mapping products that you will use during a career as a crime or intelligence analyst with a law enforcement agency. The general idea you should be thinking about now is that the audience determines what type of map is produced. You will have many different people and purposes within the audiences you will address as a crime analyst. One way to classify the maps most often produced is to place them into a few categories that align themselves with the three basic categories of crime analysis:

- Administrative
- Strategic
- Tactical

Administrative maps are often reference maps (beat maps, city council districts, etc.). They are generally less specific and often show the entire city boundaries. They generally show yearly or monthly data aggregated to some type of boundary layer and one center of focus or purpose for developing the map. They do not always have to be reference maps and can show hot spots, point data, and other incident efforts; however, the goal is to be more general in nature, and the use of these maps are more appropriate for Compstat meetings, for example.

Strategic maps are often the first step toward problem solving efforts within a police agency. They generally cover shorter time periods and focus on a specific geographic area within the city boundaries, and the overall objective is to identify problems and suggest possible solutions. These maps will evolve within a crime analyst's tenure at an agency and will probably need to be updated often to track problem areas. These maps can also be hot spot, graduated point, pin, or other types of maps, but again, the general purpose is problem solving rather than administrative depiction of data. We often see crime clusters and patterns in this stage of crime mapping.

Tactical maps involve very short time periods. Tactical maps are usually used to predict a new crime in a series or spree or identify where the suspect may live in a series of offenses. With tactical mapping products, you wait for the next offense to happen to create a new map and prediction and help catch the offender. This is an exciting phase of crime analysis, but sadly, only a few analysts across the country ever do tactical analysis. This function seems to be increasing, but it does require the analyst to work very closely with the case detectives, and that relationship has to be nurtured over time.

Within these three categories of maps, we find that a large variety of cartographic methods and geoprocessing efforts may come into play. Each of these different ways to portray data (cartography) and show relationships between different types of data (geoprocessing) can be used for any of the three mapping categories already discussed. It is not uncommon to have administrative hot spot maps lead to a strategic version of the hot spot map, and finally have a tactical set of what we could call hot spot maps to track a crime cluster, spree, series, or pattern of crimes that was originally identified at the administrative level.

As a simple matter of what looks good and what doesn't, you should think about a few things:

- Who is your audience? The person or group you are generating a map for will often govern the type of data you place onto the map. The issues involved include privacy, data accessibility for that audience, the point you are trying to make with the map, and the political ramifications.
- What type of output is going to be required? If only black and white maps on a laser printer are going to be printed, why use colors for symbols? You should consider changing the symbol for each different item rather than the color in this case.

- What size map will be printed? A large map from an E-size plotter will require having symbols that are often larger than your standard 8.5″ × 11″ map. They also print much slower, and if an aerial photo or graphic is included, the picture will pixelize a little bit and be more useful when viewed from farther away.
- Printing 22 copies of a map at 36″ × 48″ is possible but not recommended.
- Does the map have the following elements?
  - Title
  - Scale bar
  - North arrow
  - Legend
  - Disclaimer text
  - Notes or file location information so it can be found again later
  - Map with one center of attention or point being made
  - Nonpsychedelic color scheme
  - Good use of white space

It is important that the area around the center of attention is free from confusing data and allows the eye to see breaks among different items on the map. Each map you create should have only one purpose or story to tell; avoid trying to tell two or more stories in a single map. Is the map at the right scale, and are you zoomed in to the important story you want to tell with the map? Did you look at the data from several different viewpoints to make sure the type of map you are using is the best? Does your map read left to right and top to bottom as we read everything here in the United States? Using really bright colors for nonimportant items makes the map confusing and difficult to look at. Try and keep to pastel colors for the reference parts of a map, and keep the bright colors to help illuminate the key point of the map.

Before publishing the map, did you let a few coworkers review the it and let you know what they think the map is telling them? If they get the wrong point from the map, maybe a revision is needed!

The average supervisor or officer within a police department does not normally make maps. They generally have a very limited knowledge of what can be done with crime data and statistics. Both authors of this book have asked the following questions several times with new groups of students and at various presentations that provides a demonstration of this. "How many of you have taken a college-level statistics class?" Most of the group raises their hands. The second question is, "How many of you remember what you learned in that statistics class?" This question results in 75% of the hands going down. The third question, "How many of you use statistics on a weekly basis?" typically results in all hands down unless some crime or intelligence analyst sneaked into the classroom or meeting. The results of the questioning would also be very similar if you asked how many people have seen a map and then asked how many know how to make a map. If you then asked about cartography, or the art and science of making a map, chances are only one or two persons in the police group, if any, would raise their hands.

This is where most of the problems lie with crime mapping in law enforcement: (1) the analyst does not know how to make a map that speaks to the issues, or (2) the audience does not know how to read or interpret a map clearly and thus makes mistakes.

As discussed in previous chapters, crime mapping deals with administrative, strategic, and tactical map assignments. Within these general categories we know that there are reference maps, point or pin maps, hot spot maps, and maps that are derived from slightly more complicated geoprocessing or multilayer analysis. We need to remember that the ultimate goal of mapping is not to make a pretty map or one that makes you feel good about yourself but one that helps the officer or supervisor make sound decisions. It isn't a matter of making sure the specific cartographic technique you use is technically sound. After all, what good is a map if no one will read it after you've spent all day making it? The dilemma appears for the crime analyst when we begin to learn the right way to make a map based on what cartographers say and what our fellow employees can comprehend. Most analysts can tell you of at least one incident where they made a technically sound and accurate map but the audience within the police department simply looked at it once and asked, "So what does this mean?" We can only guess how many maps we've made that wound up in a trash can someplace as well. Like any other analysis product, maps are temporary and will not hang around very long. Crime is dynamic, so the maps we create must also be dynamic to keep up with the current crime trends and goals or objectives of the police department. Do not become too attached to the maps you create because they really won't last very long in the reality of the world of crime fighting.

Analysts need to make every effort to make sure the analyses they perform are accurate and reliable. They should learn about making cartographically sound maps and include things like north arrows, scale bars, legends, and titles to help the user fully understand the map. At least once in your career, you will probably experience a supervisor who asks you to remove these items from your map because "it just clutters it up!" That probably means that the elements are too large or take up too much of the map area, so reduce them in size and place them in an obscure corner, but don't take them out. If the supervisor still insists on removing them, follow orders, but add training about reading maps to your list of things to do within your police department.

You will find that most officers and supervisors are very comfortable with point or pin maps, and most of the requests you get for maps will be for these types of maps. You should always remember scale and multiple points as keys to understanding and making your audience understand your pin maps. Showing multiple points on a map at a small scale (zoomed out from the Earth's surface) can just show there is a lot of crime. At certain scales, having multiple legend categories or showing every crime as a different symbol or marker can be confusing and often a waste of time. Be comfortable with your spatial data as well as the attributes of the data behind each point. Know what it looks like when you zoom to a citywide level and how it changes as you zoom in to a specific beat or reporting district. Keep in mind that the audience does not understand the data as well as you do, and you need to use the KISS (keep it simple, stupid) method when making pin maps.

Remember that there are often multiple points under one single dot on the map. Somehow convey this message to the audience through graphs, charts, maybe an additional graduated point map, or a hot spot map that shows those concentrations at one location. Realize that the underlying geography or landscape of a place can determine why the points are arranged as they are. If you have an area that looks like crime is extremely high and everything else around it has no crime at all, be suspicious. What is probably happening is that the surrounding area includes dirt lots or agriculture areas, and the high crime area is the only residential area or apartment complex within a few miles. The best advice anyone can give you is to know your data inside and out; it saves you from making analysis errors and saves your audience from making decisions based on faulty assumptions.

When creating hot spot maps, remember that there are different ways and methods to create them. They are all based on spatial autocorrelation and distance decay theories. This simply means that things that are closer together are more likely to be related than things that are farther apart. If we have a classroom full of students and the instructor calls on a student for the answer to a question, the people in the front row are often the victims, so most people want to sit way back in the room. Many crime theories depend on the fact that this assumption is true.

You can create hot spot maps that are based on clustering algorithms (such as in CrimeStat III) where the math determines that the points are not randomly generated but are clustered together in significant groupings that are higher than the entire geographic area being analyzed. Another method uses density analysis (like spatial analyst results) where the math counts the number of points within a certain distance of an artificial grid that is laid over the area being analyzed. Those areas where the points are closer together and more numerous (within a set search radius) makes the hot spot more dense than other areas in the total geographic area being analyzed. We discussed terminology such as nearest neighbor, kernel density, spatial density, cluster analysis, STAC clusters, fuzzy mode, and several other analysis methods in other chapters. They all relate to different math calculations that attempt to find areas of abnormally high activity for the kind of crime you may be analyzing at the time. Spatial autocorrelation can be tested for using different programs and will tell you if the data is clustered, random, or dispersed. Here is where it is important to understand the landscape under those points and make some of your own decisions about the usefulness of these products. A hot spot map is usually developed to find the source of significant problems that can be addressed through enforcement activities. It isn't really a research project to find out how well hot spot methods work when compared to one another. It is a great idea to do this research and look at the results of the various methods together. Which areas of your city seem to show up with every method? Chances are these are the hot spots you should really be focusing on. The final map for your audience does not have to show all of these. In fact, it would be more useful if you just circled the most crime-ridden area for the user and explained the types of crime that are the biggest problem. It is the job of the analyst to provide the maps and information and also to do an analysis, which generally means to provide a presentation of your findings, not just regurgitate the data.

When you need to make a reference map, make sure you interview the people who will use it and find out under what conditions they will use the map. You can make a beautiful color map for daytime patrols, but the nighttime patrols cannot distinguish three of the four areas of the map because of the colors you chose. This is also true when you send an e-mail or electronic document in full color, but the units in the field only have a black and white laser printer.

Geoprocessing deals with the concept of logical steps to create data. Much more thought needs to be given to this process to make sure the analysis product is useful and gives us the right answer. What we are doing is using multiple layers of data to come to a single analytical result. Each step in the process needs to be planned out, and we need to make sure that we don't make incorrect assumptions with each individual step. Those steps could include a specific layer or analysis in the form of point data, line data, polygon data, hot spots, and more.

■ Maps for Specific Audiences

In addition to the tips that have previously been discussed, specific products can be created for various units within a police department, the city government, and citizens.

Patrol Officers

Patrol officers utilize maps for several different purposes. They are primarily a way to convey information about what is currently happening in their assigned areas and how crime and other incidents have increased or decreased from the recent past. Maps for these purposes are often distributed prior to the commencement of their shift and/or are discussed during roll call. The most important factor to consider when making maps for patrol officers is that they include real time information that they can use. Information that is several days old is not necessarily useful to them (although a historic map is helpful, it does not include the most recent data, which is needed to better inform decision making about what to do about crime and other incidents day to day). Remember, the primary purpose of crime maps at the patrol level is so officers can develop strategies to respond to crime (remember the SARA model in Chapter 5?) within their assigned areas. Another important factor to consider when making maps for patrol officers is that the mapped area should be specific to the areas the officer patrols. Along these lines, some placement of landmarks and street names helps orient the officers to where they are on the map, but placing too many landmarks and providing too many street names clutters up the map and is unnecessary because patrol officers are usually intimately acquainted with the areas they patrol. Including a table that details key elements of the crime incidents (crime, date, address, and victim or offender characteristics, for example) allows officers to refer back to important variables without cluttering up the map and provides some context to the dots. Last, using sharply contrasting colors and dark, readable fonts on the map is helpful, especially if the map will be printed in black and white or for officers who may have to refer to the maps in low light conditions. If it is difficult to see important items on the map, the map is of limited use.

Paulsen (2003) suggests that maps are underutilized by patrol officers. Creating maps that are easy to read and include up-to-date statistics that help them make decisions about their assigned patrol areas may increase patrol officers' utilization of crime maps. As examined in Chapter 5, well-informed crime maps can help officers identify trends and patterns of crime that can be relayed to citizens to reduce their victimization (as in the "garage shopping" case where residents were alerted to the increase in garage thefts and were encouraged to shut and lock their garages when not in use). In addition, maps can produce suggestions about where traffic officers might best place themselves to reduce traffic accidents caused by speeding. Finally, maps can indicate how seemingly unrelated crimes are in fact related to suggest possible areas of surveillance that might be most beneficial.

Figures 11–1 through 11–3 are examples of maps that would be useful to patrol officers. Figure 11–1 depicts an example of a briefing map that is designed for patrol officers to pull up in their patrol cars. It only includes incidents that officers are able to trend or work on and is designed primarily to help officers know what happened in their area during other shifts and for the week. It is a building map, so the first map completed for Monday shows what happened over the week, the next one completed for Tuesday shows what happened on Monday, and so on. It is broken down by shift, and the callout boxes contain information requested by the officers so that if they need more information, they can pull it up on the department's record management system.

Figures 11–2 and 11–3 depict traffic maps that use two different ways of displaying traffic accident data. Figure 11–2 uses graduated points, and Figure 11–3 uses a join between point and street segments to get a number of accidents per street segment calculation.

Investigators

There are multiple uses of crime maps for investigators. "A recurring theme is that maps often reveal a whole picture that is greater than the sum of its parts. This happens when many small and seemingly isolated and insignificant pieces of evidence take on critical importance when viewed as part of a pattern" (Harries, 1999, p. 70). Recall the case in Knoxville that was examined in Chapter 5 where a clustering of seemingly unrelated crimes were related to a drug dealer (who was also a parolee) who lived in a cul-de-sac. Upon arrest of this person, it was discovered that many of the crimes were committed by the drug dealer's clients and friends on their way to and from his house. Maps that can bring together all of the information in a coherent manner to

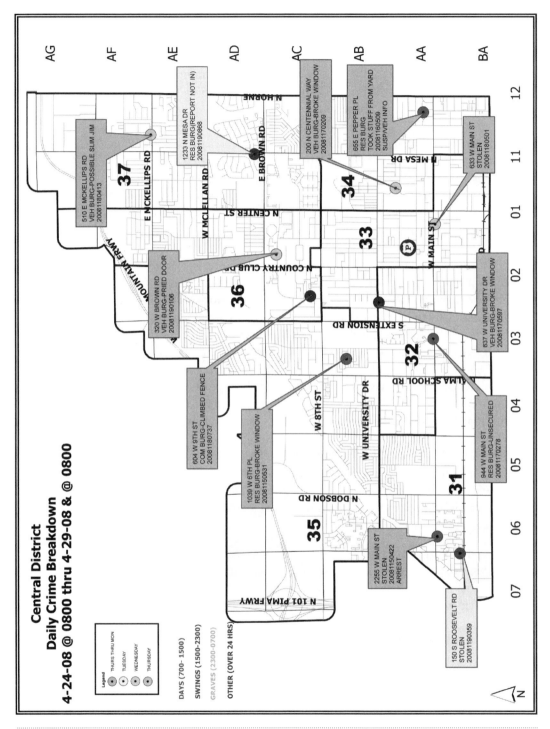

Figure 11–1a Central District Daily Crime Breakdown

Source: Jackie McClanahan, Crime Analyst, Central District Mesa Police Department

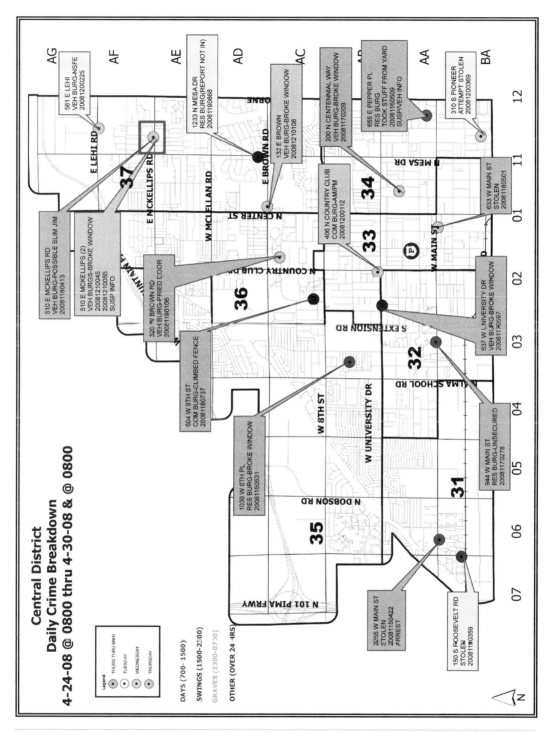

Figure 11–1b Central District Daily Crime Breakdown
Source: Jackie McClanahan, Crime Analyst, Central District Mesa Police Department

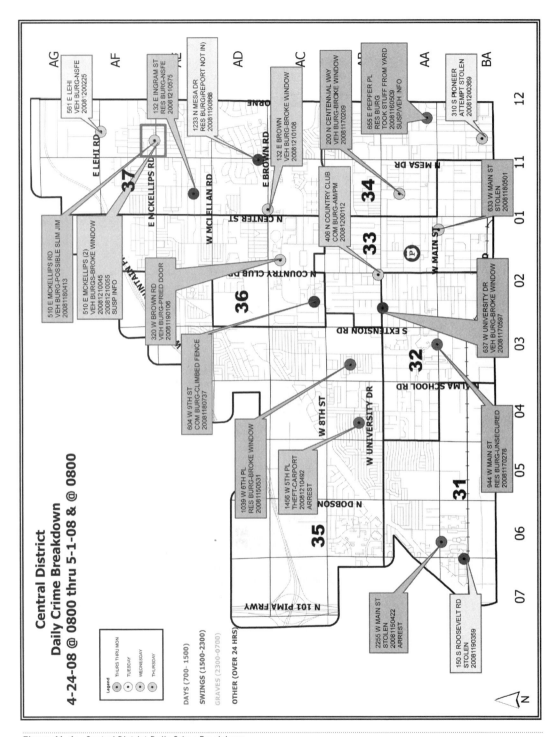

Figure 11–1c Central District Daily Crime Breakdown

Source: Jackie McClanahan, Crime Analyst, Central District Mesa Police Department

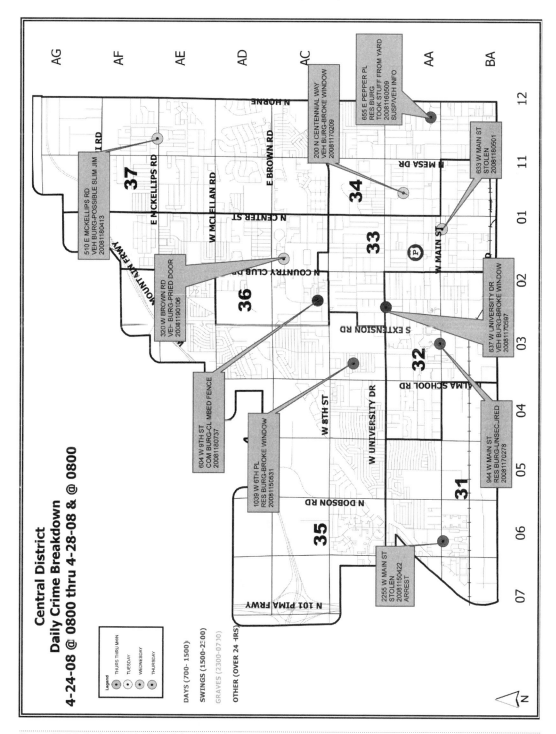

Figure 11–1d Central District Daily Crime Breakdown
Source: Jackie McClanahan, Crime Analyst, Central District Mesa Police Department

434 CHAPTER 11 Mapping for Your Audience

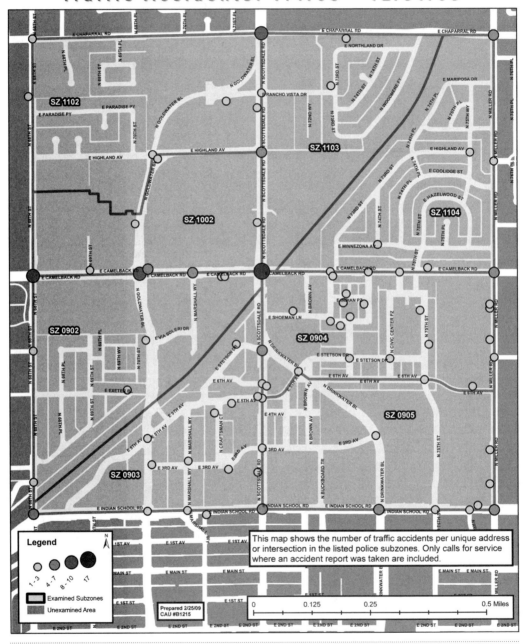

Figure 11-2 Traffic Accidents, Scottsdale, Arizona, Graduated Circles
Source: Scott Peacock, Police Analyst, Scottsdale Police Department

Maps for Specific Audiences 435

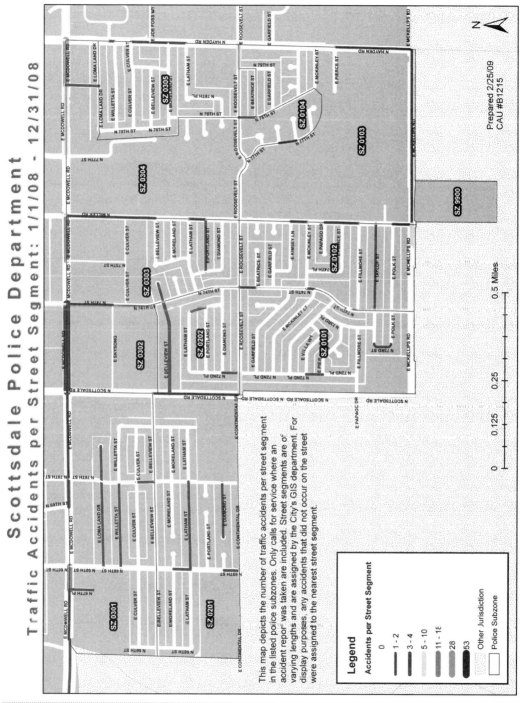

Figure 11–3 Traffic Accidents, Scottsdale, Arizona, Street Segments Colors
Source: Scott Peacock, Police Analyst, Scottsdale Police Department

provide a visual of the whole picture are most useful to investigators. Spatial analyses of crime can identify areas most likely for an offender to live, which can reduce the number of persons an investigator needs to investigate. For example, utilizing databases of last known addresses of offenders within a targeted area to perform queries based on the elements of the crimes is an important tool. Maps can then be distributed to investigators along with a prioritized list of offenders to be questioned. Recall several successes of this technique discussed in Chapter 5 in apprehending burglars and sex offenders. Maps can also be used to confirm the alibis of suspects and to corroborate eye witness accounts of crimes. Recall in Chapter 5 the mapping of the locations of victims and offenders based on cell phone transmissions.

Again, these maps need to include legible fonts for landmarks, street names, any other text that is important to the map, and contrasting colors that make the map easy to read. Including too much detail clutters a map and does not add to the investigation. Neglecting to add enough information (and the right information) to the map limits its ability to inform an investigation.

Figures 11–4 and **11–5** provide useful examples of maps for investigators. Figure 11–4 depicts a map showing the known associates of a wanted offender. Detectives searched the suspect (Percy), and they thought he was being hidden by his friends or family. Several officers and detectives who were not familiar with any of the characters assisted with the search, so the map was used to show who the parties were and where they lived. The cell phone tower hit locations indicated the areas where Percy was located when he made calls from his cell phone.

Figure 11–5 is a map that provides a timeline of a suspect's (Ruiz) activity over a couple of months. Detectives held many briefings to try to identify all of the events and places involved in their search for the suspect. The time line and map were used to keep all of the events straight during the investigation.

Management

Police managers have a difficult task in that they are responsible for having an effective, yet efficient agency as well as providing equality in the services they provide. The inherent conflicts that arise in achieving these goals are numerous. Managers need maps that present a larger picture of crime and other problems within their jurisdiction. They need maps that can inform them of recent changes in the distribution of crime and that can help them deploy resources based on predictions about crime distributions in the future. For example, one of

Maps for Specific Audiences 437

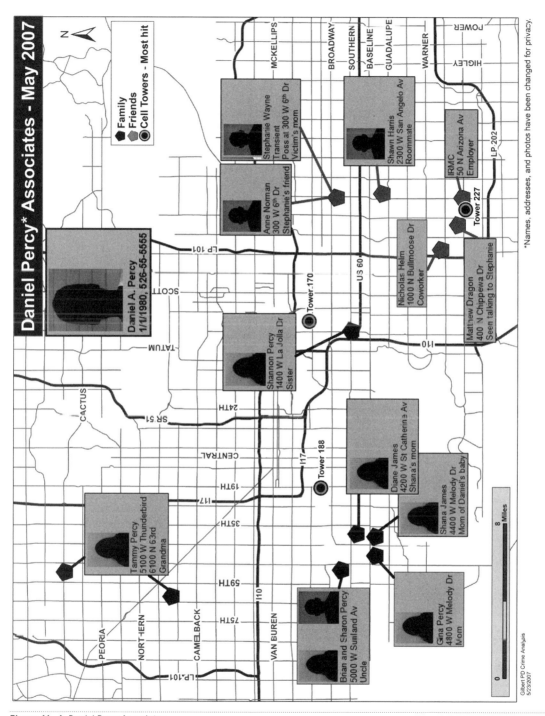

Figure 11-4 Daniel Percy Associates
Source: Judy Fernandez, Crime Analyst, Gilbert Police Department

438 CHAPTER 11 Mapping for Your Audience

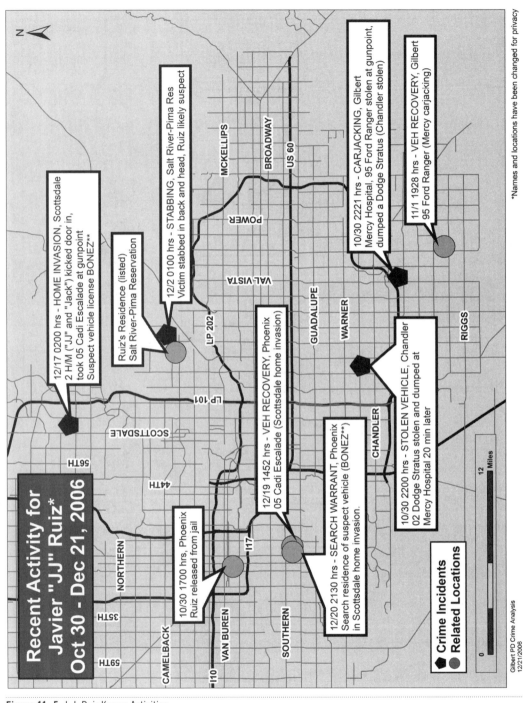

Figure 11–5 J. J. Ruiz Known Activities
Source: Judy Fernandez, Crime Analyst, Gilbert Police Department

the most common tasks of police managers is resource allocation. As discussed earlier, only a small percentage of calls for service actually involve criminal events.

Maps that depict the demands placed on law enforcement by the community can be helpful in several ways. First, calls for service can be mapped based on the severity of the call. These maps can identify areas of service that need more patrol. In addition, maps can help police managers with redistricting in attempts to equalize the workload for officers. Using techniques such as repeat address mapping (graduated dot maps) can produce maps that clearly show addresses that are responsible for a large percentage of calls for service. Second, hot spot maps can also help managers develop solutions to chronic problems (such as in the case of the Boston miracle discussed in Chapter 5) and at the same time minimize the effects of displacement. Third, maps that depict before and after levels of crime can be used to assess the effectiveness of strategies (and to show the community at large the successes) and to inform needed adjustments to crime prevention efforts. Fourth, maps are also useful as visual representations of need in grant applications for equipment and personnel. Last, in an age of cultural diversity and community policing, it is critical for police managers to have an understanding of the demographics within their jurisdiction. While access to recent data on demographic shifts can be difficult (for example, other than in large cities, the US Census Bureau only collects demographic information every 10 years), it may be useful in understanding how changes in the population may affect crime trends. For example, knowing that some communities have become increasingly more racially and ethnically diverse might help law enforcement understand some of the conflicts that may have occurred and resulted in increased numbers of crimes such as assaults. In addition, a particular neighborhood may have a higher percentage of young persons (perhaps between the ages of 12 and 24 years), which might explain higher than average crime rates within this neighborhood. These maps, if based on solid predictions about demographic and cultural shifts, may be useful in making decisions about future resources, including the hiring of officers who are bilingual.

Figures 11–6 and **11–7** are examples of maps that are useful to managers. Figure 11–6 is a choropleth map that was produced at the request of crime prevention officers when they spoke at neighborhood meetings.

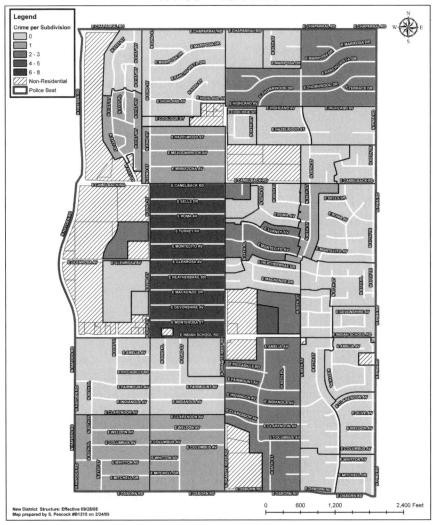

Figure 11-6 Hot Spot Map Burglaries/Thefts from Autos
Source: Scott Peacock, Police Analyst, Scottsdale Police Department

Figure 11–7 is a monthly crime density map made for each patrol district to highlight high crime density areas.

Community

The demographics of neighborhoods can change drastically over a relatively short period of time. This can create conflict amongst the persons who reside in the neighborhood and between residents and

Maps for Specific Audiences 441

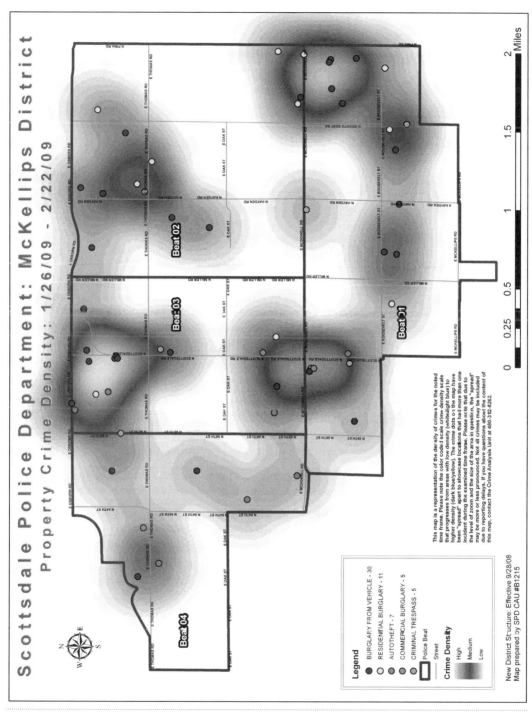

Figure 11–7 Crime Density Map, Scottsdale, Arizona
Source: Scott Peacock, Police Analyst, Scottsdale Police Department

police. Under community-oriented policing, law enforcement must work with communities to address crime and other problems that occur in their neighborhoods. One method of sharing information about crime with a community is through interactive mapping programs that are typically available online. As discussed in Chapter 6, caution should be used in providing information in this format because it can be used in a deceptive manner by persons who wish to capitalize on a neighborhood's crime problems. The maps (whether created by law enforcement and posted for community viewing or by a system that allows community members to create their own maps with the use of basic queries) should convey useful information about crime, demographics, and community resources. Access to online mapping capabilities varies from department to department. Some departments allow any user with Internet access to create detailed maps of crime at the address level. Other departments allow open access but only provide aggregate information at the block or larger level. In some cases, departments only allow certain members of community organizations to have access to mapmaking abilities. Last, some departments post static maps that community members can view but cannot change or even query.

Figure 11–8 is an example of a map for communities. It was created and submitted as part of a press conference that the Tucson Police Department/Counter Narcotics Alliance gave regarding the arrests of 39 people who were connected to a meth cell. By showing the areas where the offenders lived, the map demonstrated that meth is not a problem just in disadvantaged neighborhoods or high crime areas but that meth affects the entire community.

Courts

The main function of GIS in the courtroom is presentation. Maps can be created at the building level to reconstruct crime scenes or to illustrate the points where key pieces of evidence were found. Maps can also be useful to show the paths the offenders used to travel from place to place in committing their crime(s). For example, as discussed in Chapter 5, maps of serial offenders' movements can be mapped to provide the jury with a better picture of how a crime unfolded. In addition, maps can be used to present evidence about an offender's whereabouts in temporal and spatial proximity to a victim (such as mapping locations of offenders and victims based on cell phone call logs). The value here is the visual impact that maps can add, especially

Maps for Specific Audiences 443

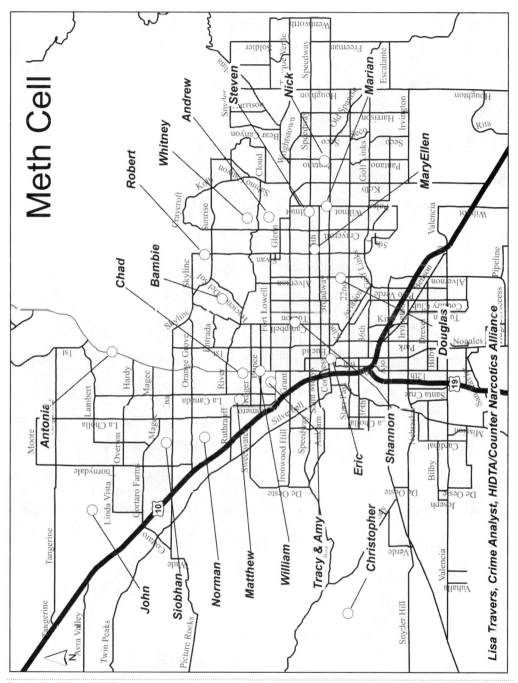

Figure 11-8 Meth Cell Map

in a long and complicated trial. Maps can synthesize a great deal of information and project the complete story in a picture. (Remember, a picture is worth a thousand words.) Many people also naturally believe that maps are always true. This can aid the prosecution of a case if the maps are well-prepared. Maps should be simple and easy to follow with bright, engaging colors that project only the information that needs to be projected (anything less minimizes a map's impact, and anything more can confuse jurors or redirect their focus to unimportant elements of the case).

Corrections

The possibilities for GIS applications in corrections are unlimited. Mapping programs can be used to identify problem areas within an institution. For example, assaults within an institution can be mapped to identify areas that need more security, or maps of gang activities within an institution can be created and used to develop strategies to minimize gang problems. In addition, GIS can be used to map the residents (and workplaces) of individuals who are on probation or parole that can be shared with law enforcement to develop strategies that improve the effectiveness of investigation and applications of problem-oriented policing. Furthermore, GIS can be used in conjunction with GPS to monitor the movements of parolees and probationers to ensure these individuals are not visiting places (and people) they are barred from by their probation and parole stipulations. Information about offenders' movements can also be compared to current cases to prioritize and investigate suspects in efforts of solving these cases. Also, government and community resources can be mapped to identify areas within large jurisdictions that are lacking in the supports necessary for parolees and probationers to successfully complete their sentences. This is critical to offenders who have limited means of transportation. Last, due to residency restrictions that have been passed in many states for sex offenders, maps can be created to identify areas where these offenders are allowed to live. (Incidentally, these maps can also discover enclaves or clusters of sex offenders living together.) **Figures 11–9** and **11–10** are examples of maps that can be used for correctional purposes. The maps depict offenders on parole for the crimes of robbery and possession of a stolen vehicle and are used by the street crime unit for knock and talk applications. These maps are updated monthly and can be made for offenders who are on parole for a variety of crimes.

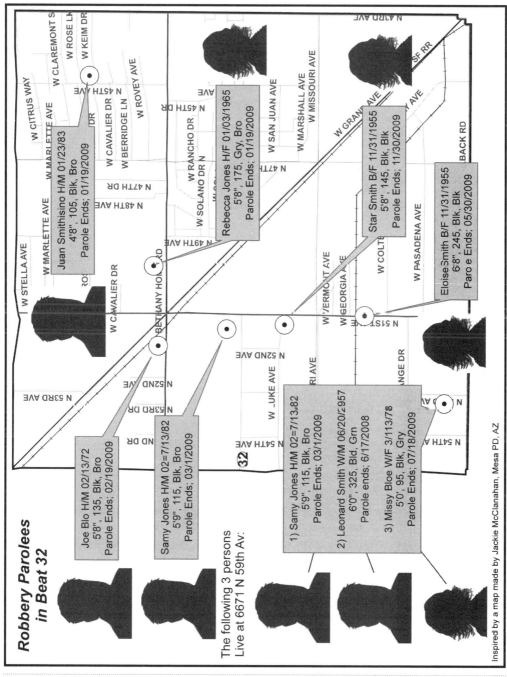

Figure 11–9 Robbery Parolees

446　CHAPTER 11　Mapping for Your Audience

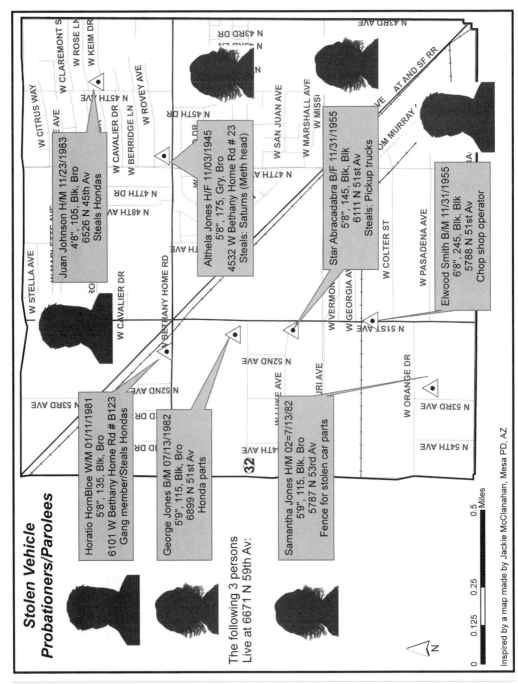

Figure 11–10 Stolen Vehicle Probationers/Parolees

Conclusion

As you can see, there is a plethora of GIS applications in the criminal justice system. However, maps must be tailored to the function(s) and audience they serve. Special attention must be paid to scale, color, text, and the data used to create useful and meaningful maps that assist criminal justice in its efforts to manage crime. Maps that are difficult to read due to scale, color, or content are problematic and will be discarded and ignored rather than yielding the intended results.

Questions for Review

1. Why must an analyst intimately understand the data he or she is working with?
2. What are the three basic categories of maps that are made in crime analysis?
3. What items should be part of any map an analyst makes?
4. What should an analyst keep in mind when making maps for:
   - Patrol officers
   - Investigators
   - Managers
   - Community members

Chapter Exercises

Exercise 24: Data Frames

Data Needed for This Exercise
- EX24.mxd contains the majority of the data you will need; however, as the lesson progresses you may find you need to add data to assist you with your project.

Lesson Objectives
- Create a layout using single data frames and multiple data frames.
- Define who the audience will be for your presentation.
- Make sure that the legend, north arrow, title, map, and disclaimer text are included as needed to help explain your map and make it more accurate.
- Move and resize graphics on the layout.
- Add a graph to the layout.
- Add a table to your layout in several ways.

Task Description

- The chief has asked for a large format map from the plotter that shows the population breakdown for the City of Glendale from the 2000 census by census block group and a hot spot map showing the total Part I UCR crimes by census block group as well. He doesn't care if the map shows rates or counts, but he wants them to be consistent. He would also like a third addition to the map that shows the relationship of Glendale to Maricopa County. A fourth section of the map should show the grids that have the most Part I crime of all crime within their individual borders or something that would show how serious Part I crimes are within each grid along with the other maps.

1. Open EX24.mxd and maximize the screen.
2. You will see a layout and project that resembles DVD Figure 24–1.
3. We are using multiple data frames in the TOC to manage the individual map displays on the layout that we intend to send to a plotter. As you can see, we only have three of the four things the chief wanted on the map at this time. We will need to make some room for another data frame (map display) by adding a new data frame to our TOC. We will also need to create the hot spot map for the Part I crimes because all we have now is the raw data in point format. We will then need to add new data to the new data frame that will help us create the last map that the chief requested (hot spot of Part I crimes compared to all crimes).
4. The audience for this map will be the chief and anyone he shows the map to. We know that we are only going to have to print one copy on the plotter, and it needs to be 36″ × 48″ in size. Knowing this information can help us plan the map. If this was going to be a 8.5″ × 11″ map, chances are the multiple frames would be very hard to see, but at the larger plotter size, it should be just fine. Nevertheless, we want to make sure that fonts are reasonable in size (five points or greater) so that the data is useable even on a large plotter sheet. Knowing your audience is key to making a good map that speaks to the issues and problems the user needs from your analysis. The audience also determines what size the map needs to be and whether or not color can be added. It would be a waste of time to create a map at 36″ × 48″ inches in full color if it were only going to be duplicated at 8.5″ × 11″ on a black and white copier, for example.

5. The layout toolbar allows us to move around the layout and gives us tools to accomplish various tasks (see DVD Figure 24–2).
6. The buttons on the left allow us to zoom in and zoom out on the layout much as we did in the map display with the standard toolbar; however, it only zooms us in and out of the display and does not change the scale of our themes.
7. The second section of buttons on the left also allows us to zoom in and out with a click rather than have to draw a box around what we want to zoom in on. The seventh tool from the left allows us to zoom out to the full extent of the map layout. This is very useful at times.
8. The next button with the blue arrow will let us go back and forth to zoom levels we have been using while working on this project.
9. The drop-down box allows us to zoom to a specific percentage of the layout's size.
10. When you have a *huge* amount of data in the map, it can sometimes be frustrating to have to wait for it to redraw each time. The third button from the right allows you to view the layout in draft mode, which does not redraw the maps and instead replaces the map display with the name of the data frame.
11. The second button from the right allows you to focus the data frame. In layout view, focusing the active data frame allows you to work with the features and elements in that data frame as though you were in data view. You can create, delete, and edit features, graphics, and text in a focused data frame. This is kind of like moving to the data view screen from the layout screen so you can change things on your map that you added in the data frame without having to actually switch to the data view to do it.
12. The last buttons allow you to choose various map templates that are premade and make map-making pretty easy sometimes. The formatting of the template will be applied to your data and can save you time. The templates can also confuse and frustrate you, but time and practice will help this. This project and layout was created with a template by pushing this button and choosing USA.mxt in the USA tab. (Do not click this now, however, or it will replace all our hard work!)
13. Practice with the zoom tools on the layout toolbar and then the zoom tools in the standard toolbar to see the difference in how they work. Remember that you can use the Zoom Previous tool to go back.

14. Can you identify the various items that are in the layout?
    a. Data frames being used
    b. Scale bars shown
    c. Text frames shown
    d. Legends being shown
    e. Graphic boxes being shown
    f. Graphic lines shown on the layout
    g. Miscellaneous pictures and graphics shown on the layout
15. You should get around 18 different items altogether.
16. Let's use what we learned in the Chapter 8 exercises to make our graduated color or shaded grid hot spot map so it works well in our layout.
17. Open the attribute table for Grids in the Part 1 Crime Hotspots data frame. You may need to right click on this data frame and then choose **Activate** from the pop-up menu to make this the active frame (so you can see it in the map display).
18. Notice that at the end of this table we have two fields:
    - P1SUM: The sum of all Part I crimes in that grid
    - ALLSUM: The sum of all crimes in that grid
19. We will create a shaded grid map using the P1SUM field as the Quantities value.
20. Right click on the **Grid Theme → Properties → General** tab. Change the name to *Part 1 Property Crimes by Grid*.
21. Click on the **Symbology** tab and choose **Quantities** from the left box, **graduated colors** as the classification type, and **P1SUM** for the values box.
22. Click **Apply**. Hopefully we have a blue to red color ramp and not some less clear colors.
23. Switch to the Data view (button at the bottom of the map display window).
24. Our map display should look something like DVD Figure 24–3.
25. Go back to the Layout view and see how far you have come.
26. Save this project.
27. Take a break for 5 minutes and come back ready to think hard!
28. We need to create a new data frame in this project and then add data to it to accomplish the last goal of creating a map that shows the total number of Part I offenses per total crimes for each grid or census block group.

29. The word "per" generally means "divide by." In this case: Part I crimes divided by all crimes results in a rate we can use in our shaded grid map.
30. We also need to make room for the new data frame on our legend and make it look halfway decent.
31. Shrink the data frame for Part 1 UCR Crime Hotspots by clicking the **minus sign** to the left of the data frame.
32. Go to the menu bar, and click **Insert → Data Frame**.
33. If you are in Layout view, you will see a new highlighted box on the layout. If you are in the Data view, you now have a blank map display.
34. Right click on the new data frame (it's likely called New Data Frame 3) and then choose **Properties** from the pop-up menu.
35. Go to the General tab and change the name of this data frame to *Part 1 Crime Rate*.
36. Click **Apply** and close.
37. Make sure you are now in the data view of the project (there will be a little world icon at the bottom left of the map display).
38. Open up the data frame called Part 1 UCR Crime Hotspots again, and right click on the **Grid** theme and choose **Copy** from the pop-up menu.
39. Right click on the **Part 1 Crime Rate** data frame, and now choose **PasteLayer(s)** from the pop-up menu. Voilà: instant copying of data from one layer to another.
40. Copy the street labels and related street outline boxes themes as you did in steps 38 and 39.
41. From step 18 of this exercise, we know that there were two fields in the grid layer: P1Sum and AllSum. Let's create a new field in the Grid layer within the Part 1 Crime Rate data frame, make it "double" type, and call it P1Rate.
42. Now right click on the new field name, and using the field calculator, calculate the value for this new field to be [P1Sum] ÷ [AllSum].
43. You should receive an error message. Look at the table and see if you can figure out what the error might be.
44. The error resulted because several records have 0 in the AllSum field, and you can't divide any number by 0.

45. We need to first select all of the records that have a 0 in the AllSum field using the Select By Attribute query that can be accessed through the Options button on the table window.
46. Choose **Options → Select By Attributes** and enter a SQL code such as *[AllSum] = 0*.
47. Run the query and close the query window.
48. Use the field calculator to calculate the value of the selected records for the P1Rate field as 0.
49. Now click on **Options → Switch Selection** to select all other records (with a number greater than 0 in the AllSum field).
50. Field calculate the value of these records as [P1Sum] ÷ [AllSum] like we tried to do in the first place.
51. Clear any selected records and close the table.
52. Right click on the **Grid** theme and choose **Properties → Symbology** tab.
53. Change the legend to a graduated color, quantities legend type. Use the new field, P1Rate, as the value field.
54. Apply your new legend.
55. Your map should now resemble DVD Figure 24–4.
56. Use the Identify tool to click on several of the red, orange, green, and blue grids and see how the values of P1Sum and AllSum affect the overall rate of Part I crimes. You should notice that very small numbers in either field can cause a grid to pop out as a hot grid, but you may dealing with just one or two offenses.
57. Go back to Layout view and add a legend (**Insert → Legend** from the menu bar), a scale bar, and anything else you think this data frame needs. Rearrange the various objects on the layout until you get an overall map that looks good and can be sent to the plotter.
58. Save your project. A possible variation of this project is shown in EX24b.mxd, which is in the same folder as EX24.mxd.
59. Save and close this project.

Exercise 25: Putting It All Together for One Project

Data Needed for This Exercise

- _____
- _____
- _____
- _____
- _____
- _____

Lesson Objectives
- Use the skills you have learned so far to create map(s) and products that will help the Gateway patrol commander determine where he should direct resources to combat auto theft in his patrol division.

Task Description
- This topic will be monitored and tracked for a period of six months. The commander wants a map that shows where vehicles have most commonly been stolen from (hot spots) in 2004. He would also like a day of week and time of day graph on the map and a hot spot map of stolen vehicle recoveries as well. He wants the initial map for the current year (2004) only, but he will want a new map with these same items every month, so the map items need to be reproduced over time. After you show him the initial map sample and the hot spots for stolen vehicle activity, he will advise you on which hot spot(s) he wants to focus on. Then he will want an additional map of the actual stolen vehicle crimes for that hot spot(s) (a pin map) with a breakdown of the make of vehicles being stolen in the hot spot and any other information you think would be useful.

1. You need to get organized to accomplish this analysis and should consider what steps you will need to take to complete this assignment. A good way to start thinking about this is to simply take out a piece of paper and start drawing a simple picture of what the final map should look like (see DVD Figure 25–1).
2. Your drawing doesn't have to look pretty. It's just a plan for making this map like any other project you might undertake.
3. From our little bit of preplanning, we know that we'll need at least three data frames in our project, some graphics and text titles, a north arrow, a scale bar for each data frame, and probably a legend for each data frame. We'll also need to add some day of week and time of day charts to the layout and be able to add tables to the layout for the make of the stolen vehicles.
4. At first this may look like a massive project, but what we need to do is break this down into parts, work on them one at a time, and not try to do the entire map all at one time. Being organized and having a plan not only helps you visually comprehend the final product, it also helps you decide what data you will need for the project. If you don't know where the needed data is, you can start looking right now. Fill in the data layers you think you will need

(I suggest using streets, stolen cars, recoveries, and maybe city boundaries).

5. Let's begin by opening a new ArcMap project.
6. When ArcMap opens, you want to add a new empty map, but don't immediately add data unless you are sure what you need for the first data frame.
7. Save this new project as EX25.mxd in your student folder.
8. Start by renaming the new data frame that ArcMap automatically adds to a new project. Name it *Stolen Vehicle Hotspots*, and save your project.
9. Find and add the point data for stolen vehicles to the data frame. You have two options: a shapefile and a personal geodatabase file, which are in the CIA folder.
10. Some of the factors we need to consider from the commander's request are:
    a. 2004 data only
    b. Stolen vehicles reported to police
    c. Crime events happened completely inside Glendale, Arizona
11. Because we are going to have to recreate these maps every so often, we need to think about how to keep current data coming from our main and updated data source. A theme definition seems most appropriate because it will limit our data to 2004 or greater, but it will always update when we reopen the project.
12. Create a theme definition (**right click → Properties → Definition Query**) with an SQL string as follows:
    [CVDATE] >=20040101
13. Apply this and close the Theme Properties dialog menu.
14. The commander was specific about wanting the hot spots by census block group, so we need to find and add our census block group theme now.
15. We know that stolen vehicles occur within the city limits, but recoveries often occur outside the city limits. Do we need the census block groups for the entire county for this hot spot map? Will it be easier than clipping the census blocks to the city right now? Will it matter?
16. Add the Countywide Census Block Groups personal geodatabase theme for now.
17. If you cannot find it, remember you can use ArcCatalog to search files in an entire folder with certain text values.

18. By opening the attribute tables of these two themes, we can see that there is no census block information in the Stolen Vehicle Theme and no stolen vehicle information in the Census Block theme. We need to *spatially* join the census blocks to the stolen vehicles so that each stolen vehicle incident will know in what census block it occurred (kind of like a cookie cutter and dough concept where the census block polygons are the cookie cutter and the stolen vehicles are the dough).

19. Select the **Stolen Vehicle** theme and right click it. Create a spatial join between it and the census block groups. If you do not remember how to do this, click on **Help** and enter *Spatial Join* under the search tab. Browse around in the help file until you find the section on creating a spatial join. Save the output file as StolenCensus.shp in your student folder.

20. The StolenCensus.shp file now has the STFID number from the census data in the table for auto thefts. The STFID field is the unique identifier field for each census block. We now need to summarize this field to find the total number of auto thefts per census block group.

21. Open the attribute table for StolenCensus.shp and right click on the field named **STFID**, choose **Summarize**, and save this table as SC1.dbf.

22. You *do want* to add the table to the project.

23. Now right click on the **Census Block Group** theme and choose **Joins and Relates → Join** to add a tabular join to the SC1.dbf table by STFID field names in both tables. You are joining the data from the SC1.dbf table to the attribute data of the census blocks by their common field STFID.

24. Right click on the **Census Block Group** theme now, and go to the Symbology tab in the Properties dialog. Create a graduated color legend by quantities using the value of the Count_STFID field you'll find toward the far right side of the attribute table.

25. You should have something that looks similar to DVD Figure 25–2 when you are done.

26. The map display is showing only those census block groups that had any stolen vehicles in 2004 by the total count of auto thefts.

27. We need to experiment a little to make sure that the red areas on the map are really logical hot spots for auto theft activity. Use the Identify tool to click on several of the census block groups and see what the counts are.

28. Remember to set the census blocks as the only selectable layer to help limit the results to just what you want.
29. Notice also that the census block groups are not all the same size. Some are rather large, and some are very, very small. When you see this sort of thing, you might want to consider looking at the data as a rate map. In this case, create a rate map by total square miles of each census block group or stolen vehicles per square mile.
30. To create this rate, we need to make a new field in the census block theme called *SVrate* and calculate it to be equal to the following:

    [SC1.Cnt_STFID] ÷ ([Census_Blk_Groups.Shape_Area] ÷ (5280^2))
31. Because the Shape_Area contains values that are square feet, we need to also convert feet to miles to get our square mileage rate, thus we include the 5280^2 part of the formula.
32. Also remember that we cannot divide by zero or null, but because the Shape_Area has no null or zero values, we are okay (sort the field to find out).
33. Make sure to make your new rate field a number data type of "double" as well.
34. Did you get reasonable values in your rate field? Did you get any error messages? Because the [SC1.Cnt_STFID] has several null values, you may get an error message, but just click **Yes** and it should go ahead and finish calculating correctly. To avoid this error message you could have first selected all records where Cnt_STFID was > 0 or was not null and then made your calculations, but in this case it doesn't matter because it does not stop ArcGIS from making the calculations.
35. Compare the rate values to the count values by sorting the [SC1.Cnt_STFID] field in descending order. That really boosts the numbers in some cases. Think about this for a moment! If we had a set of four census block groups and the square mileage for them was 10, 1, 0.5, and 0.25 miles and the count or total auto thefts within each block group was exactly 10, what actual rate or stolen vehicles per square mile would we expect to see in each block group?
    - BG1: 10 square miles, 10 auto thefts   10 ÷ 10   = _____
    - BG2: 1 square mile, 10 auto thefts   10 ÷ 1   = _____
    - BG3: 0.5 square miles, 10 auto thefts   10 ÷ 0.5 = _____
    - BG4: 0.25 square miles, 10 auto thefts 10 ÷ 0.25 = _____

    Because we are calculating auto thefts per square mile, the smaller the area, the more dense the crimes; that is, the crimes per square mile (rate) is higher. The larger the area, the lower the rate if the total count of crime remains consistent.

36. Revamp your legend to use the new field we just created and see how it changes the hot spots (if it does at all).
37. What happened to all the red areas? The legend shows that there should be some red areas on the map for 387 and above.
38. Open the attribute table of the census theme. Find the SVRate field and sort it in descending order. Select the top record that has over 600 for a SVRate value. Minimize the table and then right click on the **Census** theme again, choose **Selection**, and **Zoom to the Selected Features**. It's a tiny census block way out on the west side of town. Repeat this for the second and third highest scores to see where they are on the map. Notice that the second census block group was red in our count map. Are you getting an idea of how we could use these two maps to come up with a target area? A statement such as the following might do it for us: Census block groups where the total count of auto thefts in 2004 is highest and the rate of auto thefts per square mile is in the top five yields a good target area for continued analysis and enforcement action.
39. Choose the top 10 SVRate record values and then make a new shapefile from them, or write the STFID numbers down so you can remember them for the next step.
40. Call this new shapefile HighSVRate.shp and save it in your student folder.
41. Now sort the attribute table by the Count_STFID field in descending order, pick the top 10 census block groups by that value, and make a new shapefile called HighCount.shp.
42. You may have to select the top 11 census block groups because the 10th and 11th records both have a value of 19.
43. Make the symbology of the HighCount theme red and set its transparency value to 40% (**right click → Properties → Display tab → Transparency**).
44. Make the symbology of the HighSVRate theme blue and set its transparency value to 40% (**right click → Properties → Display tab → Transparency**).
45. The three census areas where these themes overlap and become purple are the best choices for a project site for auto theft prevention. Congratulations, you have just done sophisticated geoprocessing in a GIS environment.

46. The commander who requested this analysis is from the Gateway patrol division, so in reviewing the possible choices for the hottest auto theft census block group, we now have just one choice. The only one that is in the Gateway patrol division is STFID # 040130930001, which is located around Maryland to Glendale and 43rd Av to 45th Av.

47. We will show our entire map to the commander, but highlight these sections as ones that met our hot spot criteria. Another option is to calculate Z scores for each of these fields, create new fields with the appropriate Z score values, and then add them together and present the hot spot map to the commander in that format. That way we would have one map that shows both selection criteria and gives us an overall value for each census block group. For this exercise, we will just use the count value for the hot spot map and use some text callout boxes to identify the hottest areas.

48. DVD Figure 25–3 shows what the map looks like when you add the Z scores together for each field. Notice that the highlighted areas show where the two fields indicate areas to focus on.

49. Let's set the theme legend back to using the count field as our value field and add some callout boxes to show where the two layers overlap.

50. Change the legend back to a graduated color legend using the Count_STFID field as the value and a red to blue color ramp.

51. We need to make this map more presentable, so load the grid layer, the street labels, and the outside grid boxes if you think they are needed to explain the analysis.

52. You might also want to add the city boundaries layer.

53. Go to **View** in the menu bar and then select **Toolbars**. Make sure the Draw toolbar is turned on, and then find the little New Text (A) tool in the Draw toolbar. Click on the down arrow at the bottom right of this tool, and choose the **Callout Text** tool. Click in the center of the census block we identified in step 46, drag it left and up slightly, and then let go of the mouse. In the small text area that comes up, type *This area was the highest overall!* and press **Enter**.

54. You may have to move the text box up and away from the map display some. You can also right click it, go to **Properties**, and resize the text or change the font as needed.

55. Clean up your table of contents and delete any themes you are not using anymore for the final map.

56. It is a good idea to make a folder before you begin any project and give the folder a name that will help you remember what kind of project files are within this folder. When you create any files, summary tables, etc. always make sure they go into this folder. That way, when 30 days roll past and you need to find everything you did, it will be right there. Also, when 90 days go by and the commander has changed his mind and wants to work a new area, you can completely delete the project and any files that were created along the way and keep your hard drive clear (or zip them and archive them someplace else).
57. If your map looks similar to DVD Figure 25–4, you are ready for the next section where we will repeat all of this for recovered stolen vehicles.
58. In most cases, along with the map, you would write a memorandum or an analysis bulletin to summarize the results of your analysis. There might be some confusion with this map because the area that has the highest overall auto thefts is not red, and most people associate red with bad.
59. Go to the menu bar and click **Insert → Data Frame**.
60. Name this data frame *Recovered Vehicle Hotspots*.
61. Load the data layers you think are necessary to accomplish this analysis. (*Hint: You will perform this analysis exactly like the stolen vehicle analysis; however, recovered vehicles may be found all the way across Maricopa County.*)
62. The SQL code to limit recoveries to just those recovered in 2004 is as follows:

    [CVRECVDATE] >=20040101
63. You are also dealing with data that is countywide, and it will probably take longer to do the spatial join and other joins, so be patient!
64. When you are done with the spatial join between the census theme to the recovered vehicles theme and then the summary table, and then the tabular join between that summary table and the census theme, you can create a graduated color legend. When you get to this stage, your map display ought to look something like DVD Figure 25–5.
65. There are many things to think about with this part of the project. What is going on with the river bottoms and washes? They are all red in this version of the map. If you use the Identify tool you'll find that the washes and river bottoms are just one huge census

block group, and there were 116 recovered vehicles found somewhere across the county within one of the washes that is covered by this polygon(s).

66. Does the commander really need to know where all the vehicles were recovered in Maricopa County for his action plan? Will he care? With the exception of the STFID #040131141001 (south and east of Glendale) and the washes, not too many census blocks are red outside of Glendale, but several are orange.

67. Maybe zooming in on Glendale a little tighter would give the commander better information. Let's do this and cancel out information overload by dealing with what we know, rather than the entire county.

68. We really only have one census block group right down the middle of Glendale's downtown area that is red (the main police station).

69. Several of them are orange, however, and any one of them could be useful to the commander.

70. We can get rid of the red washes in our map by using an exclusion query in the legend editor. Right click the **Census** theme, choose **Properties**, and open the **Symbology** tab. In this menu, near the upper right, click the **Classify** button.

71. In the next dialog menu, click the **Exclusion** button.

72. In the next dialog menu enter the following SQL code under the Query tab:

    [Census_Blk_Groups.STFID] = ' '

73. Stay in the Classify dialog menu and go to the **Legend** tab, and check the box that says **Show Symbol for Excluded Data.**

74. Click the **Symbol** button, and choose a light blue fill with no outline.

75. In the Label box enter *Washes, etc.* and click **OK**.

76. Click **OK** again, and then one more time.

77. Now all those red washes have turned a pastel blue and are not confusing our map.

78. Now, what are we going to do about all those census block groups outside the city limits and the Street Labels and Outline Boxes themes that just make the map hard to read?

79. We can use a feature of the data frame properties to help us see only what we want to see.

80. Add the theme C:\CIA\Exercises\Exercise_25\AreaLimits.shp to your recovery data frame, and then just turn it off and minimize the legend for it.
81. Right click on the **Recovered Vehicle** data frame and choose **Properties**.
82. Go to the Data Frame tab and find the check box that says **Clip to Shape** and enable it. Click on the now active **Specify Shape** button, and in the Clip to Shape dialog menu, check the **Outline of Features** box and then choose the **AreaLimits** theme from the list. Change the Features box to **All** (see DVD Figure 25–6).
83. Click **Apply** and watch everything (well, almost everything) disappear outside of our desired extents.
84. If I had spent a lot more time drawing this boundary, we would not see the little blue bits around the periphery.
85. Make sure you save your project.
86. Take a 15 minute break and come back ready to continue.
87. Okay, now that you are back and refreshed, let's start actually making the layout for this new multidata frame map and learn how to add graphs and tables to our layout.
88. Click on the **Layout View** icon at the bottom of the map display.
89. We have decided that this map is going to be 24″ × 36″ with a landscape orientation, so let's start by setting up the map screen.
90. Go to **File → Page** and then **Print Setup**.
91. Uncheck the section that says Use Printer Paper Settings under Map Page Size. *This assumes you do not have a large format plotter in the classroom to be used to actually print the map.*
92. Change the paper size to Arch D and make it landscape.
93. Click **OK** and observe the very small data frames.
94. Separate the two data frames from each other. Place and resize them approximately as shown on our hand drawn rough draft.
95. Use the Draw toolbar and the Draw Rectangle tool to draw a box about 0.5 inch inside the boundaries of the layout background, and make the graphic hollow with a black eight point line.
96. Right click this box and choose **Order → Send to Back** in the pop-up menu.
97. This will move the box behind any other items on the layout.

98. Grab the rectangle tool again and draw a box along the entire bottom of the layout, just inside the previous box you drew, and make it about an inch or so high. Change this to a dark blue fill with a black two-point line surrounding it.
99. Click on **Insert Text** and add four different text items as follows:
    a. *Today's Date*
    b. *State Plane Coordinate System, NAD 83, Arizona Central, Feet*
    c. *Stolen and Recovered Vehicle Hot Spots in 2004*
    d. *Created by: GPD CAU; Source; GIS layers maintained by the CAU*
100. Put these three text items on the dark blue rectangle you created in step 98. Change the attributes for each item as follows:
    a. Item A: Far left placement, two or three lines, Arial 14, white in color, left justified
    b. Item B: Center placement, one line, Arial 52, light yellow in color, centered
    c. Item C: Far right placement, two or three lines, Arial 14, white in color, right justified
101. Add text above each of the data frames to indicate that one is hot spots of stolen vehicles and the other is hot spots of recovered vehicles. A disclaimer under the recovery hot spot map should also be added to indicate that all recoveries are not shown and are limited to the general area within and adjacent to the city boundaries. Write the wording however you think is appropriate, and the font, color, and size are up to you.
102. Add another couple of rectangles and some text to frame the graphs and tables section on the left side of the layout. Your layout so far should resemble DVD Figure 25–7.
103. To create a day of week chart for the stolen vehicle data, we need a field somewhere in our point data that has the day of week in it. We do have a field called Weekday, but it contains numbers and not the names of the days.
104. Add two look-up tables to the project:
    - C:\CIA\GIS_Data\Lookup_Tbls\DOW_LU
    - C:\CIA\GIS_Data\Lookup_Tbls\TOD_LU
105. Join the DOW_LU table to the weekday field in the Stolen Vehicle Point theme, and add the TOD_LU table to the Hour field in the Stolen Vehicle Point Data theme.

106. You need to pay attention to which field in the look-up tables you are joining to which field in the point themes (Weekday to DOWCODE and Hour_ to Hour).
107. Open the attribute table for the Stolen Vehicles theme and go all the way to the right to see our added fields.
108. To create the graph we need to summarize the data first.
109. Right click on the DOW_LU.DOWSHORT field name, and choose **Summarize** from the drop-down menu. Save the summary table to your student folder as DOWSUM and add it to the project. Repeat this process for the TOD_LU.TIME_PERIO field name, except name this table TODSUM.
110. Go to **Tools → Graphs → Create**.
111. Create a graph for the DOWSUM and TODSUM tables. I'll walk you through the first one (DOWSUM). For graph type, choose **Vertical Area**. Select the DOWSUM table you previously created. For the X field, choose **minimum_weekly** and **ascending**. For the X label field, choose **DOWSHORT**. The vertical axis should be **Left**, and the horizontal axis should be **Bottom**. Make sure you have a check mark next to **Add to Legend**. Choose a shade of red for your color and set the multiple area type to **Stacked**. The Stairs mode should be off. Check the **Show Border** box. Last, set the transparency at 16%. Click on the **Next** button to create titles for your new graph and graph components. To add the graph to the layout, make the graph active and then right click the top bar of the graph window. Choose **Add to Layout** and then close the graph. Move and resize the graph on the layout as needed. Repeat this process with the other graph.
112. Your layout should now look something like DVD Figure 25–8.
113. Notice that the previous example has a logo graphic for the City of Glendale at the bottom left. To add pictures and photos to a map, simply go to the **Insert** menu item at the top of the ArcGIS project window and then choose **Picture** from the pop-up menu. Navigate to where your photo or logo is and click **Open**.
114. Now we need to add some text above these charts to indicate that they are for citywide day of week and time of day for stolen vehicle offenses and add a table of the most stolen vehicle makes. Add the text as needed above the charts.
115. We need to find out what the count is of all stolen vehicles in 2004 by make of vehicle. We can create a new summary table to accomplish this.

116. Make sure you have named the recovery data frame to describe that it contains the data for hot spots of recoveries, and then right click the stolen vehicle data frame and choose **Activate** from the pop-up menu.
117. On the layout you'll see that a dotted line is now around the stolen vehicle hot spots data frame instead of the recovered vehicle hot spots frame.
118. Open the attribute table for the stolen vehicles.
119. Make sure that no records are selected in the table.
120. Find the field called Make or VehMakeCd, right click the field name, and choose **Summarize** from the pop-up menu.
121. Save this summary table to your student folder, and name it Makesum.dbf.
122. Add the results to the map and close the attribute table.
123. You now have a table of 51 different makes of vehicles that have been stolen. Do we really want to show all 51 makes or just the top 10, 20, or 25?
124. To show you how useless tables are in layouts that come directly from ArcGIS, click on the **Options** button in the table view and choose **Add Table to Layout** in the pop-up menu.
125. Notice the new table on the layout. Zoom in to the table, making sure to use the Layout zoom tools and not the Map Display zoom tools.
126. Try resizing the box around the table. Can you change the size of the text? While the table is open, it does preserve the sort order, but when you close it, the table reverts back to whatever sort order it was created with.
127. The table is dynamic, which means that it is directly and dynamically linked to the table itself so that if the data in the table changes, the data on the layout will also change. Other than that, tables inside ArcGIS are not pretty or easy to manipulate.
128. An alternative method to add a table to the map layout that is much better looking is to create the summary table but do not add it into the map and just save it to the hard drive. Open this DBF table with Excel, manipulate it, and change the colors, sort order, fonts, etc. until it appears like you want it (see DVD Figure 25–9). When you are finished, save the Excel spreadsheet. Highlight the table you manipulated in Excel, and press **CTRL** and then **C** to copy it. Then make the ArcMap project active and press **CTRL** and **V** to copy a graphic image of the table into

the layout. (*Hint: I prefer to use **Edit → Paste Special → Picture** because it does not add as many megabytes of data to the map project as when it is pasted directly.*) Position, resize, and move this graphic table on the layout as needed.

129. You will then end up with a graph you can add that resembles DVD Figure 25–9.
130. You can do much more with the data, such as create total columns and rows, add formulas, calculate percentages or rates, or do other statistics to make the table more useful in the map layout in Excel.
131. We can see from the table that within Glendale, the most commonly stolen vehicles in 2004 were Chevrolet, Honda, Nissan, Ford, and Dodge.
132. All we have left for our final layout it to create a new data frame for the hot spot we have decided to focus on and create a pin map of the stolen vehicles that are found within that census block. We also need to recreate the two temporal graphs and the table by make for just this hot spot area.
133. Create a new data frame and name it *Auto Theft Hotspot 1*.
134. Notice that a new, empty box appeared on your layout.
135. Position this box approximately as shown in your original drawing, and then add the stolen vehicle theme and other themes you need to show the hot spot in census block group STFID # 040130930001.
136. You will need a Select by Attributes query like *[STFID] ='040130930001'* for the census block groups layer, and then zoom in on just that selected polygon.
137. Add the streets and city boundaries to show about where it is located, and add labels for the street names.
138. Perform a **Select by location** for all auto thefts that are within (or intersected by) 0.25 miles of the selected hot spot polygon.
139. Create DOW and TOD summary tables, and Vehicle Make summary tables from these selected auto thefts.
140. You will also need to add the two look-up tables into this data frame so they will be available when you try to do the joins to get the name of the day of week and the time period for the charts.
141. Name the summary tables MakeHSSum.dbf, DOWHSSum.dbf, and TODHSSum.dbf, respectively.

142. Add the DOW and TOD summary tables into the map to make the charts, and open the Vehicle Make Summary table in Excel and make it look nice as you did in step 128 and 129.
143. You may also wish to add legends (**Insert → Legend**) for each data frame in your layout. You will need to click on each data frame in Layout view first, and then insert a legend for each one individually.
144. Save your project when completed and exit when you are satisfied with your work.
145. A completed project, called EX25.mxd, that you can use for reference is available in the exercise 25 folder. Your final map product should look like DVD Figure 25–10.

Exercise 26: Practical Administrative Exercise 1

Data Needed for This Exercise
- _____
- _____
- _____
- _____
- _____
- _____

Lesson Objectives
- Use the GIS skills you have developed to complete a city council map for patrol officers.

Task Description
- In this exercise the goal is to create a city council map that can be used by the patrol officers and detectives when they are in the field. Your map should be similar to DVD Figure 26–1.

1. Do not worry about matching the colors exactly as shown in this map or adding the photo to the lower right corner. Take this example and improve it, if you can, to come up with an easy-to-read map that patrol and detectives can use in the field to help them recognize which city council district they are working in. In the CIA260 folder there is a folder called photos_graphics that contains several graphics that may help you make the map prettier. You can add one of these graphics to a map in the Layout view by clicking on **Insert** and then **Picture**. You can then place the picture wherever you think it looks best. This is a sample of an administrative reference map.

2. When you are done, show your map to your classmates, and critically review it for what looks good and what does not so that you can learn to make better maps. Remember that all good maps always have at least the following elements:
   - Title
   - Disclaimer text, law enforcement use only, or confidentiality statement if needed
   - Scale bar
   - North arrow
   - Legend
   - Map display that shows one center of attention
   - Entire map reads left to right where possible
   - Sufficient white space between items so the map does not look cluttered
   - Good use of color, hue, tone, and spectrum
   - Legend items of seven or fewer
3. Would you change anything on this map if it were only going to be viewed by patrol officers who work the night shift? Remember: Know your audience so your maps can speak to them.

Exercise 27: Practical Administrative Exercise 2

Data Needed for This Exercise
- _____
- _____
- _____
- _____
- _____
- _____

Lesson Objectives
- Use the GIS skills you have developed to complete an analysis of crimes around hospitals in Glendale for a patrol commander.

Task Description
- In this exercise, the Foothills patrol commander has asked for an analysis of crimes around the Arrowhead and Thunderbird Hospitals. He wants you to check the crimes within 0.5 miles around both hospitals, compare them, and provide a map of each hospital. He wants you to include a table that summarizes the total UCR offenses within a 0.5 mile radius for each

hospital and a pie chart or some other graph that compares both hospitals. Many citizens and hospital staff at the Thunderbird Hospital have complained that burglary from vehicles and auto theft is much higher than at Arrowhead Hospital, and they think the hospital should have more police patrols in the area during certain hours of the night. Your job is to find the data the commander is asking for, create the maps and tables or graphs he is asking for, and respond to him with a memorandum (there is a template in C:\CIA\Documents\Word-Templates\) that discusses the differences between the two hospitals and either supports or does not support enhanced patrol activities around Thunderbird Hospital in the evening hours.

- Remember that *you are the analyst*, and your analytical effort will help the commander make a decision about increasing police patrols around Thunderbird Hospital. Keep your analysis as brief and simple as it needs to be to answer these questions, but make sure you actually answer the questions the commander has asked.

■ References

Harries, K. (1999). *Crime mapping: Principle and practice.* Washington, DC: US Department of Justice Programs. Retrieved March 2008, from http://www.ncirs.gov/htm/nij/mapping/pdf.html

Paulsen, D. (2003, March). *To map or not to map: Do crime maps have any measurable impact on patrol officers?* Presentation at the Academy of Criminal Justice Sciences, Boston, MA.

Future Issues in Crime Mapping

CHAPTER 12

▶ ▶ LEARNING OBJECTIVES

This chapter provides a brief discussion of future issues in crime mapping. These issues include technological opportunities and challenges, the future of crime mapping research and education, implementation issues, and issues related to ethics and privacy. After studying this chapter, you should be able to:

- Identify some future opportunities and challenges facing the field of crime mapping.
- Identify future directions for crime mapping research and education.

■ Introduction

This introductory text discussed several issues related to crime analysis and crime mapping. The first section examined the theoretical underpinnings of the spatial context of crime. The second section discussed the data available to crime analysts and provided a brief discussion of the limitations of the data that is used in crime analysis and in the creation of crime maps. In addition, Chapter 5 reviewed the relevant research of GIS applications in law enforcement. Section 3 provided an introduction to statistics and also included the types of analyses crime analysts perform on a regular basis.

This last chapter provides a brief examination of the academic, practical, and technological issues that crime analysts will face in the coming years.

■ General Topics

In the world of geographic information systems there are very few sources for data that remain constant over time. As the use of GIS increases in many fields, including crime analysis, there are always new data sources that will be added or modified to provide more knowledge to the decision makers. Remember that with GIS, we want to think along the lines of the business intelligence field where data becomes information and information becomes actionable knowledge. In the future, it may not be enough to know where all the crimes happened; we will need to know why they happened and what the root causes of crime really are. There are many sociological theories, psychological theories, and, as we discovered, criminological theories of why crime happens, but these are usually very general to crime and society as a whole rather than to individual geographic groups. As crime mapping progresses, we will learn how to use tools from other fields of analytical study to apply to crime analysis. We have already learned that some of the lessons learned by biologists who track animals in the wild may have some application with tracking suspects in criminal offenses; we could learn in the future that crime evolves much like a communicable disease and that epidemiological research or methods may have some use within the crime analysis field. Other related fields, such as anthropology, archaeology, biology, psychics, and geology, may provide answers to questions we don't know how to answer with today's GIS crime fighting tools. In addition, as we explore new methods and new sets of data, we'll be able to add more information behind those points, lines, and polygons. For example, a city may have been drawing sewer and water lines all along in their enterprise GIS system with attributes such as length, material, and build date. As they progress with GIS, they find that they can also do flow analyses but need to add the elevation and material flow values to the pipe segments in the GIS data so that they can do modeling to find out what happens if they add a T branch at a certain juncture. With crime analysis, as we add new elements of information, such as modus operandi (MO) data to GIS points of where and when a crime occurred, we can start using the data to find trends, patterns, series, and clusters of crime by type and MO characteristics more easily. We may want to later include US Census Bureau data, zoning data, and other elements we do not always use in crime analysis but later find might assist in our analysis or trend detection. This may lead to the desire for additional data from the health department and other agencies concerning welfare programs and teenage pregnancy levels in specific sections with high crime. Who knows, at some point we may even find out that historical soil contamination from pesticides in certain areas that are now downtown actually may have some causal effect on crime. There is no end to the data we will need when we know how to use the data we already have.

As with any technology implementation, technology itself changes over time, sometimes dramatically. Think about the fact that just a short 10–20 years ago we were talking about how huge a 20 megabyte hard drive was for a new computer. Now the storage capacity of your simple desktop PC has more capability than a whole bank of servers that filled an entire basement of a building a mere 25 years ago. We'll see more and more of our maps being used in patrol cars and other active enforcement activities. This will lead to questions concerning the accuracy of the data for some places that may have been neglected during the data creation phase or are older than some of the other data. We will find that exact accuracy may become much more important than it is now as courts begin to use GIS analysis to convict persons charged with crimes (e.g., Megan's Law and others, which require suspects to adhere to certain distance requirements from something). Presentation devices will change as they already have in the past few years within the court environment. A short time ago, bringing a PC and a movie into the courtroom was somewhat frowned upon, but now, in some modern day courtrooms, video conferencing and electronic media are status quo.

■ Education and Crime Mapping

The academic future of crime mapping looks very promising. More and more colleges across the country are offering courses that are focused on GIS in law enforcement. In a survey of 26 police agencies across the country, Paynich et al. (2007) found that one of the frequent complaints about hiring new crime analysts is that the recruits lacked hands-on experience and/or training. In addition, the new recruits were either very knowledgeable about crime or criminal justice but had very little background in statistics or crime analysis and crime mapping computer applications or vice versa. One respondent stated in the survey, "I'd add in geographic profiling, psychology, and criminology courses. I think that you need to have a working knowledge of criminology and especially... how the process flows (from dispatch to the courts). Not that this makes mapping easier, but you can't work in a vacuum." Another respondent observed, "There are plenty of skilled GIS people out there, but very few understand crime data. They often make maps that are misleading and are not really actionable for the guys on the street."

Although the number of college courses about GIS applications in law enforcement have grown, the use of crime mapping by law enforcement also continues to grow. The use of crime mapping by law enforcement agencies increased from 14.5% in 2000 to 17.9% in 2003 (Chamard, 2006). In addition, if the use of GIS by smaller agencies is to increase, more widely available training is needed in the academic realm. Smaller departments, for example, are much more likely to discontinue the use of crime mapping. This discontinuance is strongly related to staffing in technical support (Chamard, 2003). In the future, this may lead to new and simpler ways to provide mapping technology to these smaller agencies that do not have staff on hand to develop GIS for their agency. Google Earth and other similar applications seem to be on the forefront of offering such services to smaller agencies without the technical staff to develop dedicated GIS functionality, for example.

■ Crime Mapping Research

The future of crime-mapping-related research is also promising. The largest national criminal-justice-related organizations, the Academy of Criminal Justice Sciences (ACJS) and the American Society of Criminology (ASC), both host panels focused specifically on GIS applications in criminal justice at their annual meetings. In addition, the National Institute of Justice (NIJ) has to date held 9 national research conferences on crime mapping with a 10th scheduled in August of 2009. Other associations, such as the International Association of Crime Analysts (IACA), also hold annual conferences with special panels and presentations on crime mapping research. There are also many local crime analysis associations that routinely provide training on aspects of GIS with a crime analysis focus, and several private vendors specialize in various forms of GIS training outside of the academic environment to fill training needs.

In addition to conferences, several publications have emerged in recent years that are dedicated solely to GIS applications in criminal justice. ESRI, for example, publishes the *Public Safety Log: GIS for Public Safety*. In a recent issue of ESRI's *ArcNews*, several articles focused on GIS in law enforcement and highlighted how crime mapping and analysis can improve the effectiveness and efficiency of a department. For example, one article discussed how the United States Postal Service utilizes GIS to track the purchase and redemption of money orders (a common tool for criminals to launder money because individuals can purchase money orders under $3000 without providing identification). These money orders then can be redeemed anywhere in the United States. "The department has embraced the use of GIS technology to more effectively detect and in some cases deter crime. GIS maps show where suspicious activities may be occurring and link data on individuals, transactions, and/or offices to reveal potential criminal patterns. Modern mapping and spatial analysis also help USPS managers make sense of extensive transactional databases and millions of bits of data to ensure they are meeting regulatory compliance" (ESRI, 2008, p. 33).

Essentially, post offices and individuals with an unusually high number of suspicious money order transactions can be flagged and investigated. The USPS article is a prime example of many other crime-mapping-related pieces that are being published in academic and trade journals. *Geography and Public Safety* (put forth by The National Institute of Justice (NIJ), Mapping and Analysis for Public Safety (MAPS), and the Office of Community Oriented Policing Services (COPS)) is another recent publication that is dedicated to crime-mapping-related issues. In this publication, readers can find discussions of theory and practice related to crime mapping as well as helpful tips for crime analysts.

▪ Geographic Profiling and Forecasting Crime

In 1999, Harries predicted that geographic profiling and crime forecasting methodologies would continue to evolve and gain more support in the future. The research and literature that has been reviewed throughout this text indicates that Harries' prediction was correct. In geographic profiling, a number of programs, both free and for purchase, are available to help analysts make predictions about offender decision making in efforts to arrest serial offenders. In forecasting crime, analysts are developing models that include related crimes to identify emerging hot spots (for example, examining robbery, assault, and burglary to identify an emerging drug hot spot; Olligschlaeger, 2003).

With this increase in the availability of training and software to perform such analyses, we are also seeing more analysts across the country using the software and tools. There is still much disagreement about what works best, when and why it works, and what related skills are needed to make the analysis products useful to the law enforcement officials who are doing the investigations. In many cases and inside many police agencies, these predictions and analyses are still met with skepticism among the investigation personnel, but they are gaining credibility as time goes on and more analysts use these techniques effectively. Using these tools is almost a change in social perception and policies more than just a technological issue.

▪ Technology

Technological advances within the last decade have provided many opportunities in crime mapping. High resolution and 3-D applications have enabled analysts to make maps of buildings, such as apartment buildings or correctional institutions, which illustrate problem areas within these buildings. The use of GPS in police cruisers allows for enormous potential in data collection and in producing real time maps for administrators and patrol officers alike. In the future we may see that GPS, photography, and 3-D views are extended out from the officer's vehicle to handheld devices the officer can carry right to the crime scene. There are still some hurdles to cross when making accurate GPS in a small enough unit to be carried by the officer at all times during his shift, but it is on the horizon. Many aerial photography companies now do more frequent updates for locations with high growth, so the aerial photos are increasing in detail and accuracy over time as well. There are also some more widely available satellite images that show other elements such as elevation, weather, and many other types of data.

Automated programs, such as the one that Hicks (2007) designed in Minneapolis, allows for maps and analysis to be completed on a routine basis, saving valuable time and resources. Further research and development in geocoding through entity-named extraction is also on the horizon. Many analysts across the country are starting to develop multilevel geocoding processes to increase the accuracy of the points they analyze. An example is to use parcel addresses as the first pass to geocode calls for service data, then use a verified point address file as a second pass, and finally use the street centerline file to geocode the remaining rows of data that did not geocode in the first two passes. With each run through the data, the addresses that do not match at one level match at the next, so your overall data is more accurate in the long run.

Commercial applications, such as Google Street View, which is a feature of Google Maps and Google Earth, allows the user to see a 360 degree panoramic street-level view of selected cities at ground level. Launched in May 2007, only five American cities were included, but it has since expanded and includes many more US cities and locations in France, Italy, Australia, and Japan. Useful in both the practical and academic realms, Google Street View provides enormous detail and power in crime mapping. Many crime analysis product vendors are making use of this functionality to deliver maps in a new way to law enforcement agencies that purchase their products.

■ Other Issues

When it comes right down to it, there are very few limitations to what can happen in the future with GIS. We may find that aerial photos, CCTV, and 3-D modeling get so good that when a call comes out of a robbery, the officer can simply hit a few buttons on his vehicle computer, and a map will pop up that shows the quickest way to get there and where all his backup units will come from. With another click of the button, he will be able to see CCTV images of the robbery scene and watch the suspects run behind a business and get into a white vehicle, then the officer will be able to tag that vehicle electronically while he is responding. CCTV and aerial or satellite images will track the tagged vehicle, and information will be sent directly to all the units that are responding. If the suspects take hostages, a 3-D model of the buildings and access points around the crime scene will pop up to allow SWAT units to position themselves the most advantageously through line-of-site analysis and other tools right from their vehicles. Although this may sound somewhat like science fiction, some of these applications of GIS are already available but not yet widely used. We may also see crime analysis units coming to the scene of major crimes with a vehicle that is equipped with everything they have at their desks back at the office. The units will assist and run intelligence and investigative analyses for the on-scene detectives to aid the investigation in real time. This type of crime analysis activity is already happening in Riverside, California.

While crime maps are wonderful tools for crime analysts, and the future holds even more promise for new ways of using this awesome tool, there are still unresolved issues related to ethics and privacy. Modern technology allows for ease in accessing and analyzing large quantities of data. As discussed earlier in this text, if this data (in raw or map form) falls into the wrong hands, it can be used to discriminate against people or areas (such as in the process of redlining). With the advent of new data sharing agreements among federal, state, county, and local law enforcement that are either happening now or on the horizon for many states, there will be ownership and need-to-know issues with some of this massive quantity of law enforcement data that is currently or soon will become available. The analyst must be able to be discrete and keep focused on helping the investigation and not hinder the investigation through information overload. GIS enables us to access a lot of data, and sometimes we can manipulate that data enough to give the end users some information, but too much information can bury what they really need to know. That is why the analyst must analyze the data and information available and produce maps and other products that provide knowledge that can be acted upon rather than just muddy the playing field.

In addition, crime analysts need to find an appropriate balance between an individual's right to privacy and the public's right to information. In addition, agencies must also continue to address barriers to implementing crime mapping in their agencies, including (but not limited to) money, training, social changes within the department, and staffing.

Recommended Reading

Boba, R. (2008). A crime mapping technique for assessing vulnerable targets for terrorism in local communities. In S. Chainey & L. Tompson (Eds.), *Crime mapping case studies: Practice and research* (pp. 143–151). Hoboken, NJ: Wiley.

Chainey, S., & Tompson, L. (Eds.). (2008). *Crime mapping case studies: Practice and research.* Hoboken, NJ: Wiley.

Groff, E. (2008). Simulating crime to inform theory and practice. In S. Chainey & L. Tompson (Eds.), *Crime mapping case studies: Practice and research* (pp. 133–142). Hoboken, NJ: Wiley.

Monmonier, M. (1991). *How to lie with maps.* Chicago: University of Chicago Press.z

Olligschlaeger, A. M. (1997). Artificial neural networks and crime mapping. In D. Weisburd & T. McEwen (Eds.), *Crime mapping and crime prevention* (pp. 313–347). Monsey, NY: Criminal Justice Press.

Paulsen, D., & Robinson, M. (2004). *Spatial aspects of crime: Theory and practice.* Upper Saddle River, NJ: Pearson.

Ratcliffe, J. (2002). Damned if you don't, damned if you do: Crime mapping and its implications in the real world. *Policing and Society: An International Journal of Research and Policy, 12*(3), 211–225.

Sorensen, S.L. (1997). SMART mapping for law enforcement settings: Integrating GIS and GPS for dynamic, near-real time applications and analysis. In D. Weisburd & T. McEwen (Eds.), *Crime mapping and crime prevention* (pp. 349–378). Monsey, NY: Criminal Justice Press.

Chapter Exercises

Exercise 28: Practical Strategic Exercise: Vandalism Unveiled

Data Needed for This Exercise

- _____
- _____
- _____
- _____
- _____

Lesson Objectives

- Use the GIS skills you have developed to complete an analysis of vandalism offenses along Grand Avenue within the Ocotillo district.

Task Description
- A city council member for the Ocotillo district has advised the police chief that she is getting numerous complaints about vandalism/criminal damage offenses that are occurring along Grand Avenue within her district. She believes that a gang has begun to terrorize the businesses along Grand Avenue and wants to know if the number of vandalism offenses "around" Grand Avenue, within the Ocotillo district, increased between 2003 and 2004 and what the total numbers are. The police chief asks you to create a map to answer this question and also asks you to advise him, if possible, where along Grand Avenue the number of vandalisms are the greatest or where they increased between 2003 and 2004.
- For this analysis you will need to create a buffer around Grand Avenue that is within the Ocotillo district, and then use this buffer to select offenses from 2003, and then 2004, and then compare them. You will want to create a map or a table that shows if the vandalisms increased, and show the chief where on the map the offenses increased the most (if you can figure this out). The results can be e-mailed to the chief, so a formal memorandum or letter is not needed; however, you need to write out an informal e-mail to the chief to go along with your map or table that explains what you did and your conclusions. Again, be brief but detailed enough to advise the chief of the situation and any problems he should be aware of. Remember that sometimes the data cannot answer the questions, and you can say no, but make sure you can explain why the question cannot be answered when you say no. You may find that if you do not know your data well, or if you have not thought about the problem enough, someone else will be able to answer the question when you said it could not be answered. Therefore, saying no is acceptable sometimes, but just make sure you are right.

Exercise 29: Practical Strategic Exercise 2: Intelligence-Led Policing

Data Needed for This Exercise
- _____
- _____
- _____
- _____
- _____
- _____

Lesson Objectives
- Use the GIS skills you have developed to complete a complicated GIS analysis through preplanning, knowing your data, and flexing the GIS skills you have already learned to develop focus areas or high intensity enforcement areas that would best serve the community and the enforcement abilities of patrol.

Task Description
- The patrol commanders have asked that the crime analysis unit provide them with a focus area within the Foothills division and the Gateway division (sectors). They have specified that they want to find a quarter square mile area (grid) or similar sized area (census block) where the number of calls for service are the greatest, the Part I person (violent) UCR offenses have the most density, and the Part I property crimes also have the highest density. They would ideally like to have three grids or census tracts to choose from and would like a summary table and charts for each of the three grid choices within each division that advises them of the following:
    - What are the top 10 calls for service types in each grid?
    - What is the most prevalent violent crime reported in each grid?
    - What is the most prevalent property crime reported in each grid?
    - What are the most active days of week and times of day for the top 10 calls for service?
    - What are the most active days of week and times of day for the most prevalent violent and property crimes in each grid?
- The commanders each want a map of just their division that shows the hottest three grids and also how the surrounding grids scored. (Provide a hot spot map of each division with the top three grids or areas identified on the map. Each grid has a total score based on geoprocessing of the three data types previously explained.)
- Remember that square mileage can be a large factor in density maps if all of the grids or census tracts are not equal in size. You may want to substitute a rate map based on the total population, total number of reported crimes or calls, or some other reasonable factor to determine the Part I crime densities. You could also consider providing data (if available) on the average response time for calls for service or the average work time

for all calls within the areas as another measurement of areas that would benefit most from a focused enforcement program within each division. This is your analysis, so take a stand and tell them what you think.

1. Break this project up into several steps and ask yourself the following questions:
    a. What data do I need to make these maps and do this analysis?
        i. Calls for service
        ii. Part I UCR offenses
        iii. Streets
        iv. Grids or census blocks
        v. _____
        vi. _____
    b. What tools do I need?
        i. ArcGIS
        ii. Excel
        iii. _____
    c. What steps should I take and in what order?
        i. Create a new project
        ii. Add my data
        iii. Aggregate data to the grid or census block polygons as needed
        iv. Add any other data I need to get a good density map by something that is reasonably useful
            1. Square mileage of grid
            2. Total crimes per grid summary
            3. Total population by census block
            4. _____
    d. What maps should I make and how will I produce them?
        i. During analysis
        ii. For the final product
        iii. Who is my audience?
        iv. What printer or plotter can I print these to?
        v. Will the maps be in color or black and white?

Exercise 30: Practical Tactical Exercise 1: Robbery 101

Data Needed for This Exercise
- This exercise starts out as an Excel spreadsheet, which can be found in the C:\CIA\Exercises\Exercise_30 folder and is named Robbery.xls.

Lesson Objectives
- Use the GIS skills you have developed to complete a tactical analysis of a robbery series.

Task Description
- You are a crime analyst working for the City of Glendale and you have received a robbery crime series. The investigator who is assigned to the case wants a map showing where the next crime may occur so he can arrange for surveillance units. He would also like to know the next probable date, time, and day of week. If you have time, he would like to know if you have any opinion of where the offender may live in relation to the offenses and would like any leads you can offer.
- The suspect is a white male subject who is between 20 and 35 years of age, has a moustache and beard, and is believed to be balding on top of his head. The suspect wears a dark colored ski mask and brandishes a chrome semiautomatic handgun in all of the robberies. He always says words to the effect, "Give me the money or I'll kill you!" In 10 of the crimes he brought his own plastic grocery-type bag to the crime scene and had the attendant place the money in the bag. In the other two incidents the suspect pistol-whipped the attendants, knocking them to the ground, and then took the money from the register himself. He has worn gloves in at least six of the incidents, and in the others the attendant cannot remember if he was wearing gloves. No useable prints or video have been recovered; however, in the second robbery the suspect asked for and received the surveillance tape, and he asked for a surveillance tape in at least half of the incidents.

Exercise 31: Practical Tactical Exercise 2: Burglary 101

Data Needed for This Exercise
- This exercise starts out as an Excel spreadsheet, which can be found in the C:\CIA\Exercises\Exercise_31 folder and is named Burglary.xls.

Lesson Objectives
- Use the GIS skills you have developed to complete a tactical analysis of a business burglary series.

Task Description
- You are a crime analyst working for the City of Glendale, and you have noticed that a business burglar seems to be operating in a part of Glendale. His main MO is that he enters the business

by cutting the alarm/phone lines and then prying open the rear door. The offender has taken expensive photo equipment and supplies from specialty photo stores.
- You need to generate a map and analysis bulletin for the patrol bureau to be aware of the crime series and this offender. Include the next probable date, time, and day of week of the next hit. If you have time, do a journey-to-crime analysis as well to see if any suspects can be identified.

Exercise 32: Practical Tactical Exercise 3: Pick a Crime

Data Needed for This Exercise
- This exercise starts out as an Excel spreadsheet, which can be located in the C:\CIA\Exercises\Exercise_32 folder and is named Massage.xls.

Lesson Objectives
- Use the GIS skills you have developed to complete a tactical analysis of a massage parlor burglary series.

Task Description
- You are a crime analyst working for the City of Glendale, and a city council member sent you an e-mail from a concerned businessman in the city who owns a massage parlor. He indicated that several massage parlors have been burglarized late at night (early morning hours) in Glendale and would like an enforcement response to assist in helping them capture the offender. You need to create an internal bulletin for law enforcement and also create an e-mail response to the city council member regarding what you have found and what actions the police department will probably take with this crime series.

■ References

Chamard, S. E. (2003). Innovation-diffusion networks and the adoption and discontinuance of computerized crime mapping by municipal police departments in New Jersey. PhD Dissertation, Rutgers University, New Brunswick, NJ.

Chamard, S.E. (2006). The history of crime mapping and its use by American police departments. *Alaska Justice Forum, 23*(3), 1, 4–8.

ESRI. (Summer 2008). USPS fights crime with GIS: GIS pinpoints potential money laundering. *ArcNews, 30*(2), 1, 33.

Harries, K. (1999). *Crime mapping: Principle and practice.* Washington, DC: US Department of Justice Programs. Retrieved March 2008, from http://www.ncirs.gov/htm/nij/mapping/pdf.html

Hicks, D. (2007). *Spatial analysis of probationer contacts and automated email alerts.* National Institute of Justice Workshop. Washington, DC: US Department of Justice Programs. Retrieved February 2009, from www.ojp.usdoj.gov/nij/events/maps/pittsburgh2007/papers/Hicks%20Workshop.pdf

Olligschlaeger, A. (2003). Future directions in crime mapping. In M. R. Leipnik & D. P. Albert (Eds.), *GIS in law enforcement: Implementation issues and case studies* (pp. 103–109). New York: Taylor & Francis.

Paynich, R., Cooke, P., & Mathews, C. (2007). *Developing standards and curriculum for GIS in law enforcement.* Presentation at 9th NIJ Crime Mapping Conference, Pittsburgh, PA.

Microsoft Excel: A Generic Primer for Crime Analysis

APPENDIX

> This appendix is a tutorial for using Microsoft Excel 2003 and earlier. The majority of these functions will also work in Excel 2007; however, the location of the menus, buttons, and functions may be in a different place on the ribbon menu that is part of Excel 2007. This tutorial will go over basic Excel skills that the analyst may need on a daily basis and allow the student to learn more about Excel through practical exercises.

■ Introduction

Crime analysts often need a variety of software tools to create a final analysis product. One of the most widely used tools for day-to-day analysis by analysts across the country is the Microsoft Office suite of products. Excel probably tops the analysts' most wanted list in any crime analysis unit across the United States and is used in a variety of ways. Analysts can collect intelligence information on persons, do analytical studies of crime phenomena, or simply use it to balance their checkbooks. One obvious use for Excel is to create administrative, strategic, or tactical analytical products and bulletins. Practicing analysts could use Excel for creating the bulletin itself or they could use it to clean and prepare their data, make calculations, and then cut and paste that information into Microsoft Word or other applications.

This tutorial does not intend to suggest what an analyst should use. It is only a how-to manual for Excel that contains some of the lessons and processes we've learned over the past 15 years or so. There are often many other ways of doing the same things as shown in this tutorial, and it is not intended to be a definitive source about Excel.

This tutorial is divided up into sections that cover specific topics and examples in Excel, and the included Excel exercises also follow the section numbers. You may also find the following Web sites to be useful for Excel hints and tricks:

        http://www.ce.memphis.edu/1112/excel/xlstart.htm
        http://spreadsheetpage.com/
        http://www.ozgrid.com/Excel/default.htm
        http://www.xlhelp.com/Excel/help.htm
        http://www.meadinkent.co.uk/excel.htm
        http://homepage.ntlworld.com/noneley/

■ Section 1

Introduction to Excel and Some General Information

When you first open Excel, the spreadsheet application should look something like DVD Figure App 1–1. Your menu bars and items may be slightly different depending on who has used your computer last and how they left Excel, but generally, this should be the way it looks. If you are using Excel for Windows XP, you will also see a window to the right that shows some shortcuts for using basic Excel functions and creating new spreadsheets. I typically turn this window off because it takes up my working space, but that is up to you.

I will not spend a great deal of time showing you the Excel interface, but I will point out a few things so we can find them easily later. The Data menu item provides you with several functions, but this is where we find the Pivot Table item. The Tools menu item holds a special menu for Add-Ins and allows you to turn some exciting features of Excel on or off as you need them. One of the most useful add-ins is the Analysis Toolpak (for Excel 2007, please check the help menu for how to turn on the Analysis Toolpak). This set of tools adds several new statistical functions for the user and enables you to do some slightly more advanced work. As we go through this process, I will show you how to turn this on and give you some things to consider about add-ins that could prove useful for you. For the moment we will just discuss the basic application window of Excel.

Excel uses some odd terms at times to describe differences in various things you may see on the screen. For example, the entire Excel application, whether you are using one or more worksheets, is called a workbook. If you close Excel and it asks if you want to save your work, you are saving an Excel workbook. At the bottom of the Excel application window, you will see one or more tabs with wording such as Sheet1, Sheet2, etc. DVD Figure App 1–1 has one tab in the workbook, and it is labeled sheet 1. You can create a whole bunch of these individual sheets, and Excel refers to them as worksheets. Think of an Excel file as a book that contains multiple sheets.

Within each sheet, the working surface is separated into cells. Each cell is represented by the gray squares (gridlines) on the worksheet surface. These gridlines can be turned on and off so you do not see them when viewing a worksheet, but typically we want them there to help us organize our projects. These gridlines are really only for the user, and unless you set Excel to print them, they will not show up on the printed page. To make this somewhat more confusing, the makers at Microsoft decided that users would want to put other borders and such around these cells and groups of cells, so instead of calling them "printed gridlines" or something else equally useful, they called them borders. The width, line color, and style can be changed for each cell, or a group of cells, by selecting the cells you want to change and then pressing CTRL+1 or by selecting Format → Cells from the menu bar. The background color, text color, and a variety of other things can be changed for each individual cell, or groups of cells, to make the final product more eye-catching and appealing. You can practice this whenever you want to, and any basic Excel course should have introduced you to these items.

Another useful tip is knowing what Excel Help screens mean when they indicate that you can make *relative cell references*, or *absolute cell references*, in a worksheet. This should not be confused with R1C1 reference style, which is discussed next.

Click on the **Tools → Options** menu items at the top of the Excel application (see DVD Figure App 1–2), and in the dialog that pops up (DVD Figure App 1–3), click the **General** tab. Under the General tab, there is a Settings area that you can check to have Excel change the normal way it refers to cells (A1 for the cell in column A, row 1) to a slightly different row 1, column 1 (R1C1) reference to the same cell (columns and rows both use numbers). Users who are familiar with Multiplan, Lotus 1-2-3, and some other spreadsheet programs may find that this setting makes everything easier to understand, but most Excel users currently prefer the A1 references instead of the R1C1 references. When you have R1C1 checked, instead of seeing columns labeled A, B, C, D, etc., the rows and columns are both labeled with numbers (1, 2, 3, 4, etc.). This means that you need to type in R1C1 instead of A1 when you refer to the same cell.

You will also notice that when you have the Tools → Options dialog open and the General tab selected (view right), you can change the default type style and size, default file location, user name, how many sheets will be created when you open a new session of Excel, and how many Excel spreadsheets will be displayed in the recently used files list. A word of warning about the Alternate Startup File Location item: If you enter a directory name or an Excel file name here, Excel will open that spreadsheet, or every single spreadsheet in the specified directory, when it opens. You should therefore be cautious if you use this feature, and you may want to read the Help file before using it.

The other kind of cell referencing pertains to how we relate one cell to another cell in the worksheet we are working with. If I have cell A1 and it contains the value of say, 31, and I want to add it to cell B1, which has a value of 40, I would want to enter a formula in cell C1 that says something like "=A1+B1". This would give me the value of 71 in cell C1. This is called relative cell referencing because you are saying that cell A1 should be added to B1, and they are "relative" to each other for the value you want in cell C1. You will always take whatever is in column B at that level and add it to whatever is in column A at the same level, or in the same row. Now expand this thought to data you have entered into cells A1 through B25. You have two columns worth of data (columns A and B), and in column C you want to use the previously described formula, but you want it to change as it moves from row 1 to row 2, to row 3, etc. By simply entering the formula as previously shown in cell C1 and then double clicking the bottom right-hand side of that cell (black plus sign), you can copy the formula into all of the cells below it down to the end of your data (*open Section1.xls to practice this*). You can also drag the cell and formulas down by clicking on the bottom right-hand corner of the cell with the formula, holding your left mouse button down, and dragging it to the cells below. In the individual rows in column C, you see that the formula also changed because it was a relative formula. Cell C2 has the formula =A2+B2, cell C3 has the formula =A3+B3, and so on. This is the use of the relative cell reference function in Excel.

On the other hand, if we want an absolute reference to a particular cell, we need to modify our formula a little to make Excel calculate it the way we want it to. Now think of the rows of data as the following examples show, where column E contains some text value, column F contains a numerical value, and we want column G to calculate the total percentage of all the different types. We will want to use an *absolute cell reference* in our formula to accomplish this.

As you can see from the formula used columns for the two types of referencing (see DVD Figure App 1–4), the absolute cell reference (column H) uses the "$" sign before the column and row numbers to make it refer to the total crimes cell for each of the calculated percentages. The first operator, which refers to column F, is a relative reference, so the formula changes as it is copied down, but the last operator is an absolute cell reference with the $F$11 type of notation, so it always looks "absolutely" to that cell for the second half of the formula.

Practice finishing the Section 1 sheet in Section1.xls, and then recreate this sheet (see DVD Figure App 1–5 and 1–6) with the data displayed under the Student Exercise tab in Section1.xls.

This concludes the general information about Excel because you should already have basic abilities with it, so we will approach some harder topics and uses of formulas and functions in Excel for crime analysis.

■ Section 2

Working with Standard Deviation, Average, and Other Statistics

Start this exercise with the file named Section2.xls (see DVD Figure App 1–7).

You want to be able to give your command staff a prediction of how many crimes will probably be reported for your jurisdiction by the end of 2004. We deliberately left out any crimes that have already been reported in 2004 so that we can use the previous 19 years to predict 2004 levels. Typically, the more data you have, the more likely your prediction will be accurate, and most of the time you will want at least 20 months, days, weeks, or years to predict the next period you are aiming at with any degree of certainty. This will change depending on the variability of your data from year to year and you may use only 5 years, or 35 years as needed. Keep in mind that this prediction is really only reasonably accurate for the predicted next year, week, or month, and the reliability of the prediction decreases as you try to use your data to predict further into the future. Inferring what may happen based on the past is always a shaky proposition, and with crime data, it is often not as accurate as we may want it to be. Combining a description of what has happened in the past with an inference of what *may* happen helps your analysis to be understood and should at least be considered each time you try this.

As an analyst you should be able to look at the data and be able to get a feel for what your prediction will most likely look like, especially with this example.

Based on the data from 1985 through 2003, what would you expect the total number of crimes to be in 2004 by simple observation of the

Count field? (Will it be higher than 2003, lower than 2003, or about the same as 2003?)

The data appears to be steadily increasing from year to year for the most part, and except for 1994, each year has always been higher than the previous year. You should expect that if you use the mean and standard deviation, or the slope and standard error, to predict 2004's count, either method will most likely show 2004 higher than 2003. This is reasonable for this data set, so why do we need Excel? Law enforcement agencies have been using "gut feelings" and "by the seat of their pants" types of analyses very successfully for many years. Most of that past analysis is based on pure common sense. Make sure that when you start applying scientific methods and statistics to crime analysis you keep common sense!

Let's test this theory by calculating the following statistics for our set of numbers.

Enter the formulas as shown in column F of the Section 2 worksheet (see DVD Figure App 1–8) and match your results to those in the table (DVD Figure App 1–8).

If you get any numbers different than those listed in the table, make sure you entered the formulas correctly and make sure that your data matches the data in DVD Figure App 1–8.

Now that we have used Excel's statistical functions to calculate the mean and standard deviations of our yearly data, what does that allow us to say about this data?

We could say that we average at least 94.3 crimes per year in this city. We can also say that 68% (mean plus/minus one standard deviation) of the years experienced between 55.5 and 122.1 crimes, or that 95% (mean plus/minus two standard deviations) of the years experienced between 38.8 and 149.9 crimes. We cannot really predict what the total will be for 2004 using the mean and standard deviation, and we can only *describe* the crimes per year for the previous years.

To predict what 2004 is most likely to experience, we have to use Excel's **Forecast**, **Slope**, and **Steyx** (standard error) functions. These will allow us to know if the linear trend is up or down and give us an estimate of what might be expected in 2004. The statistics we will need to calculate are shown in DVD Figure App 1–9.

Enter the formulas in column F in the Section2.xls/Section 2 worksheet.

This shows that with no standard error calculated, the forecasted number of crimes in 2004 should be 143.1 crimes. If you add in the standard error (or standard differences between numbers from year to year), then within one standard error, the predicted range for 2004 is between 138.5 and 147.6 crimes. Extending this to two standard errors brings the range to between 133.9 and 152.2 crimes. If you look at the Slope of the linear trend line (see DVD Figure App 1–10), it is a positive number, meaning that the trend over the past 19 years has been a positive one, or upward.

The chart of the forecasted information that we will create in the following steps may help you to understand it better. I have found that the linear trend line works about as well as any other trend or forecasting tool with crime data. There are more sophisticated methods, such as polynomial, logarithmic, moving average, and many others, that can be used, and you should take the effort to look at these other methods at some point in your career and empower yourself to be able to use them as the data lends itself to those analysis methods. The only one covered in this tutorial will be the linear trend line.

To create this chart in Excel, follow these steps:

1. Highlight cells B1 through C20 with your mouse.
2. Click on the **make graph** icon (see DVD Figure App 1–11) to bring up the Chart wizard. (Charting is much different in Excel 2007, so please refer to Microsoft Help to learn how to do this in Excel 2007.)
3. In the Chart wizard, select the area chart in the top left section of the area charts, and then click **Next** (see DVD Figure App 1–12).
4. When the next screen pops up, click **Finish**.
5. A new chart will appear in your Excel worksheet.
6. Position the chart wherever you want it on the sheet by dragging and dropping it to that location.
7. To add a trend line, right click the blue area and pick **Add Trend Line** from the drop-down menu.
8. Select **Linear Trend Line** from the drop-down menu that appears, and click **OK** to add the black trend line.
9. You can reposition the legend to the top by right clicking it, selecting **Placement**, and checking **Top**.

You can clearly see that the statistics we calculated for this data certainly seem to match up to the chart/graph that we created of the data. Combining these visual objects with the analytical data allows the analyst to not only make sound analytical statements; it also allows the command staff users to understand the significance of the statistics you used through a very simple chart or graphic of the data.

The tab in the Excel exercise spreadsheet labeled Section 2 Example from Glendale shows how this type of analysis was used on an actual analysis effort that tracked calls for service information for several apartment complexes in Glendale that were hot locations. The chart (see DVD Figure App 1–13) is a visual representation of the data and should be reviewed so you can decide what conclusions you would make in a written document to describe this data.

Based on this information, provide a short description of the data presented for the addresses shown on the chart, describe how the data has changed over the past months shown, and provide information on what the predicted values will be for the next month, which will be August 2004 (200408). Refer back to the various fields in the Excel spreadsheet and the calculated data for reference when needed. An example of the descriptive text for our exercise is as follows:

> For all apartment complexes listed, the trend appears to be slightly upward for all calls for service over the last 31 months. 5701 N 67th Av also shows a slight increase in CFS, however 6565 W Bethany Home Rd and 5801 N 67th Av show a slight downward trend over the past 31 months. The monthly average for all apartment complexes is 85 CFS and the forecasted number of calls to these apartment complexes in August 2004 should be between 77 and 107 within one standard error of estimate.

For further exercise using these functions in Excel, go to the spreadsheet labeled Section 2 Student Exercise and complete the formulas for the data listed. Create a short memorandum that explains the data, what the trend is with each of the crime types listed, what the forecasted number of crimes should be for August 2004 for each type of crime, and any other information you think would be appropriate for a crime analysis effort requested by your chief for Part I UCR offenses in your jurisdiction. For a more in-depth review of these processes and some other considerations, review Chapters 15–20 in Gottlieb, Arenberg, & Singh (1994).

The following summary is an example of information you may want to include when you describe this data to your chief.

> Based on the information included in the following tables and charts, it appears that the number of homicides have been steadily increasing over the past 31 months. There have been a total of 278 homicides in the past 31 months, with an average of 9 per month across that entire time range. If this basic trend holds true, there will be between 16 and 30 homicides in August 2004, or greater than a 162% increase in homicides from the July 2004 count of 6. From past statistical reports, the reason for the sharp increase in homicides in recent months may be linked to gang violence erupting in the city's downtown area.

Students should be able to review the information in the Section2.xls/Student Exercise worksheet tables, insert the correct formulas and functions in the cells that are blank, summarize the information in a similar fashion for each crime type, and potentially make several generalizations about the data and the status of violent crimes and property crimes in the city as a whole in the past 31 months.

Remember that the time period you choose to analyze may often be as important as the data you are analyzing. If you use a short time period in your analysis, your prediction and forecast may be highly inaccurate. If you use a time range that is too long, you may see a trend that does not exist in more recent months or may be changing rapidly from month to month. This is where common sense comes into play when doing these types of analyses using Excel. You can only describe the data you choose to analyze and forecast for that one next time period with any degree of certainty. Make sure you understand these limitations, and make sure that the persons for whom you create your analysis understand the limitations of a linear trend. You may also want to consider seasonality in your trend analyses (crimes go up or down naturally with the weather, school getting out, etc. every year despite the current trend), and look for potential causes to explain severe upward spikes or low troughs in the data. Often these really high spots and low spots may be due to a comprehensive enforcement project your agency was involved in, a change in policy on how certain crimes are recapped, or some other situation that was not normally present (such as a new computer system installation, a statewide auto theft task force or DEA focus in your city, etc.).

When it is possible to list some of these situations as potential confounding variables on your analysis product, you will usually provide a better product than if you did not look for them and reported the information simply from the charts or graphs. Of course, as with any other analysis, the peaks or valleys may not be explainable, and that may need to be a statement in your analysis as well. Do not be afraid to state your opinions in your analysis products, but be intimate enough with your data to support those conclusions and wise enough not to argue too much if they are not accepted or considered by personnel for whom the analysis was meant.

Now that we know how to describe trends in singular datasets, what happens if we want to know how one item, like population, affects an increase in crime? This is called a correlation, and I have heard this term used incorrectly many times. The most incorrect use of this term is when someone says that population increases "caused" crime to rise because they are highly correlated. Although this statement may have some truth behind it, a correlation is simply looking at the relationships between two sets of data or numbers and whether or not they are going up or down in the same manner over time, not if one "caused" the other. This incorrect usage should be avoided.

There are three basic types of correlation you may deal with when analyzing crime data:

1. Two variables increase over time together (positive correlation).
2. Two variables decrease over time together (positive correlation).
3. As one variable increases, the other decreases over time (inverse or negative correlation).

Items 1 and 2 are typically called a "positive" correlation because they both move in the same direction. The positive part of this is that both sets of data move in the same direction: up or down. Item 3 is typically called a "negative" or "inverse" correlation because as one moves up, the other moves down. You can find positive and negative correlations in just about any two columns or rows of data. The common sense part of this entire calculation comes in what two variables you choose to analyze. For example, if your commander asked you to check on whether the color of the dirt in his sector affected the crime rate, your response should be, "Are you kidding?" On the other hand, if he advised you that he has been doing a 90 day increased enforcement activity for traffic violators in his precinct, and he wants to know if it has affected the number of traffic collisions, you may have a different opinion of his request. In this example, you would want to find a negative correlation; as the number of citations went up, the number of crashes went down.

Open the Section2.xls workbook and the worksheet labeled Correlation Exercise. You will see our old data from previous exercises along with some population data for each year and the chart shown on DVD Figure App 1–14. By using your common sense and viewing this chart, you can see that as population increased, the number of crimes also rose.

The degree to which these elements changed over time is not exactly the same, but we can use common sense and conclude that there is a positive correlation between population rate and crimes reported. When we use the Correl function or the Pearson function in Excel, we find that the correlation coefficient (with either function) between the population rate and the reported crimes is 0.99. For both functions, we specify that array 1 equals the population data column (D2:D20), and array 2 equals the total number of crimes (E2:E20). Because the "r value" or degree of correlation (values of +1 to −1 with a value closer to +1 or −1 indicating a very strong correlation, with +1 indicating a strong positive correlation, and −1 indicating a strong negative or inverse correlation) is very close to 1, we can say that there is a *strong positive correlation* between the population rate and the reported crimes. This does not mean that one causes the other, but as population increases, we could expect that crimes will also rise (*pure and simple common sense in action*). The lower the value of the correlation coefficient (or the closer to 0), the less likely that we will see a change in one variable when we see a change in the other. If the value is −0.99, then the relationship would be a strong, negative correlation. What you are looking for is a value that is closest to 1 or −1. A correlation coefficient of 0 would mean that there was no correlation of any sort between your two columns of data. The table in DVD Figure App 1–15 explains what we might say about the strength and direction of a correlation, depending on what the value is.

If you use the function Correl (this function is only visible when the Analysis Toolpak is on) or Pearson, you should get the same result for the correlation coefficient or r value with most crime datasets. Just knowing whether or not you have a strong or weak correlation and if it is positive or negative is part of the process, but you also need to have an idea of how likely it is that you could arrive at that r value by chance or accident. The simplest way to get a handle on this possibility is to either square the r value or use the RSQ function, which does the same thing. The RSQ function needs the same data columns as your Correl functions, but you can also just square your r value obtained from the Correl or Pearson function by using the following formula in a new cell:

=(R_ValueCell)^2

For our example, we had an r value of 0.99. If we square it using either RSQ or the preceding formula, we get a value of 0.987. We can then make a statement such as, "When the population goes up, 98.7% of the time the crimes reported will also go up." This doesn't tell us how much it will go up, of course, but that is life in statistics and math.

Using the Section2.xls/Student Correlation Exercise worksheet, enter the data from DVD Figure App 1–16 and create the correct functions and formulas so that your charts and other data come out correctly. Use the correlation sample sheet as your guide to the correct formulas to enter.

- Is there a correlation between these two sets of data?
- What is the strength of the correlation and the direction of it?
- When the number of traffic citations increase or decrease, what happens to traffic crashes?

- What percentage of the time does this occur?
- Create a summary statement to explain this information to the commander who asked you to check into these statistics for him.

■ Section 3

Converting Dates and Times

Start this exercise with the file named Section3.xls.

Dates and times can often be one of the most difficult pieces of data that you may run into when working with Excel. ESRI's ArcView 3.x or 9.x products often do strange things to dates and times, and many database products, records management systems (RMS), and computer aided dispatch (CAD) software applications store and retrieve date and time data much differently. In this very long section we will discuss some of the basic date and time data you may run into and how to utilize the nifty date and time handling functions inherent in Excel to make your job as an analyst much easier.

Let's start out by deciding what we want to do with a date or time. Next to the "where" category or location (geography of crime), the "when" (date and time or temporal factors) question seems to pop up most often during an analyst's normal routines. Even in the examples in Section 2 we used year and month as factors in our analysis. If we have a date, we can determine several things about an incident, or we can aggregate or simplify this information. Excel has functions that deal with dates and can calculate all of these different variables when needed. But before we can deal with the data, we first have to tell Excel that the data we are using or have entered is in date and/or time format. If Excel treats the data as a number or text field, then the date and time functions that are built into Excel cannot be used with the data.

Several of the things we may want to know about a date on which an incident occurred may be as follows:

- Date or time
- Day of the month
- Hour of the day
- Day of the week
- Month
- Year
- Week

Of course, we may also want to combine several of these factors into some time period to make it easier to understand, make the data work better for a chart or table, or explain some aspect of the data being analyzed. We may also want to use the information to describe when something happened or predict when it may happen again based on prior occurrences.

To be able to do all of these things, we first have to provide Excel with data that is in date and/or time format. To enter dates into Excel, you should normally enter the information as follows: 08/24/2004 17:34. If you neglect to enter the time, Excel will normally use 00:00 (midnight) instead of the actual time of the incident, so you should be careful of this when entering data. When you convert data that is only available as a date, such as 08/24/2004, without a time, Excel will store the time as 00:00 for this event. If you need to capture the second that something happened, add the second at the end, for example, 08/24/2004 17:34:42. If you fail to do this, Excel will also make the seconds :00, and any calculations you do could be in error at the seconds level.

Using the Section3.xls spreadsheet as a guide, enter the following dates and times starting in cell A5.

- 01/01/1900 12:00:01
- 01/01/1998 12:01:01
- 12/31/2003 13:14:01
- 06/21/1994 06:05:03
- 08/31/2000 00:00:01

As you enter each of these dates and times in the spreadsheet, the days, hours, year, etc. fields should automatically calculate. The Week field may show Number? or Value? instead of a number. This is because the WeekNum function is part of an add-in for Excel that has to be turned on before you can use that function. To turn on add-ins for Excel, click the Tools menu item. In the drop-down box, click on Add-Ins. If the person who installed Excel on your computer installed all of the Excel add-ins that come with Microsoft Office, you should see several add-ins in the add-in drop-down dialog box. Find the one that says Analysis Toolpak, and check the box to the left of this item. Click the OK button, and this will load the Analysis Toolpak functions for this session of Excel and any future sessions if you save this spreadsheet at any time during this lesson. There is also an Analysis Toolpak for Visual Basic that you do not need to check, but it is helpful if you are going to write any Visual Basic code in Excel that may use a function that is installed with the Analysis Toolpak. If for some reason these add-ins are not available to you, then they will have to be installed before you can use the WeekNum function and others that may be discussed in this tutorial. To find out how to do this, open the Help menu and enter *How do I install add-ins?* in the dialog, then follow the directions to install the add-ins on your PC.

You will need to refresh the calculations on the screen by clicking in any row under the week column and then pressing enter without changing anything in the field. This will recalculate the WeekNum, and values should now appear in each of the Week cells. You should now have a table that looks like DVD Figure App 1–17.

For each date and time you entered, Excel calculated the year, month, weekday (with 1 representing Sunday), day of month, week of the year, hour, minutes, and seconds. Now click on the cells in each column and see how the formula is entered using the date and time functions of Excel. These have all been entered for you for this part of the exercise. You can refer back to this simple spreadsheet as the lesson progresses for information on how to use each formula.

I often create all of these fields in a set of data I am going use for analysis even if I don't think I will use them. It helps me to remember these functions, and if I decide to do some date and time analyses, I already have everything I need. Some additional fields I often create are concatenations (combinations) of several of these fields. I often create a field called YrMnth, which is the year and month of an incident. This is how the data was analyzed in Section 2. I may also create a field called YrMnthWeek and use it for trend charts by week for some analysis projects. The only trick to these fields is to observe that some months have only one digit in them, and others have two. To make your data sort correctly, you need to make all of your data the same when you combine or concatenate fields. See the table in DVD Figure App 1–18 for an example.

If I use the standard formula for concatenating two fields, the cells under YrMnth Before Edits would contain a formula that resembles =B12 & C12. It refers to the first cell (B12), the ampersand advises Excel that a new cell or function is coming, and then C12 refers to the second cell to be combined. This formula takes two number fields (right justified), combines them, and converts the data to a text field (left justified), but each cell is not the same length. If you sort them, the long list of data will not sort properly. To keep this from happening, change the formula to the following:

=B12 & IF(LEN(C12)=1,"0" & C12,C12)

The LEN statement stands for length, and it finds out whether or not the value in C12 is one or two digits long. If it is one digit, it adds a 0 in front of it, and if not, it accepts the value of C12 and combines it to the year. On the Section3.xls spreadsheet, use this same logic to create the YrMnthWeek formula so that each of the weeks is also exactly two digits long in the final result. Remember that you already have the year and month combined correctly in the YrMnth (Corrected) field, so you do not need to concatenate all three fields in this formula. On the exercise spreadsheet you will also see that I added the Excel function INT (integer) to convert the text value of the combined year and month back to a number (right justified). This is sometimes useful when sorting large sets of data because Excel sorts numbers much faster than text. An alternative to using these functions individually is the use of the TEXT function. The data you wish to convert must be a date and time field and you will use the TEXT function to get the parts of the date or time you wish to use. Some examples are included below.

- = TEXT(A14, "yyyymm")—provides 200408 from the date and time in A14
- =TEXT(A14, "HH")—provides the military hour of the time in A14

For more help with the TEXT function refer to the Excel help, or visit the Microsoft Web site and search for the Excel Text function there.

■ Converting Data into Date Format

There are several ways that temporal information can get into your Excel spreadsheet, and it comes in several different types. The table in DVD Figure App 1–19 shows data in several different formats that refer to a date or date and time. The easiest way to tell numeric data from textual data (most of the time) is that if the data is left justified it is probably text, and if it is right justified it is normally a number. You can change the formatting of every cell in an Excel spreadsheet, however, using the Format → Cell item in the menu bar (or press CTRL+1) or by right clicking the cell and then selecting Format Cell from the drop-down box, so this is not always the case.

These are just a few of the examples that I have seen, and your data may come in a variety of formats and combinations when you are working as a crime analyst. You will need to become intimately familiar with the data your agency collects and know what odd collection methods, programming, or policies make the data come in those various ways to your desktop. An example I like to use is from my days with another police department. When we were creating a crime analysis unit database, we pulled data from our RMS system, which was built in-house. The date and time were collected in two different fields in the RMS system for most incidents, and they were not combined. Going through the data, we noticed that several date fields had "0" instead of a blank or an actual date value. We assumed this was exactly midnight. The fields that did contain dates had text such as 19990101, which represented 01/01/1999. Every time we brought this information from our RMS into Excel, it converted the 19990101 to a number. This seemed to work fine, and we logically concluded that "0" in the date field meant that the date had not been known and was not entered for this incident or was null. Unfortunately we did not conclude the same thing for the time field. The time was also a text item in the crime analysis data tables, but it was entered such that 1 represented 00:01, 23 represented 00:23, 123 represented 01:23, etc. When we brought this data into Excel, it changed the data into a number. For almost a year we did analyses with this data and converted these two separate fields into a concatenated date and time field in Excel through some manipulations of several Excel functions. During a meeting on a totally different subject one day, a programmer mentioned that the value of 0 in our RMS system anywhere represented a null value or something that had not been entered. This meant that we had been translating 0 in the time field to be 00:00, or midnight, incorrectly. Luckily, there were not too many incidents where this information had been left out, but it still caused some problems because we had not known enough about our data and how it was collected in our own RMS system. In this case, the RMS system could not capture midnight as a time because programmers had used 0 instead of 999, 99, or some other common number as a null or missing value. We could only record from 1 minute past midnight to one minute before midnight, or 2359, because, when it was converted, 0 didn't represent 00:00; it represented a null value. Therefore, nothing could happen exactly at midnight in that RMS system.

A side note is that if you convert 0 to a date, Excel will show the date as 01/00/1900 because that date is represented by the number 0 in Excel's storage of date and time information (see Excel's Help file for information about how Excel stores data and time data). All other dates are serial numbers counting up from that date or the number 0. For example, 08/24/2004 is 38,223 days from that date, so 08/24/2004 can also be represented by the serial number 38223.

There are several examples of data conversions under the Excel Hints and Tricks tab in Section3.xls that may be helpful to you for the remaining lessons in this tutorial and for future use. Feel free to copy, print, or use them as needed.

In the Section 3 Exercise tab, the data in DVD Figure App 1–19 is included in a table along with several suggested formulas for converting the date and time data into one field containing both the proper date and time. I canot tell you how many problems I have solved by simply advising an analyst to combine separate date and time fields into one date/time field and then make the calculations. When you enter a formula into the Results column and a number such as 38233 appears in the cell, remember that this is the way Excel stores date/time information. If you want to see this information in a standard date/time format (such as 01/01/2004 15:34) you will need to format the cell so that it shows the date and time the way you want to see it (see Formatting Cells in Excel Help).

The general rule for converting data into dates that Excel understands is to do whatever it takes to make Excel think of the information as if you had entered it as in the first part of this lesson: 08/24/2004 17:34. You want to convert the data that was given to you with whatever functions Excel needs so that it interprets the data in that format. For example, let's use the following data and see what we need to do to it to make Excel see the data as a date:

Cell A21: 01/01/2004, Format: text
Cell B21: 00:01, Format: text
Target cell for formula: C21, Format: mm/dd/yyyy HH:mm

This information is almost in the format that Excel likes to see; however, it is contained in two different fields. We need to use the DateValue and TimeValue functions of Excel to look at this text information, see it as a date and time, and then just add them together into a new field. The formula for doing this would be as follows:

= DATEVALUE(A21) + TIMEVALUE(B21)

This would give you the result of 01/01/2004 00:01 in cell C21, where you entered this formula. This kind of date and time manipulation is easy and is what you hope for in life. You may also run into data that actually looks like "01/01/2004 12:34:05" that may come from ArcView or another application, and it appears to be a date already. When you attempt to do calculations between dates, however, you will get back an error message like Error or Value. This is because Excel is looking at this data as a text field and not a date. You can sometimes simply format the field as "mm/dd/yyyy HH:mm:ss" using the custom formatting function under the Number tab in the Format Cells dialog, and Excel will do your calculations correctly without actually converting the data into a date and time. Formatting simply displays the information in the way you want it, but it does not change the underlying data type in the sheet in most cases. You can also create a new field and use the following function to look at cell C2 and convert the "01/01/2004 13:42:13" data into a date and time field:

=DATEVALUE(left(C2,10)) + TIMEVALUE (RIGHT(C2,8))

If you just used the DateValue function, you would only get the date part of the text data and not the time as well. If you only used the TimeValue function, you would only get the time part of the text. Excel will only do what you tell it to do and will not guess what you meant to do.

Complete the cells in the table under the Section 3 exercise for the different date formats. Remember to use the Mid, Left, Right, DateValue, and TimeValue functions in your formulas to arrive at combined date and time fields that reflect the correct dates and times for the data provided. Use the Excel Help dialogs to find out more information about these functions if you need to. The format for using each of these functions can be found in the Help menu; however, a very short description of their use and a few other helpful functions are provided in DVD Figure App 1–20.

Some other useful text functions taken directly from Excel Help are as follows:

- **Char**: Returns the ANSI character specified by the code number.
- **Clean**: Removes all nonprintable characters from text.
- **Code**: Returns a numeric code for the first character in a text string.
- **Concatenate Concatenate (text1,text2,...)**: Joins several text items into one text item.
- **Dollar**: Converts a number to text using currency format.
- **Exact**: Checks to see if two text values are identical.
- **Fixed**: Formats a number as text with a fixed number of decimals.
- **Lower**: Converts text to lowercase.
- **Proper**: Capitalizes the first letter in each word of a text value.
- **Rept**: Repeats text a given number of times.
- **Substitute**: Substitutes new text for old text in a text string (this can be used for much more complicated operations than the Replace function).
- **T**: Converts its arguments to text. You can sometimes use this to see if a field is actually text or some other type of data. It returns an empty/blank cell if the field does not contain text.
- **Text**: Formats a number and converts it to text.

- **Upper**: Converts text to uppercase.
- **Value**: Converts a text argument to a number.

In the Section3.xls spreadsheet, there is a tab labeled Section 3 Example from Glendale. This tab shows a robbery series that occurred in Glendale and shows that the dates and times were in two separate fields and not usable with Excel's date functions as they were. I used a formula that combined several of the functions previously mentioned to calculate the dates and times into one field so that this information could be used for doing day of week, time of day, and next date of hit predictions as shown. In Section 5, we will go over how this is done with a tactical crime series bulletin.

A very useful function is the **text** function when you already have the data in date and time format. For instance, if you want the day of the week from a date, and the date and time are in cell A2, you can use the text function to get it in several ways. For the following examples, let's assume the date 03/14/2009 13:14:55 is in cell A2.

=TEXT(A2,"d") returns 13
=TEXT(A2,"ddd") returns Fri
=TEXT(A2,"dddd") returns Friday

Although we do not use this function in this tutorial, you should at least be aware of it for later use in your own analysis efforts.

Using what you have learned from this discussion and from reviewing the spreadsheets provided, click on the Section 3 Student Exercise tab and create a formula for the columns that do not have anything in those cells. These are RptNum, DateTime1 (use the Date_Time field to calculate), DateTime2 (Use CvDate and CvTime to calculate), Day of Week, Hour of Day, Month, YrMnth, and Days Between Calls columns. Then drag your formulas down to fill the remaining cells for every call listed. Make sure they all calculate correctly and there are no problems with your data or formulas for all of the records. Consider that all of the data will need to be sorted by order of occurrence to calculate properly for the Days Between Calls column, and make sure you sort the data correctly prior to entering a formula for this field.

After you have completed entering these formulas, see if you can calculate the average number of calls per day of week and hour of day. You might consider placing this information in a graph to make it easier to understand. You should also calculate the average number of days between calls. Consider how this information could be used in your daily crime analysis assignment and what value it could be to your agency. Write a short memo describing the conclusions you came up with concerning this data and your temporal analysis results complete with the charts, tables, and graphs you think most appropriately support your conclusions. After you have completed your analysis, review the Word document entitled Section3.doc as an example of the analysis that was done with this information. The very simple date and time charts were added for this tutorial only.

■ Section 4

If you are using Excel 2007, you will need to search for a paper on the ESRI Web site (www.esri.com) to find out how to work with Excel and Access 2007 data sources. Instead of trying to export to a DBF table from Excel, you will save as a comma delimited text file (CSV) for use in ArcGIS.

Notes on Working with ArcView and Excel

It is usually very easy to get data from ArcView to Excel, or from Excel to ArcView, but there are a few tricks that make this process much simpler. One of the first bits of advice I can give you pertains to an ArcView shapefile and the format in which it is stored on your hard drive. When you create a shapefile in ArcView, it actually creates and stores three interconnected files that make up a shapefile. The files in the following list are the three files that are absolutely necessary to a shapefile and cannot be changed or modified independent of one another without corrupting the shapefile and making it totally useless:

- SHP: The shape geometry of a shapefile (polygon, line, point, and related information)
- SHX: The locations and spatial elements of a shapefile (where it is)
- DBF: The data behind the shapefile (the specific rows and columns of data)

In many cases, you will take some time in querying a set of data in ArcView and create a shapefile of the selected data so that you can use it in Excel or some other application. Many instructional materials teach you that you can use the DBF table that ArcView creates to view and manipulate data in Excel, Access, SPSS, and other products that read DBF files. The problem comes in when they forget to advise you to avoid editing the DBF table and saving it under the same shapefile name, same shapefile folder name, and as the same shapefile DBF table where you obtained it. If you write over the shapefile's DBF table with edits you have made in SPSS, Excel, or Access, you will totally corrupt the shapefile, and it will no longer be useful in ArcView as a shapefile. This can cause regeocoding of a lot of data and some unhappy moments. Therefore, one of the first lessons in this section is that if you edit data that comes from a DBF table that is part of an ArcView shapefile, *do not* save it under the same name or same directory as the shapefiles DBF source table you took it from. This will save you hundreds of hours of work.

Suppose that you have data in ArcView and want to do some calculations and create some new fields in Excel so that you do not have to try to use ArcView's limited database tools. There are a couple of situations that could occur in which it would make more sense to pull your data into Excel to create and calculate some new fields rather than try to do it in ArcView. The first situation is when you deal with dates or times. ArcView handles dates much like any other application, but manipulating dates and calculating dates and times can be a bit rough unless you have some programming experience or really like to work slowly. We can do everything we might think of or want to do in Excel, but we can't do it as easily in ArcView, and sometimes it may just be too much of a headache to try.

The first, most obvious, way to get data from ArcView into Excel is to simply pull the data into Excel from the ArcView shapefile's DBF attribute table. Remember that if we edit the shapefile's DBF table in Excel and save it, we will corrupt the shapefile. There are two options to work around this problem. The first is to give every incident in your shapefile an absolutely unique ID number or record number and then bring only the data you will need from this table into Excel, add your new fields, and then using that unique record number, join it back to your shapefile so the data is useable in ArcView. The second method is to add the X and Y coordinates of your data to the shapefile, bring the DBF file into Excel, create, edit, and add the new fields, then bring this data back into ArcView as an event theme, and display all the information that way in your map. ArcGIS 9x has a tool in the ArcToolbox that will add the X and Y coordinates to the attribute table. I will show you an example of both methods in the next few pages, and you can decide which would work the best for your project.

In example 1, we will use a shapefile called RobSeries1.shp (SHX and DBF). This is a set of data of a robbery series in the City of Tazzieville, and we have geocoded the addresses and created a shapefile of that information from our RMS system. The data will be located on this companion DVD under CIA/Exercises/Excel/RobSeries1.shp (SHX, DBF). We need to pull the DBF table into Excel from this shapefile to calculate some common tactical analysis fields we will need. This particular table does not have X and Y coordinates in it. However, the

DR_NO field is a unique identifier for each incident in this series. To start using this data in Excel, open Excel and follow these steps:

1. Open Excel and click on **File**.
2. In the drop-down menu, click on **Open**.
3. In the dialog box, navigate to the drive that contains the CIA course data, and find the subdirectory CIA/exercises/excel/.
4. Change the Files of Type box at the bottom of this dialog box to Dbase Files (*.DBF) to display the RobSeries1.dbf file. (This works for all versions of Excel.)
5. When you have found this file, click on it once, and then click **Open**.
6. The data will now be in your Excel spreadsheet as shown in DVD Figure App 1–21.
7. Click on the small box in the upper left-hand side where the rows and columns come together to select all the rows and columns.
8. Now click on **Format → Columns → Autofit Selection** to resize all of the columns to fit the data. (In Excel 2007 this is under the Home→Ribbon→Format.)
9. For the data we need to calculate, we only need the RPTNUM, DATE_, and TIME fields, so delete all of the other columns in this spreadsheet by selecting the entire column and then pressing the Delete key.
10. The table you now have should look like DVD Figure App 1–22.
11. Create five new fields as shown in DVD Figure App 1–23, and enter the formulas as shown to calculate these items.

If you entered all of these formulas correctly, your Excel table should have the values as shown in DVD Figure App 1–24 in each of the calculated cells.

You should notice that the DateTime field is displaying the serial number for the date and time. This is the default way that Excel stores dates and times. If you want to see this as a date and time, use the Format → Cells (or CTRL+1) dialog and format this to be "mm/dd/yyyy HH:mm." Excel does not need the data to look like this, and it still calculates the days between hits, cumulative days between hits, hour, and day of week correctly. Format the DateTime field how you would like to see it in ArcView. You will see that ArcView seldom cares how you saved it and will typically make it into a text field when you bring it back in (depending on the version of ArcView you are using.)

Now make sure you Save As this data as an Excel spreadsheet in a folder of your choice. Make sure you *do not* save it as a DBF file under the same name as the shapefile, or that shapefile will no longer be useful to you. Select **Save As** again, and save this Excel spreadsheet as a DBF table (in Excel 2007, save as a Comma Separated Values [CSV] file instead) with a new name (RSLU1.dbf [or csv]). This will enable us to use the table in ArcView and join it to our shapefile. This is the hard part: *Remember where you saved these files so you can find them again!*

After you have saved this spreadsheet as an Excel workbook and a RSLU1.dbf table, you can close Excel. You must close Excel or you cannot add the data to ArcView. Open ArcView and add the RobSeries1 shapefile. Add the RSLU1.dbf (or csv) table to your project. Occasionally Excel plays a trick on you, and during the DBF/CSV export it only exports the first field, or it may export other fields but leaves them blank or truncates the data in some fields. You need to open the table in ArcView to make sure that everything is there before proceeding. If it is not, there are a few guidelines you should follow:

1. Open the Excel spreadsheet with your data. Select the entire range of data.
2. Click on **Insert > Name > Define**. (In Excel 2007, right click on the range and choose Name a Range from the pop up menu.)
3. Call the whole sheet you have selected (all your data) "Database".
4. Select **Add**.
5. Rename all your field names so they are a maximum of 10 characters long.
6. Rename all your fields so there are no spaces or other characters from the Shift + number keys (1–0) in the field names. Underscore can be used in the field name.
7. Save the spreadsheet again.
8. Save As RSLU1.dbf (or csv) again.
9. Replace the bogus version.
10. Bring the file into ArcView again.

Now you should have the entire table complete with the data you calculated. Now open the attribute table of the RobSeries1 data, and Join the RSLU1.dbf table to the RobSeries1 attribute table by the field named RPTNUM. The RSLU1.dbf table will disappear and will now be joined to the far right-hand side of your shapefile. You can now use it to do some temporal displays of the data, show the ones that occurred on Saturdays or between certain hours in another color, etc.

If you were really observant during all this, you will have noticed that my data had more than 12 characters in the RPTNUM field (yours will not). I have found that it is much easier in the long run to use an DSN/ODBC/OLE DB connection to bring this data from Excel into ArcView so my field names and data will not be truncated (especially in Excel 2007).

Using similar steps, with one slight variation in the beginning, makes my life a bit easier. Before I bring the data into Excel, I open it in ArcView and use an ArcView tool or extension to add the XY coordinates to the shapefile's attribute (DBF) table. Then I bring the data into Excel as shown in the previous example, don't delete any fields or data, add the fields and calculate them, make it pretty in Excel, and do everything else I need to do for my crime bulletin.

If you have to get this data back into ArcView, use the following steps:

1. Select all the rows and columns that make up your series of incidents.
2. Insert a name for that data, and save this workbook as an Excel workbook.
3. Close the Excel workbook.
4. Open ArcView.
5. Create an database connection to the Excel sheet.
6. Select the named range to add.
7. Voilà! The complete data from your Excel spreadsheet is in ArcView. (See ArcView help for creating and adding a new DSN/ODBC/OLE DB database connection for further assistance or refer to Chapter 2 exercises.)

Because the XY coordinates were part of the data, you could then add this table to your view as an event theme to get points on the map again, and all the fields you calculated in Excel will come along. It is then a very simple matter to convert the DBF table you added to a shapefile with ArcView complete with all of your new Excel fields and calculations.

In summary, my advice to you is:

- When dealing with ArcView and dates, retreat slowly and use Excel.
- If you really want to bring the data back into ArcView after you have made several calculations and added data, then name the entire data range (all cells in your worksheet), save it as an XLS sheet, and use a database connection to bring the data back into ArcView as an event theme (XY coordinates in the table).

- *Do not* edit a DBF file that is part of an ArcView shapefile, or you will corrupt the file!
- When all else fails, read the Help instructions carefully.

■ Section 5

Tactical Prediction

This section deals with doing date and time predictions in Excel using three different methods, and it mentions another paper and how-to publication I created to do the next date prediction in CrimeStat III. Some of the ideas behind different temporal calculations will be briefly discussed as well, such as days between hits and dollars per day. These are my own observations and not part of scientific research, although some testing has been done with a multitude of crime series I have analyzed over the past 15 years.

■ Decimal Time Predictions

Decimal time is simply the process of taking a date and time, converting it to decimals, and then doing basic math calculations on it. This is the method taught by Gottlieb et al. (1994).

We are really doing two things here: (1) predicting the next date and (2) predicting the most likely time for a new hit in a crime series. A third option of predicting the most likely day of the week is also possible from this same data.

For the first part, we are going to use the days between hits (DBH) calculation to figure out the average number of days between hits and the standard deviation. We can use this information to provide a range of dates the offender most likely will strike again *based on his prior behavior*. In Excel, this is a very simple calculation when you have the data in a date and time format. We have already seen that getting the data into date and time format is the most complicated part of this process. Open the Section5.xls exercise spreadsheet and view the data in the first tab, S5 Exercise. You will see our old table from the robbery series with all the data calculated and the fields we added correct and ready to go. To predict the next date of hit based on the mean and standard deviation of the DBH data, we first have to calculate the average and standard deviation of that column of data, so in the cell to the right of the word Average, type in =AVERAGE(E3:E13). You should get an average of around 26 days if you rounded up. Now type in =STDEV(E3:E13) next to the StDev, and you should get a value of 14 (rounded). Now calculate the mean minus one standard deviation, mean plus one standard deviation, mean minus two standard deviations, and mean plus two standard deviations in the SD–1, SD+1, SD–2, and SD+2 fields. If you calculated these correctly, you will find that for the row called 68% range, the prediction is that the suspect will hit between October 25 and November 22. The row called 95% range indicates that the next hit will be between October 11 and December 6. Because October 11 is prior to the last hit in the series, which occurred on October 16, we would change this in any analysis bulletin to between now and December 6.

As with any predictions based on mean and standard deviation, you have to really look at the data to see if it describes this offender's behavior adequately. The closer the standard deviation is to the mean, the more likely your prediction is going to be less useful and so wide that the offender could hit just about any time. In this case, the mean is 26 and the standard deviation is 14. I would call this date prediction very "loose," and although I do think the offender will hit in the 95% range time frame, nailing down which of those 54 or more days he will choose to hit is much harder. From looking at just the last five or six hits, I can see that he has not hit any earlier than 21 days between a previous hit. Prior hits were much more widely distributed. Perhaps as this criminal gets better at what he does, he settles down into that 21 day or so pattern that can help you do a better best guess for the investigators. By using just the last six hits, you would get a much smaller date range for a new hit, and depending on how confident you are of that (based on modus operandi information and reading the reports), you may want to use the shorter time range for your prediction on your bulletin. It is these types of decisions you are being paid for and should be willing to make to help narrow the focus of the investigation to the best chances of nabbing this criminal in the act of committing a new crime. Another factor we may want to look at is if the "tempo" of the hits are increasing or decreasing over time. In other words, is the offender seeming to be hitting in fewer days from his last or in more days as time passes?

Now we will work on predicting the time range using decimal time. First we need to convert regular time into decimal time to do average and standard deviation calculations on it. To accomplish this we will need to create two or three new fields to work with. We already have a field called Hour that tells us what the hour of day was for each hit. We need to know the minutes, converted to decimal, and then combine the two into a decimal time number field.

In Excel, look at the two new fields at the end of the Excel spreadsheet: Minutes and DecTime. Now we need to calculate these fields.

Take a look at the formula entered in cell I2 (=MINUTES(D2)). Drag that formula down to the other cells under the Minutes column. To convert minutes into decimals, we need to divide by 60, so the formula we want in the DecTime field (cell J2) should look something like =G2 +(I2/60) or a decimal equivalent time of 23.41667. When you have this entered correctly, drag this formula down to the other cells. Because we have already calculated the mean and standard deviation for the DBH column, we can select all of the cells between Average and SD+2, press CTRL+C to copy them, and then go to cell J15 and press CTRL+V to paste the formulas. We will have to change the range of cells from J3:J13 to J2:J13, but that is fairly simple to do for the Average and StDev cells. By doing this, we can see that the average for the decimal time is 21.75, and the standard deviation is only 1.06. Therefore, the 68% time range is between 20.68 and 22.81, and the 95% range is between 19.62 and 23.88. That doesn't make any sense, does it? How can you have a time like 22.81? To avoid confusing friends, loved ones, and coworkers, we have to convert these decimal times back into regular times. *For further detail on how to do this, please review Chapter 18 in Gottlieb et al. (1994).*

To convert the decimal times you calculated back to regular time, use a formula that looks like the following:

=INT(J17)&":"&INT((J17-INT(J17))*60)

Basically you take the integer part of the decimal time (20, from 20.68) and then take the decimal remainder and multiply that by 60 to convert it back to minutes, so 0.68 × 60 = 40.8. Because we don't want the 0.8 part, we then take the integer of 40.8, which is 40, and then separate the 20 and the 40 with a colon, and voilà, instant regular time! The spreadsheet automatically calculates this for you, but you should practice it several times on your own to remember it, understand it, and be able to repeat it later when you need it.

■ Simple Graph Predictions

Another way to accomplish the same thing in a slightly easier way is to simply graph your data by hour of day, and use the graph to tell you when investigators should do their stakeout activities. Keep in mind that surveillance units typically still work an 8 or 10 hour day and we just need to give them a time span that gives them the best likelihood of catching the offender in the act. We are probably not going to be able to predict the actual minute the offender strikes with most crime series. You will come to about the same conclusion; however, you will not be accurate down to the minute, of course. Using the S5 Exercise sheet again, create a new spreadsheet and enter the data from DVD Figure App 1–25 into it.

We will use the table in DVD Figure App 1–25 to create a chart of the time distribution of this crime series. In the Count field (C2), enter the following formula: =COUNTIF('S5 Exercise'!$G$2:$G$13,A2). Now drag this formula down through the other 22 cells. You should see that all of them are 0 except for 2000 to 2359 hours. To calculate the percentage, create a cell with the name TOTAL in it under the 2300–2359 cell (B26), and then calculate C26 to be =SUM(C2:C25). You should get a total of 12 incidents in cell C26. Now in cell D2 enter the following formula and copy it down to the other cells: =C2/$C$26. (Remember absolute referencing from section 1.)

Format (CTRL+1) column D as a percentage with one decimal place, and you should now see the percentage of incidents that occurred during each hour of the day.

Now select cells B1 through C25 and create a graph like the one in DVD Figure App 1–26 using the Graph tool as explained previously in this tutorial. I prefer area charts over bar charts because they seem to be simpler and not as confusing or prone to misconceptions.

By simply viewing the chart you can see that the best time to deploy undercover personnel for stakeout activities is between 2000 and 2359 hours for this crime series. This, of course, falls right into line with the decimal time predictions, and both are likely to be very accurate. In many cases I have found that if multiple methods predict the same day of week, time of day, or next date of hit, then we can have much greater confidence in the prediction.

You can use a similar approach to predict what day of the week a serial offender will hit next. Count up the number of hits by day of week instead of hour of day as we did with the hour of day calculations. For this offender, based on the chart in DVD Figure App 1–27, the prediction would be based on those days with the highest number of hits in prior incidents.

I would make a statement such as, "*The most likely day of week for a new hit in this series is between Wednesday and Saturday (75%), with Friday being the most likely single day of the week of those four days.*" When you use this simple graphing method, you need to pay attention to the sequence of hits to see if the offender's pattern has changed a lot lately. This is also true with other methods for calculating these items. In the preceding case, you could also reasonably say that the next day of the week for a new hit will be on Friday or Saturday because both of those days of the week had more hits than any other day so far; however, it is the consecutive nature of the days or hours being high that is actually the signal that a pattern may exist, so what you really want to look for in these charts is days or hours where the largest number of hits occurred next to each other in time.

■ Split Time or Midpoint and Weighted Averaging Predictions

Weighted averaging is a way to conclude what time an incident may have occurred when you don't have an exact time. For the examples discussed so far, we have had specific times for which crimes happened. This is almost always the case with robberies, sexual assaults, and other violent crimes where a person is attacked. With property crimes, like burglary or auto theft, victims may see their property when they go to bed and wake up to find it missing.

The normal advice on this topic suggests that any times over 24 hours should not be used in calculations for any series (Gottlieb et al., 1994). This is mainly due to the fact that we don't really have any idea when the crime occurred and thus should probably not use it in our scientific analysis. This is also true with split time or midpoint time calculations, and you would get a major migraine trying to find out what time you wanted to use for your best guess split time. When you do not have one time for all incidents in your series, you can use split time to calculate these date and time differences. It is still suggested that with time periods longer than 24 hours, you should throw out that incident during your calculations. When doing standard decimal time calculations and graphs, I agree with this statement. When doing weighted averaging, I think you could either throw it out or use it if you are able to justify its use or addition to your analysis. The entire concept of weighted averaging dictates that you could weight this incident correctly and still have it be a valid part of your analysis because you don't really know when it happened, and you can only use an estimate. All you really need to know is when the time range starts and ends and how long is it between them.

First, let's discuss the easier split time calculations and how to do them in Excel. Open the Section5b.xls spreadsheet and take a look at the data. In this spreadsheet you have 25 burglaries that have been committed by the same offender. For most of these incidents there is not an exact minute that the crime occurred, and you have two date fields and two time fields to deal with. We can use Excel to subtract dates and times for us and calculate the midpoint or split time, but we need to convert these all into dates first. We therefore create two new fields called DateTime1 and DateTime2 and use the functions we previously learned to calculate these two new fields correctly. Format these fields to be mm/dd/yyyy HH:mm style (press CTRL+1). Keep in mind that because your time fields are not all four characters long, you will have to change the formula slightly to accommodate each of the time field lengths to make them come out right. You can also use an If statement, but this gets a little more complicated and may not be worth the effort for 25 incidents. Your formula should look something like the following for 3 characters in a time field:

=DATEVALUE(MID(B2,5,2)&"/"&RIGHT(B2,2)&"/"&LEFT(B2,4))+TIMEVALUE(LEFT(C2,1)&":"&RIGHT(C2,2))

I often create a new field called TimeLen to make using the If statement a little easier. I use the formula = LEN(A2) (where A2 is the cell that contains my time in number format) to get the length of the time field first. Then in my date–time formula I use If statements to add 0s or not, depending on the length. This looks something like the following:

=DATEVALUE(MID(B2,5,2) &"/"& RIGHT(B2,2) &"/"& LEFT(B2,4)) + IF(LEN(C2)=4, TIMEVALUE(LEFT(C2,2) &":"& RIGHT(C2,2)), IF(LEN(C2)=1, TIMEVALUE("00:0" & RIGHT(C2,1)), IF(LEN(C2)=3, TIMEVALUE("0" & LEFT(C2,1) &":"& RIGHT(C2,2)),"00:" & RIGHT(C2,2))))

Note that C2 contains the time field, B2 contains the date information in text format, etc.

You have to embed an If statement inside another If statement, making sure to account for each of the possibilities in your data. Excel limits how many times you can embed an If statement inside another one, typically seven times. Now that we have converted the date and time fields to real date–time data, we can calculate the split time or midpoint between these two dates and times. This formula is quite simple: =DATETIME1 + ((DATETIME2–DATETIME1)/2). Create a new field called Midpoint, and enter this formula, replacing the Datetime1 and Datetime2 above with with cell references for your Date-Time1 (F2) and DateTime2 (G2) columns. You now have a date and time that occurred directly between or halfway between DateTime1 and DateTime2. As you can see, this doesn't tell you when the incident really occurred, but it gives you an estimate. You can then use this estimate like a regular date (convert it to decimal time, etc.) and work with it to get your calculation and prediction. Some analysts choose to disregard all times that have a spread greater than a few hours in their final analysis. This is entirely up to you, and it is one of those judgments you make and live by as your experience and knowledge grows in this profession. One thing you may want to watch out for is if there are any single times or time ranges that are very narrow in your crime series. For instance, let's say you have five burglaries and three have a time span of 32 hours, but the other two show exactly when the alarm activated. You would probably use the time the alarms activated instead of trying to calculate a weighted averaging estimate as it just makes sense. I try to use all the data I can for most analysis products, even if it means using data that is questionable at times, as long as I'm sure it does not adversely affect the analysis. Most of the data we collect as a police department has many chances for human error when it is collected, so I try to view my data as a challenge and make use of as much of it as I can that makes good sense.

You can visually inspect your data and figure out which incidents will have more than a 24 hour date and time difference. In workbook Section5b.xls, if you have calculated all of the dates and time correctly, you will see that a few of time ranges are more than 24 hours apart. You can verify this by looking in the column labeled Dec Hrs Between, which simply subtracts DateTime1 from DateTime2. The equivalent decimal number is the difference between the first and second date serial numbers. The integer value (left of the decimal point) represents the total number of days between the two dates. The decimal portion represents the total number of hours, minutes, and seconds. Basically, I have a bunch of information here that I could use, but the key item is that if the integer is 1 or more, then there is a 24 hour or more gap between the two dates and times. With this field you could decide which incidents to ignore when doing your calculations for mean, standard deviation, etc. when you convert the midpoint time to decimal time.

Another way of handling these incidents is to create a weighted averaging table of your incidents and calculate the total for each 24 hour period based on your data. Although it seems a little more complicated the first few times you use this approach, it actually makes quite a bit of sense when you really sit down and look at it. It also is where I got the idea to use the simplified graphing method of determining the next date, most likely time period, and most likely day of the week. I also like to avoid math at times, and being able to look at a chart, see where the offender has been hitting the most frequently in the past, and then advise the investigators of this information seems simpler at times. Dr. Jerry Ratcliffe has a slightly different approach to weighted averaging he has termed Aoristic analysis. You can review papers he has taught on the subject at his Web site. His Web site also has a top ten tips and tricks and other publications that could be of interest to crime analysts.

Weighted averaging uses the idea that if a crime could have occurred, say, between 0800 and 1200 hours, the total possible time span is four hours. Each hour from 0800 to 1200 will get a weighted score for the total possible hours, or one-fourth. As you add up the possible scores per hour for each of the incidents in your series, you get an overall weight assigned to that hour of day of the chance that the crimes occurred during that hour, and thus, a way to predict the next chance for a new crime to occur.

Now take a look at the Weighted Averaging spreadsheet for the burglary series we have been working with in Section5b.xls (it will be blank if you have not done the first part of this exercise). You will see that for each incident there is a column of data based on the 24 hour periods of a day. I have calculated these based on the dates we calculated in the first steps of this exercise, so if there are any errors we need to go back and fix them. Each incident is evaluated on how many hours are between the start and end times, and you also have to know the start times and end times. If your time period was between 12/31/1998 23:30 hours and 01/01/1999 13:15 hours, we know that there are 15 hours inclusive when the hit could have happened. We score 2300–2359, 0000–0059, all the way to 1300–1359, with a score of 1/15th or 0.067. When we tally up all of the incidents for each of the 24 hour periods, we can see how much weight that hour and incident had on the overall crime series.

In addition, when a crime can be identified as happening at a specific time or within an hour, that hour gets the entire weight of 1. The entire weighting for that incident cannot equal more than 1. I have heard of analysts who complicate the issue and score incidents they think are weakly associated with the series with a value of less than 1, depending on how strongly they think that crime is part of the series. This gets way too complicated for me and my brain, but it is a thought on how we can rank and weight different crimes in our series depending on how well they reflect the normal behavior of the offender.

When we compare the hour of day chart from the weighted result to the midpoint result, which counted all of the crimes, we can't see much of a difference for those peak hours we are looking for in the first place (see DVD Figure App 1–28). Note that the weighted result tossed out 8 of the 24 incidents. I left them all in for the split time calculation, but in theory we should exclude the same eight incidents from both the weighted and midpoint results.

The peaks are about in the same place, although the 2200–2359 peak is much higher with the midpoint analysis results and may be a bit deceiving. You should expect to see a more level graph with weighted averaging because that is really what you are doing with the data: leveling it off and making each hour the crime could have occurred in less valuable in the overall result, or less weighted. I find that for most analysis efforts, the midpoint is fast, very practical, and works as well as the much longer and more complicated weighted averaging endeavor, even if I leave in the incidents where the time spans more than 24 hours. Because the graphing helps me understand the hourly distribution of the crime series very well, I also prefer doing the simple graph method for predicting the next time range and day of week range. The midpoint analysis and weighted averaging can both be done rather quickly when you know how to convert and use dates in Excel.

If you wanted to use all of the incidents in the weighted method, you would simply figure out how many possible hours are between the start and end times. Every hour would then get 1/nth of the weight. For example, if there were 48 hours between two times, a crime may have occurred. The start time was midnight, and the end time was also midnight 2 days later. Each of the 24 hour periods would be weighted at 2/48 (the same as 1/24) because they have two chances of being the correct time for the actual offense. If the time between the start and end was 26 hours, then every hour would get 1/26, but 2 hours would get 2/26 of weight. This gets confusing, and although you can set this calculation up in Excel, I have never been very successful at creating a spreadsheet that would do this automatically for any crime series. However, I have done this in Access, and it works quite well for most analysis efforts and does not change the outcome very much.

The methods shown are not the only ways dates and times can be manipulated in Excel to come to these answers, of course. You may develop your own method and become an expert at manipulating Excel into making your life easier. The idea is that we want to create an analysis that is operationally useful to the investigators, is not confusing to them, and has the best chance of nabbing the offender the next time he tries to commit a crime in this series. When you develop that easier or faster method, share it with other analysts who also want to make their jobs and analysis products better.

■ CrimeStat's Correlated Walk Prediction

Because this tutorial is getting fairly lengthy, I will not go into great detail about the Correlated Walk Analysis method in CrimeStat III®.

Please review the CrimeStat III manual for instructions on how to create and use the correlated walk analysis routine. I have found that the next hit date prediction it calculates is very reliable for most crime series and could be useful to you once you learn how to use it.

Take a look at the Correlated Walk Analysis (CWA) Exercise sheet in the Section5b.xls workbook. This sheet does the date calculations I use when doing a tactical prediction. If I use the days between midpoints as my base for calculations, you will see that for this series, the average days between hits is 2.5, and the standard deviation is 2.8. This should tell you that the data are widely variable, and there is no real discernable pattern in hits by date. The last hit was 10 days from the previous one, and this could mean that the offender had been jailed on another offense or you don't have all of the crimes in this series identified. Maybe the offender was sick for a few days or on vacation! Whatever the case, the prediction based on the average and standard deviation has the next hit happening prior to the last one and up to March 7. That is about as good as your next date prediction may get.

Using the three date predictions from CrimeStat III CWA routine, we see that it is predicting a new hit on March 2, 4, or 6. So far in my analysis, the median prediction has been the most accurate, therefore March 4 seems like a good date. I typically then just subtract 2–3 days and add 2–3 days to this prediction to get a range. I find that when the standard method of days between hits predicts a similar time period as the Correlated Walk Analysis routine, then it is a pretty sure bet the suspect will hit during the 7-day span that was calculated with the CWA routine (see DVD Figures App 1–29 and 1–30).

It is a good idea to try several different types of temporal analysis with your tactical crime series. I find it very validating to know that the CWA routine, my simple graphing method, and the calculated decimal time all come up with pretty much the same prediction for a lot of crime series. When they do not come out about the same, I begin to worry that I have done some calculation incorrectly, so I make a habit of checking my work for errors. If I find none, then the suspect has no discernable temporal pattern and may be very hard to predict. I may also be missing cases that happened in another jurisdiction nearby or simply missed them in my own review of potential cases from my RMS.

■ Why Do So Many Calculations?

The comment I made before about several methods all arriving at the same conclusion is the primary reason why you may want to do all of these things on every tactical prediction. For me, the fact that the decimal time calculation, days between hits calculation, Correlated Walk Analysis, and dollars per day calculation all say the same date, time, and day of week are most likely for a new hit in a crime series means that my prediction is very likely to be true, and we have the absolute best chance to apprehend the offender before he can do any more harm. When they differ a great deal, it can mean that your prediction is based on data that is just not reliable enough to make a good prediction, and you can tone down your confidence in the prediction. The other possibility is that you have made a data entry or calculation error someplace and should fix it prior to sending out that tactical bulletin. Picking the days that overlap from the two to four methods you may have tried can narrow a prediction that has a very wide time span for the prediction. As an analyst, the first thing you have to do is give the most likely estimate for a new hit in the series. The second thing you should do is state how confident you are in that prediction. You should also not stop doing analysis if you happen to be wrong now and again.

One of the methods I learned many years ago from Dale Harris, of Corona Solutions, was based on a robbery analysis. He advised that his agency in Colorado had done well with using an average dollars per day calculation for predicting when a robbery offender would hit next. This was based on the thought that if an offender was a drug user and had a $300-a-day habit, he would need to commit enough robberies to keep him in drugs. If you have a robber who is shown to have a $300 habit based on the dollars per day you've calculated, then if he made $1200 on his last robbery, you have about 4 days before he hits again. This is a very simple thought process and piece of logic, but it often works very well. If you think about your own habits, like drinking milk, for example, you may see this thought process a bit clearer. My family of four used to drink 2 gallons of milk per week when my sons were little, so every week I went shopping for milk at the local grocery store to replenish my supply. One weekend I bought 4 gallons instead of my usual 2 gallons (it was on sale) so I didn't go shopping or need milk again for 2 weeks. If I were a robber, and money to feed my drug habit were my motivation, do you think this pattern would be predictable? Of course, I would have to measure the number of gallons of milk the robber received for his crime, but you get the idea.

The formulas for doing all of these calculations are included in the exercise spreadsheet and are just very simple formulas. Review them as needed to understand how this all goes together. An example for the CWA worksheet is shown in DVD Figure App 1–31.

Section 6

Pivot Tables and Administrative, or Strategic, Analysis

A pivot table is a function of Excel that allows you to take rows and columns of data and create summary tables for the fields in your dataset. You can make pivot tables from fields that contain day of week (row) and time of day data (column) and a value field (count of RptNum) to help in describing and analyzing data more quickly.

■ Pivot Table Web Resources

The Web pages in the following list have great tutorials on creating pivot tables. You can refer to them to learn how pivot tables can be used and leveraged for crime analysis. Pivot tables and pivot charts are a crime analyst's friend, and the Excel Help file is also useful in learning how to use these powerful summary tools. You can find many other tutorials on pivot tables by searching on the internet for "Excel pivot tables."

> http://www.exceltip.com/excel_tips/Excel_Pivot_Tables/32.html
> http://www.mathtools.net/Excel/Pivot_tables/
> http://www.cpearson.com/excel/pivots.htm

■ Pivot Table Example

The following pivot table example uses some data you have already seen. Open Section6.xls and review the calls for service data. Follow these steps to create a pivot table of this data to determine what radio code type was the highest for this area. *(Note: In Excel 2007 the functions are similar, but these instructions were written for Excel 2003. Please review the pivot table help in Excel 2007 for directions on how to do it in that version.)*

1. Click on **Data** in the menu bar and select **Pivot Table and Pivot Chart Report** from the drop-down box (see DVD Figure App 1–32).
2. In the next dialog box, select from a Microsoft Excel list or database and then click **Next**.
3. In the Where Is the Data You Want to Use box, click the small box to the right of the field with the red in it (see DVD Figure App 1–33). This will fold up the dialog box so you can look around it. Now highlight all of the rows and cells in your spreadsheet (A1:K1681), click the box again, and click **Next** (see DVD Figures App 1–33 and 1–34).

4. In the next dialog box, check the box next to **New Worksheet**, and then click on the **Layout** button
5. A new dialog box will pop up where you can design your pivot table and add all of the fields you would like to be part of the pivot table (see DVD Figure App 1–35).
6. Drag the RptNum field from the right side and drop it on the Data part of the pivot table on the left. Drag the Radio_Code field from the right and drop it on the Row part of the pivot table.
7. Click the **OK** button (see DVD Figure App 1–36).
8. Click on the **Options** button on the pivot table dialog, and enter *0* in the For Empty Cells, Show: field and click **OK** (see DVD Figure App 1–37).
9. Click **Finish** and a new Excel spreadsheet should appear that looks like the spreadsheet shown in DVD Figure App 1–38.
10. This is your pivot table, and it summarizes the number of calls for service by radio code type.

Click in the **Total** column, and then you can use the sort buttons to sort this data in descending or ascending order as needed. You can also drag other fields to the pivot table, close the pivot table dialog box, move the row data to columns, and any number of other things. I modified this pivot table and made the DOW (day of week) field be the column heading and the TOD (time of day) field be the row heading The resulting spreadsheet looks like DVD Figure App 1–39.

Right clicking on the pivot table allows you to refresh the data from your table, go back to the layout wizard, and a variety of other things (see DVD Figure App 1–40). Clicking on the wizard will allow you to go back to the dialog boxes you saw when you created the pivot table, and they may be more familiar and easier to use than simply moving fields around on the spreadsheet. If you change anything in your source data sheet, click Refresh to update the data in the pivot table.

If, for example, you wanted a list of all of the details of the incidents that occurred at 2100–2159 hours on Sunday (the spreadsheet shows 19 records), you can double click on this cell in the pivot table and a whole new worksheet will be created that has a drill-down version of those 19 records and all of the details that are in the original source data, as shown in DVD Figure App 1–41.

This is a super feature for creating subsets of data you may want to use for other analyses, like date and time, trend over time, top addresses, and others based on the analytical product you want to create.

This has been a simple introduction to pivot tables for crime analysis. Attempt to create your own pivot table with the data in

Section6.xls and change it, refresh it, and make several different pivot tables to summarize the data in this table.

■ Section 7

Now open the spreadsheet called Practice.xls, which is located in the same folder as the other spreadsheets for this tutorial. The file contains a set of 12 different tactical crime series from the City of Tazzieville. I have also included four administrative/strategic forecasting projects in this spreadsheet.

The projects will not be Geocodable, so you cannot create a map; however, you can create forecasts, charts, and brief summaries of the data as if you have been assigned them during your work as a crime analyst.

You should create all of the fields you will need to do a date and time analysis and calculate split time where needed. Also complete the next hit date, most likely weekday, and most likely time range for a new hit in the series provided. You can work on these crime series individually or in groups as would be best assist you in learning how to compute the statistics and information you will need to know in the real world of crime analysis. Try to forecast the total number of Part 1 UCR crimes for 2004 in worksheet Admin1. Calculate what the total UCR crimes will be in each of the 5 grids in Admin2. Calculate the correlations between variables in the strategic worksheets.

■ References

Gottlieb, S., Arenberg, S., & Singh, R. (1994). *Crime analysis: From first report to final arrest.* Montclair, CA: Alpha.

Ratcliffe, J. (n.d.). *Research and spatial questions (past and current).* Retrieved February 2009, from http://jratcliffe.net/research/index.htm

Index

Abandoned buildings, 59, 194
Academy of Criminal Justice Sciences (ACJS), 472
Access control, 103, 106
Activity space, 111–112, 150–151, 156
Activity support, 103
Adams, T. M., 194
Adding tables (exercise), 82–94
Addresses. *See also* Repeat address mapping
  locators and find addresses (exercise), 77–82
  tables and maps (exercise), 67–77
Administrative crime analysis, 10–11
Administrative maps, 422–423
Age and crime trends, 114, 203, 256–257, 259, 266
Aggregating data for shared grid maps (exercise), 320–325
Alcohol, 204, 263, 267–268
Ambusher, attack method, 157
American Society of Criminology (ASC), 472
Analysis data. *See* Data for crime analysis
Angel, Schlomo, 99
Anselin, Luc, 16
Antecedent variable, 304
Applications. *See* Research and applications
ArcCatalog (exercise), 35–43
ArcGIS, 16, 21. *See also* Geographic Information Systems
ArcMap, 20, 21
  exercise, 23–34
ArcNews (ESRI), 472

ArcView, 16
Arson, 154, 194, 260
Assault
  hot spots of, 194
  journey-to-crime trip and, 154–155
  lifestyle exposure approaches and, 113
  motivations for, 261
  in public housing projects, 188–189
  race perception of offenders, 260
  reporting of, 265
  Routine Activities Theory and, 109–110
  on university campuses, 191
  urban vs. rural, 271
  victimization risk of, 258, 259
  victim precipitation or provocation and, 114, 266
  weapon use in, 264
Assumptions and errors in data analysis, 227, 289, 299, 301–306, 309
Attack methods, 157
Attribute table, 242
Audience considerations, 15, 421–447
  community, 440, 442, 443
  corrections, 444–446
  courts, 442, 444
  GIS mapping and, 422–428
  investigators, 429, 436–438
  patrol officers, 428–435
  police managers, 436, 439–440
  types of audiences, 226
Autodialer System, Baltimore County, 200

Auto theft
  buffering exercise, 141–143
  Compstat and, 203
  displacement and, 116–117
  journey-to-crime for, 154
  lifestyle exposure and, 113
  motivations for, 260
  reporting of, 265
  routine patrol prevention of, 229
Average. *See* Mean
Awareness space
  in Crime Pattern Theory, 111
  crime series and, 156
  mental maps and, 151–152

Baltimore County's Autodialer System, 200
Bandwidth, 375, 378
Barnes, E., 194
Barr, R., 117
Beardsley, K. C., 202
Beccaria, Cesare, 54
Behavioral learning theory, 102
Bell curve. *See* Normal curve
Benign displacement, 116
Bentham, Jeremy, 54
Bike paths, buffering exercise, 144–145
Bimodal distribution, 292
BJS. *See* Bureau of Justice Statistics
Black, Jane, 106
Block, C. R., 54, 192–193
Block, R., 192–193
BLS, 7
Boba, R., 9, 12, 371
Boston Gun Project, 202–203
Bounded rationality, 108
Boyz in the Hood (movie), 61–62
Braga, A. A., 285
Brantingham, P. J., 6, 155
Brantingham, P. L., 6, 155
Break-ins. *See* Burglary
Briefing maps, 429, 430–433
Britt, D., 194
Broken Windows Theory, 59–64
Bruce, C., 9, 10
Buerger, M. E., 371
Buffer analysis and queries, 341–342, 343

Buffers (exercise), 136–145
Buffer zone, 155
Bureau of Justice Statistics (BJS), 237, 254, 265
Bureau of Labor Statistics, 7
Burgess, Ernest, 57
Burglary
  awareness space and, 152
  of businesses, 190
  classification in mapping of, 310–313
  community policing and, 199–200
  in crime funnel model, 238
  defensible space and, 102
  geographic profiling case example, 160–163
  journey-to-crime trip for, 154–155
  near schools, 191
  in public housing, 189
  Routine Activities Theory and, 110
  routine patrol prevention of, 229
  socioeconomic status and, 258, 260–261
  surveillance based on cluster analysis of, 190
  tactical exercise, 480–481
  urban vs. rural, 272
Businesses and calls for service, 190–191
Business intelligence, 422

CAD systems, 235
Calls for service (CFS), 190–191, 200, 202, 235
Canter, David, 158–159
Canter, P., 116, 200
Cao, L., 108
Car accidents. *See* Traffic
Carjacking, 116–117
Cartographic School of Criminology, 56
Cartography, 13, 423
Casady, T., 198, 199
Casinos, 98
Catalano, P., 109, 303
Categorical data, 286–287

Causal relationship, 303, 315–316
CCTV (Closed-circuit television), 106–107, 116, 474–475
Cell phones, 197–198
Census. *See* US Census Bureau
Central tendency. *See* Measures of central tendency
CFS. *See* Calls for service
Chainey, S., 116
Chamard, S., 12
Chicago School of Criminology, 56–58, 63
Child abuse, alcohol and, 204
Child protection services (CPS), 204
Choropleth mapping, 288, 310, 375, 439, 440
  exercise, 247–252, 325–333
City council maps, 466–467
Clarke, R., 7, 107, 110, 117
Classical School of Criminology, 54–55, 107
Classification in mapping, 309–313
Clean data, 227–229, 236
Closed-circuit television (CCTV), 106–107, 116, 474–475
Cluster analysis, 379–383, 427
Coefficients in regression, 306, 308
Cognitive images, 150–152
Collective efficacy, 58
Community building, 106, 107
Community Oriented Policing Services (COPS), 17, 473
Community policing, 59, 61, 199–200
  audience considerations and, 440, 442, 443
"Commuter." *See Poacher*, target search method
Compstat, 203
Computerized geographic profiling, 158–160
Concentric zone theory, 57
Confidence interval, 382
Contextual data, 238–241
  local social agencies, 240–241
  US Census Bureau, 239–240, 254
Contingent variable, 304
Continuous variables, 290–291, 301, 302, 306

Convex hull representation, 383, 386
Coordinate systems, 14
COPS (Community Oriented Policing Services), 17, 473
Corrections personnel as audience, considerations for, 444–446
Correlations, 307. *See also* Regression analysis
Count (exercise), 347–351
Courts as audience, considerations for, 442, 444
CPS (Child protection services), 204
CPTED. *See* Crime prevention through environmental design approach
Crime analysis, defined, 9–12
Crime displacement, 115–117, 160
Crime forecasting, 473
Crime funnel model, 237–238
Crime fuse, 62
Crime mapping
  basics of, 13–16
  discontinuance of, 471–472
  history of, 4, 12–13
  increased use of, 471
  map data, 15–16
  map information, 13–14
  for patrol commander, 467–468
  types of maps used by crime analysts, 14–15
Crime Pattern Theory, 6, 111–113, 156
Crime prevention through environmental design (CPTED) approach, 6, 100–107
  Newman's defensible space, 100–102
  theory and research, 102–107
Crime Prevention Through Environmental Design (Jeffery), 102
Crimes as criminal events, 4, 7–9, 341
Crime series, 156, 161, 339, 342, 344–346
CrimeStat, 16, 159, 337

Crime trends, 254–268. *See also specific crimes*
　demographics and, 439
　macro-level forces driving, 62
　motivations for offending, 260–262
　offenders, 259–260
　offending patterns, 263–265
　online resources, 272
　places/locations, 268–272
　Rational Choice Theory and, 108
　victim-offender relationships, 262–263
　victims, 254–259
Criminal Geographic Targeting algorithm, 158
Criminal intelligence analysis, 11
Criminal investigative analysis, 11
Criminology
　defined, 4
　history of, 54–56
Critical intensity zone, 193
Crowe, T. D., 103
Cultural issues, 58, 62, 104. *See also* Race

Dangermond, Jack, 11–12
Data cleaning, 227–229, 236
Data DVD (exercise), 22–23
Data for crime analysis, 225–252. *See also* Crime trends
　contextual, 238–241
　　local social agencies, 240–241
　　US Census Bureau, 239–240, 254
　crime funnel model and, 237–238
　data sharing issues, 243–244
　official crime, 231–236
　self-report, 236–237
　tools, 241–243
　　Microsoft, 241–242
　　Statistical Package for the Social Sciences (SPSS), 242
　types used in statistics, 284–292
　usage issues, 226–230
Data frames (exercise), 447–452
Data sharing issues, 243–244
Davies, G., 189–190

Day, G., 198
The Death and Life of Great American Cities (Jacobs), 99
Defensible space model, 100–102, 188
Demographics
　in contextual data, 238–240
　crime rates and, 256, 439
　crime theory and, 54
　drugs and, 195
　in geographic profiling, 160
　journey-to-crime analysis and, 154
　robbery and, 189
　school-related crime and, 192
Dempster, J. M., 189
Density analysis, 375, 427, 441
　exercise, 347–351, 406–411
Density calculation, 377
Dependent variables, 303, 306–309
Descriptive statistics, 292–297, 306
Determinism, 55
Dichotomous variables, 286
Differential Association Theory, 58
Diffusion of benefits, 117–118
DiIulio, J., 314
Discrete variables, 290–291, 301, 302, 306
Disorganization theory. *See* Social disorganization
Dispersion, measures of, 295–297
Displacement, 115–117, 160
Distance analysis, 335–369
　buffer analysis and queries, 341–342, 343
　distance between hits, 342, 344–346
　journey-to-crime, 339–341
　mean center, 337–339
　mean center (exercise), 351–359
　methods of calculating distance, 335–337
　online resources, 345
　spider, 341
　time and, 344–345
Distance between hits analysis, 342, 344–346
Distance between hits analysis (exercise), 351–359

Distance decay, 155, 426
  exercise, 359–368
Distance to crime. *See* Journey-to-crime trip
DNA swabbing and analysis, 162, 163, 198
Domestic homicide, 153, 341
Dragnet geographic profiling system, 158–159
Drugs
  abandoned buildings and, 59
  distance to crimes involving, 154
  offender motivations and, 261
  perpetrator displacement and, 117
  Rational Choice Theory and, 108–109
  research on drug-related crimes, 195, 200, 202
  transit stops and, 193
  trends in alcohol and drug abuse, 260, 263, 267–268
  victimization risk and, 115
DUI offenses, buffering exercise, 137–141
Dummy variable, 286
Durose, M. R., 254
Dyreson, D. A., 6

Eck, J., 7, 54, 64, 371–372
Ecological fallacy, 63
Economic factors. *See* Socioeconomic status
Education. *See also* Schools
  in field of crime mapping, 471–472
Elias, G., 313–314
Ellipses, 297, 383
Environmental criminology, 6, 97–148
  background on environmental design, 98–99
  CPTED approach, 6, 100–107
  Crime Pattern Theory, 6, 111–113
  diffusion of benefits, 117–118
  displacement, 115–117
  in GPA training program, 160
  lifestyle exposure approaches, 113–115
  online resources, 118
  Rational Choice Theory, 6, 107–109
  Routine Activities Theory, 6–7, 109–111
Environmental Criminology (Brantingham & Brantingham), 6
Equal area classification, 311
Equal interval classification, 309–310
Errors in data analysis, 227, 289, 299, 301–306, 309
ESRI, 11, 472
Ethics and statistics, 284
Ethnic minorities. *See* Race
Euclidean distance, 335–337
Event dependency, 115
Event planning, 199
Exhaustive variable categories, 287
Expected crime pattern, 152

Fagan, J., 189–190
Federal Bureau of Investigation (FBI), 231–233, 255, 260, 264
Feidler, A., 195
Felson, M., 107, 110
Filer, Scott, 160
Freisthier, B., 204
Frequency distribution, 289–290, 312. *See also* Histograms
Functional displacement, 117
Future issues, 469–482
  education in field of crime mapping, 471–472
  geographic profiling and crime forecasting, 473
  GIS, use of, 469–471
  related fields and, 470
  research, 472–473
  technological change, 470–471, 473–474
Fuzzy mode analysis, 380–381

Gangs
  abandoned houses and, 59
  crime trends and, 264
  problem-oriented policing and, 201–203

qualitative and quantitative analysis of, 285
in schools, 270–271
victimization risk of, 115
Gartin, P. R., 371
Gau, J., 64
Geary's C, 305, 382–383
Gender, 265, 266. *See also* Women
Gentrification, 57
Geocoding, 14, 235, 241, 474
GeoDa, 16
Geographic Information Systems (GIS)
discussed, 11–16
future trends, 422–428
research and applications. *See* Research and applications
Geographic profiling, 11, 155–163. *See also Crime Pattern Theory; Rational Choice Theory; Routine Activities Theory*
case example, 160–163
computerized, 158–160
future trends, 473
hunting methods, 156–157
investigative strategies, 157–158
training in, 160
Geographic Profiling Analysis (GPA) training program, 158, 160
Geography and Public Safety (NIJ), 473
Geoprocessing, 423, 427–428
Geoprofile, 158
Gersh, J. S., 202
Gill, M., 106
Glendale, Arizona, 21
Global Positioning System (GPS), 197, 204, 473–474
Goldstein, Herman, 201
Google Earth, 472
Google Street View, 474
Government agencies, 240–241
GPA training program, 158, 160
GPS (Global Positioning System), 197, 204, 473–474
Graduated color mapping, 375

Graduated points (exercise), 387–389
Graphics (exercise), 43–49
Grid cell mapping, 375–379
Grocery stores, 98, 196–197
Groff, E., 301–302, 336–337, 340–341
Group offending patterns, 263–264
Groves, W., 58
Grube, T., 203
Guardians, in problem analysis triangle, 7–8
Guerry, A. M., 12, 56
Gun violence, 200, 201–203, 285

Habitual geography. *See* Activity space
Handlers, in problem analysis triangle, 7–8
Harries, K., 12, 309–310, 473
Hartford, Connecticut, drug calls, 200
Hate crimes, 262
Hedonism, 107
Hendersen, K., 191
Heterogeneous populations, 56, 102. *See also* Cultural issues
Hicks, D., 242, 474
Hijacking of persons, 198
Hill, Bryan, 20, 198, 201, 303
Histograms, 301–302
History of crime mapping, 4, 12–13
History of criminology, 54–56
Holzman, H. R., 189
Homicide
in crime funnel model, 238
distance to crime for, 153, 155, 340–341
gang related, 264
geographic profiling and, 158
GIS applications for, 196–197, 198, 203
hot spots of, 194
skewed distribution of data, 301–302
trends in, 255, 260, 262, 267
"Hotspot matrix," 196

Hot spots, 371–387
  analysis techniques exercises, 390–406
  cluster analysis, 379–383
  of crime in public housing projects, 188
  defined, 371–372
  fuzzy mode analysis, 380–381
  graduated color map or choropleth map analysis, 375
  grid cell mapping analysis, 375–379
  manual or eyeball analysis, 373–375
  online sources, 384–385
  point pattern analysis, 379–383
  repeat address mapping (RAM) of, 375, 439
  research on, 193–195, 202, 427, 439
  routine patrol of, 230
  standard deviation analysis, 383–385
  techniques and classification (exercise), 325–333
  transit station robberies, 192–193
  types of analyses, 373–383
Housing projects. *See* Public housing projects
How to Lie with Statistics (Huff), 313–314
Hubbs, R., 200
Huff, D., 313–314
Human space, 103
Hunter, target search method, 157, 341
Hunting methods, 156–157, 158, 160
Hyatt, R. A., 189
Hypotheses, 8–9

IACA (International Association of Crime Analysts), 9
IACP (International Association of Chiefs of Police), 163
IBRS (Incident-Based Reporting System), 10, 232–234
ICPSR (Inter-University Consortium for Political and Social Research), 233, 241
Image, 100–101
Incident-Based Reporting System (IBRS), 10, 232–234
Incident data, 231–236
Indecent exposure, 196, 197
Independent variables, 303, 306–309
Individual decision making, 149–183
  geographic profiling, 11, 155–160
    case example, 160–163
  journey-to-crime trip, 153–155, 166–170
  mental maps and awareness space, 150–152
  online resources, 164
  target backcloth, 152–153
Inferential statistics, 297–309
Informal social control, 8, 59, 99, 189–190
Institutional review boards (IRBs), 243
Intelligence-led policing (exercise), 477–479
International Association of Chiefs of Police (IACP)/ChoicePoint Award for Excellence in Criminal Investigations, 163
International Association of Crime Analysts (IACA), 9, 472
Interquartile range, 295–296
Inter-University Consortium for Political and Social Research (ICPSR), 233, 241
Interval-level data, 288–290, 296, 301, 306
Intervening variable, 304
Intimate partner violence, 263, 264
Introduction to the Principles of Morals and Legislation (Bentham), 54
Inverse relationship, 305

Investigative strategies. *See also specific crimes*
 applications for, 196–198
 geographic profiling, 157–158, 160–163
Investigators as audience, considerations for, 429, 436–438
IRBs (Institutional review boards), 243
Irvine Police Department, 160–163

Jacobs, Jane, 99
Jeffery, C. Ray, 100, 102–103
Jeopardy surface, 158
Joins and relates (exercise), 247–252
Journey-to-crime trip
 computerized geographic profiling and, 158, 159
 described, 153–155
 distance analysis, 339–341
 exercise, 166–181
 paths and, 111
Juveniles. *See* Schools; Youth crime

Kaplan, H., 103
Karmen, A., 113–114
Kelling, G., 59, 61
Kennedy, D. M., 285
Kernel density calculations, 378
Kidnapping, 198
Knoxville Police Department (Tennessee), 200
Kushmuk, J., 103

Land use, 6, 156, 160, 194
Langan, P. A., 254
Larceny, 155, 189, 260
Leipnik, M. R., 196–197
Levels of measurement, 286–290
Lifestyle exposure approaches, 113–115
Likert scale, 287
Lincoln Police Department (Nebraska), 199
Linear relationship, 306
Literature review. *See* Research and applications

Local social agencies, 240–241
Location, crime trends and, 268–272
Lockwood, D., 194
Long, B., 303
Lowell, R., 191

Macro-level theories, 5, 7, 62
Malign displacement, 116, 117
Managers
 as audience, considerations for, 436, 439–440
 in problem analysis triangle, 7–8
Manhattan distance, 335–337
Manual or eyeball analysis, 373–375
Map, defined, 13
Map data, 15–16
MapInfo, 16
Map information, 13–14
Mapping and Analysis for Public Safety (MAPS), 17, 473
Mapping Crime: Principle and Practice (Harries), 309–310
MAPS (Mapping and Analysis for Public Safety), 17, 473
Martin, D., 194
"Maurader." *See* Hunter, target search method
McCullough, M. J., 194
McEwen, J. T., 301–302, 336–337, 340–341
McEwen, T., 12
McGuire, P. G., 203
McKay, H., 5–6, 57–58
Mean, 293–295, 296–297, 298–302, 314. *See also* Measures of central tendency
Mean center analysis, 294–295, 337–339
 exercise, 351–359
Measures of association, 305–309, 314–315
Measures of central tendency, 154, 240, 291–295, 298–302, 314
Measures of dispersion, 295–297
Media, working with (exercise), 273–279

Median, 293, 295, 298 302. *See also* Measures of central tendency
Mental maps and awareness space, 150–152
Merry, S. E., 102
Methamphetamines (meth), 195
Micro-level theories, 5
Microsoft, for data analysis, 241–242
Microsoft Excel tutorial, 483–520
 calculations, reasons for doing so many, 516–517
 converting data into date format, 497–502
 converting dates and times, 494–497
 CrimeStat's correlated walk prediction, 515–516
 decimal time predictions, 507–509
 introduction, 484–487
 notes on working with AcrView, 502–507
 pivot table, 518–520
 simple graph predictions, 509–511
 split time or midpoint and weighted averaging predictions, 511–515
 standard deviation, average, and other statistics, 487–494
Milieu, 101, 382
Minority groups. *See* Race
Missing persons cases, 158
Mobility, residential, 271
Mode, 292–293, 295, 298–302. *See also* Measures of central tendency
modus operandi, 156, 161, 198
Money laundering, 472
Moran's I, 305, 382
Motel crime, 190–191
Motivation reinforcement, 103
Motivations for offending, 260–262
Multicollinearity, 63–64, 304, 307
Murder. *See* Homicide
Mutually exclusive variable categories, 286–287

National Archive of Criminal Justice Data (NACJD), 233–234, 237
National Crime Victimization Survey (NCVS), 236–237, 238, 254–260
National Incident Based Reporting System (NIBRS), 10, 232–234
National Institute of Justice (NIJ), 16, 472
Natural breaks classification, 311–312
Natural surveillance, 100–101
Nearest neighbor hierarchical (Nnh) clustering, 380, 382
Ned Levine & Associates, 16
Newman, Oscar, 100–102, 188
New York Police Department, 12
Nodes and paths, 111, 112
Nominal-level data, 286–287, 306
Normal curve, 297, 298–302, 312

Offender patterns and trends, 259–265. *See also specific crimes*
Official crime data, 231–236
OLS (Ordinary least-squares) regression, 306. *See also* Regression analysis
"On Crimes and Punishments" (Beccaria), 54
Operationalizing variables, 287–288
Opportunity. *See also* Crime Pattern Theory; Rational Choice Theory; Routine Activities Theory
 lack of and crime, 304
 opportunistic vs. planned offense, 152, 261
 principles of, 107
 target availability and, 153
Ordinal-level data, 287–288, 290, 296, 306
Ordinary least-squares (OLS) regression, 306. *See also* Regression analysis
Otto, A. C., 200–201

Overall, C., 198
Overcrowding, 63
Overland Park, Kansas,
 construction sites, 190, 199

Park, Robert, 57
Park DuValle, 188
Parolees, 444–446
Patrol, routine and preventative,
 229–230
Patrol officers as audience,
 considerations for, 428–435
Patterns. *See* Crime trends
Paulsen, D. J., 194, 429
Pearson's correlation, 307. *See also*
 Regression analysis
Pease, K., 106, 117
People involved in criminal
 incidents, 254–268
Percentage changes, 292
Perpetrator displacement, 117
PGM. *See* Probability grid method
Physical design. *See* Crime
 prevention through
 environmental design
 (CPTED) approach;
 Environmental criminology
Piehl, A. M., 285, 314
Planned vs. opportunistic
 offense, 152
Poacher, target search method, 157,
 162, 341
Point pattern analysis, 379–383
Police managers as audience,
 considerations for, 436,
 439–440
Police operational analysis, 11
Population and sample in data
 analysis, 226–227,
 297–309, 314
Positive relationship, 305
Positivist School of Criminology, 55
Poverty. *See* Socioeconomic status
Pratt, T., 64
Privacy issues, 475
Private spaces, 101
Proactive policing, 199–204
 community policing, 199–200
 Compstat, 203

event planning, 199
problem-oriented policing,
 201–203
Probability grid method
 (PGM), 198
 exercise, 412–417
Probablity sample, 227
Probability surface/map, 158, 198
Probablity theory, 298
Probationers, 444–446
Problem analysis triangle, 7–8
Problem-oriented policing
 Broken Windows Theory and,
 59, 61
 proactive policing applications
 for, 201–203
Profile map, 158
Projection, 14
Property crime rates, 255–256,
 269. *See also specific property
 crimes*
Prosecution, applications for,
 196–198
Prostitution, 59, 115, 153
Public housing projects, 100–102,
 188–190
*Public Safety Log: GIS for Public
 Safety* (ESRI), 472
Public spaces, 101
Public transportation and robbery,
 192–193
Putting it all together for one
 project (exercise), 452–466

Qualitative research, 284–286
Quantile classification, 310–311
Quantitative research, 284–286
Queries and buffer analysis,
 341–342, 343
Quetelet, Adolphe, 12, 56

Race. *See also* Cultural issues
 Chicago School of Criminology
 and, 57
 data analysis and, 228–229
 drug and alcohol abuse and, 267
 hate crimes and, 262
 as nominal-level variable,
 286, 287

of offenders, 259–260, 264
profiling, 228–229
reporting of crimes and, 265
robbery and, 192
victimization risk and, 189, 255, 258, 263, 270–271
victim precipitation and, 266
RAM. *See* Repeat address mapping
Random chance and data distribution, 304–305, 335, 348
Random sampling, 190, 296, 298, 300
Range of data distribution, 295–296
Rape. *See also* Sexual assault
  distance to crime for, 154–155
  drugs and alcohol in, 267
  GIS applications for investigating, 196
  location of, 270
  motivations for, 261
  race perception of offenders, 260
  reporting of, 265
  victimization risk, 258, 259
  victim-offender relationships, 262
  victim precipitation and, 266
Raptor, attack method, 157
Ratcliffe, J. H., 194, 196
Rate, computing for crime, 291
Ratio-level data, 289–290, 306
Rational Choice Theory, 6, 107–109, 156
Rational offender model, 54–55, 103, 108–109
Ratios, 291
Records management systems (RMS), 228–229, 243
Reference maps, 422–423, 427
Regression analysis, 305–308
Reliability of variables, 288
Reno, S., 191
Renters, 271
Repeat address mapping (RAM), 202, 375, 439
Repeat victimization, 115
Reporting crime. *See also* Incident-Based Reporting System; Uniform Crime Report
  data analysis issues, 227
  drug calls, 200
  environmental design and, 99
  self-report data, 236–238, 314
  trends in, 264–265
  on university campus, 191
Reppetto, T., 116
Research and applications, 187–224. *See also* Data for crime analysis
  on businesses, 190–191
  community policing, 199–200
  Compstat, 203
  on drug-related crime, 195
  event planning, 199
  future trends, 472–473
  on hot spots, 193–195
  investigation and prosecution, applications for, 196–198
  online resources, 204
  proactive policing, 199–204
  problem-oriented policing, 201–203
  on public housing, 188–190
  on schools, 191–192
  on transit stops, 192–193
Residential mobility, 271
Rieckenberg, E. J., 203
Rigel geographic profiling system, 158, 162
Risk adjusted Nnh, 382
Risk heterogeneity, 115
Risk of apprehension, 152, 155
RMS (Records management systems), 228–229, 243
Robbery
  awareness space in, 152
  in crime funnel model, 238
  displacement and, 116, 117
  drugs and alcohol in, 267
  GIS applications for, 197, 198
  hot spots of, 194
  journey-to-crime trip for, 154
  lifestyle exposure approaches and, 113, 114, 115
  location of, 270
  motivations for, 260–261
  predicting, 303
  proximity to rapid transit stations, 192–193
  proximity to schools, 191–192

in public housing, 189
race perception of offenders, 260
Rational Choice Theory and,
 108–109
reporting of, 265
routine patrol prevention of, 229
socioeconomic status and, 258
tactical exercise, 479–480
victim-offender relationships
 and, 262
victim precipitation and, 266
weapon use in, 264
Roman, C. G., 110–111
Romig, K., 195
Roncek, D. W., 189, 191
Rosenbaum, D. P., 106
Rossmo, D. Kim, 111, 158, 160, 336
Routine Activities Theory, 6–7,
 109–111, 156
Routine patrol, 229–230
R-square in regression models,
 307–308
Rural vs. urban victimization,
 268–269, 271–272

Safe Streets Act of 1968, 100
Sample and population in data
 analysis, 226–227,
 297–309, 314
Sampson, R., 58
Santiago, J. J., 116
SARA (scan, analyze, respond, and
 assess) model, 201–202
Satisficing, 108
Scale, 14, 229–230, 315–316, 426
Schmitt, E. L., 254
Schools
 buffering exercise, 143–144
 CPTED and, 102
 crime trends and, 270–271
 drug and alcohol abuse at, 268
 research on crime at, 191–192
 Routine Activities Theory
  and, 111
SDEs (Standard deviation ellipses),
 297, 383
Search base, 159–160
Search methods, target, 157, 162
Search radius, 375, 378

"Seed," 379
Selecting features (exercise),
 120–135
Self-protective measures of crime
 victims, 266–267
Self-report data, 236–237
Semiprivate spaces, 101
Semipublic spaces, 101
Sex offenders, 196, 200, 236
Sexual assault. *See also* Rape
 drugs and alcohol in, 267
 GIS applications for
  investigating, 196
 location of, 270
 race and, 258, 260
 reporting of, 265
 victimization risk, 113, 259
 victim-offender
  relationships, 262
 victim precipitation and, 266
Shared grid maps, aggregating data
 for (exercise), 320–325
Shared responsibility, 113–114
Shaw, C., 5–6, 57–58
Shaw, J., 197
Sherman, L. W., 230, 371
Shoplifting, motivations for, 261
Significance, tests of, 304–305,
 314–315
Simon, H., 108
Skewed data distributions, 294, 295,
 301–302, 314
Skogan, W., 63
Social agencies and contextual data,
 240–241
Social control. *See* Informal social
 control
Social disorganization, 53–96
 Broken Windows Theory, 59–64
 Chicago School of Criminology,
  56–58
 image and, 101
 online resources, 64
 in public housing projects, 189
 social efficacy and, 58–59
Social Disorganization Theory, 5–6,
 58, 63
Social efficacy, 58–59, 64
Socioeconomic status

crime hot spots and, 188, 194–195
in history of crime mapping, 12
as motivation for offending, 260–261
poverty and social disorganization theory, 5, 56, 57–58, 62
poverty and social efficacy, 58–59
robberies and, 192
spurious relationship between crime and, 304
victimization rates and, 258–259
Sociological criminologists, 55–56
Software options, 16
Spatial autocorrelation, 305, 337, 382, 426, 427
exercise, 359–368
Spatial deviation (exercise), 205–215
Spider distance analysis, 341
Spriggs, A., 106
Springfield Police Department and Illinois State Police experience with gun-related incidents and gang-related shootings, 201
Spurious relationships, 304
Stalker, attack method, 157
Standard deviation, 296–297, 299–301, 383–385
Standard deviation ellipses (SDEs), 297, 383
Standard deviation maps, 311–313
Statistical Package for the Social Sciences (SPSS), 242
Statistics, 283–333
classification in mapping, 309–313
descriptive, 292–297
inferential, 297–309
online resources, 313
sins to avoid, 313–317
types of data, 284–292
Strategic crime analysis, 10
exercise, 476–479
Strategics maps, 423
Suitability of target, 152
Supermarkets, 98, 196–197

Suresh, G., 188
Surveillance
enhancement efforts, 106
GIS analysis and, 197
natural, 100–101
physical design and, 103
robbery and, 193
Sutherland, Edwin, 58

Tables, addresses, and maps (exercise), 67–77
adding tables (exercise), 82–94
Tactical crime analysis, 10
burglary exercise, 480–481
independent exercise, 215–216
robbery exercise, 479–480
Tactical displacement, 116–117
Tactical maps, 423
Target backcloth, 152–153, 155, 156, 160, 161
Target displacement, 116
Target hardening, 59, 106, 117
Target search methods, 157, 162
Target suitability, 152
Taylor, R. B., 59, 102, 193–194
Technological change, 470–471, 473–474
Television, closed-circuit (CCTV), 106–107, 116, 474–475
Temporal analysis
of business crime, 191
distance analysis and, 344–345
in geographic profiling, 160, 161–163
Routine Activities Theory and, 110–111
target backcloth and, 153
in theory of crime and place, 8, 9
using GIS for, 16, 197, 198
Temporal displacement, 116
"The Ten Commandments of Crime Analysis" (Bruce), 245
Territorial displacement, 116
Territoriality, 100–101
Tests of significance, 304–305, 314–315
Theft, 6, 154, 189, 190. *See also* Auto theft; Burglary; Robbery

Theoretical explanations of crime and place, 4–9
Three-D approach, 103–105
TIGER, 239–240
Time and distance analysis, 344–345
Topologically Integrated Geographic Encoding and Referencing system (TIGER), 239–240
Traffic, 105, 199, 203, 236
 maps, 429, 434–435
Training in geographic profiling, 158–160
Transit stops, research on crime at, 192–193
Trapper, target search method, 157
Trends. *See* Crime trends
Troller, target search method, 157

Uniform Crime Report (UCR), 10, 231–234, 254–255, 260
Unimodal curve, 298–299
United States Postal Service, 472–473
Units of analysis, 306. *See also* Scale
University GIS data, 241
University of Chicago, 12, 56
Urban vs. rural victimization, 268–269, 271–272
US Census Bureau, 226, 239–240, 254
Utilitarianism, 54

Validity of variables, 288
Vandalism, 59–64, 102, 229, 260
 exercise, 476–477
Variables in statistical analysis, 285–292, 303–309
Variance, 295–296
Victim facilitation, 114

Victimization risks, 113–115, 188–189, 254–259
Victim–offender relationships, 262–263
Victimology (exercise), 216–219
Victim precipitation or provocation, 114, 266–267
Violent crimes. *See also specific violent crimes*
 rates, 255–256, 268–269, 301–302
 weapon use, 264, 271
Virtual repeats, 115
Vito, G. F., 188

Warden, J., 197
Weapon use patterns, 264, 271
Weisburd, D., 12, 117, 230
Wernicke, S., 190
Whittemore, S., 103
Wilson, J., 59, 61
Wilson, S., 102
Women
 crime trends and, 260, 262–263, 265, 266, 267
 Routine Activities Theory and, 7
 victimization risks for, 189, 256, 258, 259
Wood, Elizabeth, 99
Workplace violence, 114–115

Youth crime. *See also* Gangs
 Boston Gun Project and, 202–203
 drug and alcohol abuse and, 268
 juvenile delinquency, 5, 57–58
 motivations for offending, 260
 Routine Activities Theory and, 110–111
 trends in, 260

Zedlewski, E. W., 314
Zimbardo, Philip, 61